Genome Evolution

This book is dedicated to the memory of Susumu Ohno (1928–2000)

Genome Evolution

Gene and Genome Duplications and the Origin of Novel Gene Functions

Edited by

Axel Meyer
University of Konstanz, Germany

and

Yves Van de Peer
Ghent University, Belgium

KLUWER ACADEMIC PUBLISHERS
DORDRECHT / BOSTON / LONDON

A C.I.P. Catalogue record for this book is available from the Library of Congress

ISBN 1-4020-1021-4

Published by Kluwer Academic Publishers,
P.O. Box 17, 3300 AA Dordrecht, The Netherlands

Sold and distributed in North, Central and South America
by Kluwer Academic Publishers,
101 Philip Drive, Norwell, MA 02061, U.S.A.

In all other countries, sold and distributed
by Kluwer Academic Publishers,
P.O. Box 322, 3300 AH Dordrecht, The Netherlands

Printed on acid-free paper

The articles in this book are reprinted from *Journal of Structural and Functional Genomics (JSFG)*
Volume 3 Nos. 1–4 (2003).

Cover picture: Heike Haunstetter

Printed in the Netherlands

Table of Contents

Case studies

Gene networks and evolution

A. Meyer, Y. Van de Peer (eds.), Genome Evolution, vii–x.

Foreword

'Natural selection merely modified while redundancy created': Susumu Ohno's idea of the evolutionary importance of gene and genome duplications

The year 2000 marked the 30th anniversary of the publication of Susumu Ohno's seminal book *Evolution by Gene Duplication*. This book dealt with the idea that gene and genome duplication events are one of the major mechanisms by which the genetic raw material is provided for increasing complexity during evolution. This hypothesis was not forgotten during the last 30 years. On the contrary, today, in the age of genomics, it is more often discussed and tested than ever before in its history and it continues to be expanded, and refined.

Ohno believed that the major advances in evolution such as the transition from single-celled organisms to complex multicellular animals and plants could not simply have been brought about through processes such as natural selection based on existing allelic variation in populations. He suggested rather that novelty in evolution is most often based on genomic redundancy as substrate for subsequent natural selection to act differentially on the two copies of a single initial copy of a gene. In a statement that brought his conviction to a point he postulated that 'natural selection merely modified, while redundancy created'. Ohno believed that mechanisms such as gene and genome duplications provide the raw material for divergent selection and would result in novel functions of gene copies. In his opinion, natural selection would be relegated to the back seat of evolution to doing its rather conservative job of only fine-tuning those novel functions of duplicates. These had, through duplication, the chance to accumulate a sufficiently large number of what were termed 'forbidden' mutations to bring about a change in function of duplicated genes. This surely looks like a controversial, almost anti-Darwinian, idea.

One might expect that in a fast-moving field such as molecular evolutionary biology, that 30-year-old ideas would be long forgotten. Ohno himself seemed concerned that his book would be already outdated by the time it was published. To quote from the introduction of the book 'In this golden age of biology, a book faces the danger of becoming obsolete before its publication. It is my belief that in order to avoid early obsolescence, the author, judging on the basis of the scant evidence available, is obliged to anticipate future developments and paint a picture with broad strokes of his brush. This I have done rather freely in this book'. Ohno succeeded brilliantly; far from being obsolete, Ohno's basic tenets remain current and hotly debated still today, strongly suggesting that there must be some important truths and insights here.

An impression of the importance that the scientific community is placing on studying the evolutionary significance of the decades old ideas of gene and genome duplications can be gleaned also from the number of publications that are devoted to these topics. Figure 1 summarizes this trend for the last 27 years. The graph depicts the results of a search with the ISI Web of Science with the keyword 'gene duplication'. Until about 1990 only about 10–15 publications on the topic of gene duplication were published per year. Far from being forgotten, 20 years after the publication of Ohno's book, in 1991 the number of publications jumped to more than 80 per year and continued to steadily increase until it reached a level of about 200 yearly publications during the last 4–5 years. Several new kinds of data became available in 1990/91 that might be responsible for the renewed strong interest in gene and genome duplications. An incomplete list of catalytic advances of the early 1990s includes: Piatigorsky's pioneering studies on the evolution of novel functions of crystallin genes, Peter Holland's work on the evolution of Hox clusters in deuterostomes, research on the evolution of the immune system by several researchers and also Tomoko Ohta's theoretical work on the evolution of genes in multi-gene gene families. Several of these leading researchers contributed chapters to this book.

A survey of the patterns in the publication of research on genome duplications shows a similar trend to that on gene duplications (Fig. 1). But, according

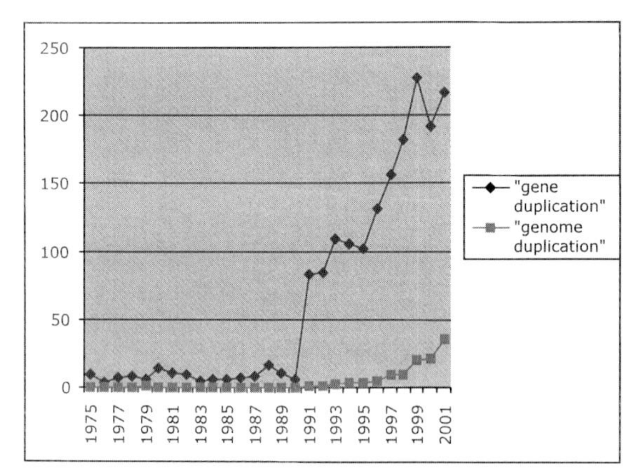

Figure 1. Numbers of publications per year from 1975 to 2001 on 'gene' and 'genome' duplications as reported by the Institute for Scientific Information (Web of Science).

the Web of Science, only since 1995 did publications on 'genome duplication' really begin to appear in the scientific literature. Also here the comparative genomic and developmental work on Hox gene and Hox cluster evolution, in particular the important work of Frank Ruddle, seems to have ignited the research on genome duplications and influenced the thinking of many additional researchers who entered the field in the last decade.

The 30th anniversary of Ohno's book was the main impetus for us to bring together researchers who are interpreting genomic data in the theoretical framework of gene and genome duplications. In the year 2000 our planning with Hervé Philippe began for a C.N.R.S.-sponsored Jacques Monod conference on 'Gene and genome duplications and the origin of novel gene functions'. It took place in April 2001 in Aussois in the French Alps. We are thankful to all who came to Aussois and made it an intellectually stimulating and overall most enjoyable endeavor and we are particularly appreciative of those who contributed their work and ideas to this book. We were overwhelmed by the extremely positive reception by our colleagues for our idea to bring together, for the first time under one book cover, contributions of many of the leading researchers on gene and genome duplications. To this day, as it has been and should be for the last three decades, the relative importance of gene and genome duplications for evolutionary progress, however defined, remains disputed. The duplication/redundancy hypothesis is rivaled most strongly by the idea that 'regulatory evolution' is what 'drives' evolution (e.g, see recent summaries by Carroll *et al.*, 2001; Davidson 2001). We are happy to have also adherents of this viewpoint represented in this collection of papers. Obviously, the two viewpoints on the relative importance of redundancy or regulatory evolution are not mutually exclusive and, as is so often the case, the truth will probably lie somewhere in the middle. In our assessment, much more comparative genomic data are needed to be able to decide which of these two alternative mechanisms might have been, in a particular case, more important in bringing about a major evolutionary transition.

Ohno's ideas were brought forth during a time where the documentation and quantification of genetic variation within populations and between species was largely restricted to scoring allelic variation in enzymes through starch gel electrophoresis and the microscopic inspection of karyotypes. Methods for effectively measuring genetic variation at the level of the gene or to even sequence DNA had still to be invented. Ohno's ideas, while expanding on Darwin's principles of evolution by natural selection, also clearly relegated it in a narrow interpretation of natural selection to only second place after duplication. The fact that many of Ohno's ideas were brought about at a time when it was not possible to check them technically is rather reminiscent of the situation that Darwin had faced. Darwin's proposition that evolution and natural selection worked because offspring resembled their parents more than the average member of the population was done, of course, when genetics had not been invented. Nonetheless, naturally, it still lies at the core of evolutionary biology.

This is the age of genomics in biology. Only since the complete genomes of organisms can be determined did it become possible to rigorously test some of the ideas related to Ohno's first bold proposals on gene and genome duplications. As the phylogenetic sampling of complete genome sequences becomes denser, it will become possible to better evaluate through what mechanisms of genomic evolution advances and major transitions in evolution are triggered and what forces shaped the evolution of genomes themselves. Many whose ideas and empirical data advanced the investigation of the evolutionary importance of gene and genome duplication during evolution for the last three decades contributed to this book. We hope that others will be attracted to working on this set of questions in the future by being exposed to complete genome sequences and wondering what evolutionary forces shaped genomes or by reading some of the research that is contained in this set of publications.

We greatly appreciate the help we received in organizing the Jacques Monod meeting by Dominique Lidoreau, C.N.R.S. We are grateful to Christiane Ehmann, University of Konstanz, and Martine De Cock, Ghent University for their editorial help. We acknowledge the generous financial support by the C.N.R.S. and the University of Konstanz.

Axel Meyer and Yves Van de Peer
Konstanz and Ghent, May 2002

References

Carroll, S.B, Grenier, J. and S.D. Weatherbee. (2001) *From DNA to diversity*, Blackwell Science, Malden, MA.

Davidson, E. C. (2001) *Genomic regulatory systems*, Academic Press, San Diego, CA.

Ohno, S. (1970). *Evolution by gene duplication*, Springer Verlag. New York, NY.

A. Meyer, Y. Van de Peer (eds.), Genome Evolution, xi–xv.

Contributors

Rami Aburomia
Department of Pathology
Stanford University Medical Center
Room 248B
300 Pasteur Drive
Stanford, CA 94305-5324
USA
aburomia@standford.edu

Chris T. Amemiya
Department of Molecular Genetics
Virginia Mason Research Center
Benaroya Research Institute
1201 9th Avenue
Seattle, WA 98101
U.S.A.
camemiya@vmresearch.org

Sébastien Aubourg
URGV
Unité Mixte de Recherche INRA-CNRS
2 rue Gaston Crémieux
CP 5708
F-91057 Evry
France
sebastien.aubourg@evry.inra.fr

Pierre-Olivier Barôme
D.E.P.S.N., UPR2197
Institut de Neurobiologie A. Fessard
C.N.R.S.
Avenue de la Terrasse
F-91198 Gif-sur-Yvette cedex
France
barome@iaf.cnrs-gif.fr

Joseph P. Bielawski
Department of Biology
Galton Laboratory
University College London
Darwin Building
Gower Street
London WCIE 6BT
United Kingdom
j.bielawski@ucl.ac.uk

Nathalie Boudet
Institut de Biotechnologie des Plantes
Unité Mixte de Recherche-Centre National de la Recherche Scientifique 8618
Université de Paris-Sud
Bâtiment 630
F-91405 Orsay Cedex
France
boudet@ibp.u-psud.fr

Jürgen Brosius
Institute of Experimental Pathology/Molecular Neurobiology
Center for Molecular Biology of Inflammation
University of Münster
Von-Esmarch-Strasse 56
D-48149 Münster
Germany
RNA.world@uni-muenster.de

Christine Brun
Laboratoire de Génétique et Physiologie du Développement
IBDM
Parc Scientifique de Luminy
CNRS
Case 907
F-13288 Marseille Cedex 9
France
brun@lgpd.uni-mrs.fr

Andre R.O. Cavalcanti
Departamento de Quimica Fundamental
Universidade Federal de Pernambuco
Brazil

John S. Conery
Department of Computer and Information Service
University of Oregon
Eugene, OR 97403
USA
conery@cs.uoregon.edu

Héctor Escrivá García
Laboratoire de Biologie Moléculaire et Cellulaire
Ecole Normale Supérieure de Lyon
46 allée d'Italie
F-69364 Lyon cedex 07
France
hector.escriva.garcia@ens-lyon.fr

Robert Friedman
Department of Biological Sciences
University of South Carolina
Columbia, SC 29208
USA
friedman@biol.sc.edu

Zhenglong Gu
Department of Ecology and Evolution
University of Chicago
1101 East 57th Street
Chicago, IL 60637
USA
zgu@midway.uchicago.edu

Alain Guénoche
Institut de Mathématiques de Luminy
Parc Scientifique de Luminy
CNRS
Case 907
F-13288 Marseille Cedex 9
France
guenoche@iml.univ-mrs.fr

Isabelle Gy
Institut de Biotechnologie des Plantes
Unité Mixte de Recherche-Centre National de la Recherche Scientifique 8618
Université de Paris-Sud
Bâtiment 630
F-91405 Orsay Cedex
France
gy@ibp.u-psud.fr

Finn Hallböök
Department of Neuroscience
BMC
Uppsala University
S-751 24 Uppsala
Sweden
finn.hallbook@neuro.uu.se

Karsten Hokamp
Department of Genetics
Smurfit Institute
University of Dublin
Trinity College
Dublin 2
Ireland
khokamp@tcd.ie

Peter W.H. Holland
School of Animal & Microbial Sciences
The University of Reading
Whiteknights
PO Box 228
Reading RG6 6AJ
United Kingdom
p.w.h.holland@reading.ac.uk

Austin L. Hughes
Department of Biological Sciences
University of South Carolina
Columbia, SC 29208
USA
austin@biol.sc.edu

Steven Q. Irvine
Department of Molecular, Cellular, and Developmental Biology
Yale University
PO Box 208103
New Haven, CT 06520
USA
steven.irvine@yale.edu

Bernard Jacq
Laboratoire de Génétique et Physiologie du Développement
IBDM
Parc Scientifique de Luminy
CNRS
Case 907
F-13288 Marseille Cedex 9
France
jacq@lgpd.uni-mrs.fr

Chris Jozefowicz
Committee on Evolutionary Biology
The University of Chicago
Chicago, IL 60637
USA
chris.jozefowicz@hotmail.com

Marika Kapsimali
Deparment of Anatomy and Developmental Biology
University College London
Gower Street
London WC1E 6BT
United Kingdom
m.kapsimali@ucl.ac.uk

Oded Khaner
Department of Bio-Medical Sciences
Hadassah College
Jerusalem
Israel
odedk@vms.huji.ac.il

Chang-Bae Kim
Genetic Resources Center
Korea Reserach Institute of Bioscience and Biotechnology
Taejon 305-333
Korea

Martin Kreis
Institut de Biotechnologie des Plantes
Unité Mixte de Recherche-Centre National de la Recherche Scientifique 8618
Université de Paris-Sud
Bâtiment 630
F-91405 Orsay Cedex
France
kreis@ibp.u-psud.fr

Dan Larhammar
Department of Neuroscience
BMC
Uppsala University
S-751 24 Uppsala
Sweden
dan.larhammer@neuro.uu.se

Vincent Laudet
Laboratoire de Biologie Moléculaire et Cellulaire
Ecole Normale Supérieure de Lyon
46 allée d'Italie
F-69364 Lyon cedex 07
France
vincent.laudet@ens-lyon.fr

Stéphane Le Crom
Stephen Michnick Laboratory
Département de Biochimie
Université de Montreal
C.P. 6128, Succ. Centre-Ville
Montreal, Quebec H3C 3J7
Canada
lecrom@wotan.ens.fr

Alain Lecharny
Institut de Biotechnologie des Plantes
Unité Mixte de Recherche-Centre National de la Recherche Scientifique 8618
Université de Paris-Sud
Bâtiment 630
F-91405 Orsay Cedex
France
lecharny@ibp.u-psud.fr

Wen-Hsiung Li
Department of Ecology and Evolution
University of Chicago
1101 East 57th Street
Chicago, IL 60637
USA
whli@uchicago.edu

Lars-Gustav Lundin
Department of Neuroscience
BMC
Uppsala University
S-751 24 Uppsala
Sweden
lglundin@neuro.uu.se

Michael Lynch
Department of Biology
Indiana University
Bloomington, IN 47405
USA
mlynch@bio.indiana.edu

James McClintock
Committee on Developmental Biology
The University of Chicago
Chicago, IL 60637
USA
jmmccli@midway.uchicago.edu

Aoife McLysaght
Department of Genetics
Smurfit Institute
University of Dublin
Trinity College
Dublin 2
Ireland
a.mclysaght@europe.com

Axel Meyer
Department of Biology
University of Konstanz
D-78457 Konstanz
Germany
axel.meyer@uni-konstanz.de

Anton Nekrutenko
Department of Ecology and Evolution
University of Chicago
1101 East 57th Street
Chicago, IL 60637
USA
anton@uchicago.edu

Joram Piatigorsky
Laboratory of Molecular and Developmental Biology
National Eye Institute
NIH
Bethesda, MD 20892-2730
USA
joramp@nei.nih.gov

Victoria Prince
The Committees on Evolutionary Biology and Developmental Biology
Department of Organismal Biology and Anatomy
The University of Chicago
Chicago, IL 60637
USA
vprince@midway.uchicago.edu

Jeroen Raes
Department of Plant Systems Biology
Flanders Interuniversity Institute for Biotechnology
Ghent University
K.L. Ledeganckstraat 35
B-9000 Gent
Belgium
jeroen.raes@gengenp.rug.ac.be

Jonathan P. Rast
Division of Biology 156-29
California Institute of Technology
Pasadena, CA 91125
USA
jprast@its.caltech.edu

Marc Robinson-Rechavi
Laboratoire de Biologie Moléculaire et Cellulaire
Ecole Normale Supérieure de Lyon
46 allée d'Italie
F-69364 Lyon cedex 07
France
marc.robinson@ens-lyon.fr

Frank H. Ruddle
Department of Molecular, Cellular, and Developmental Biology
Yale University
PO Box 208103
New Haven, CT 06520
USA
frank.ruddle@yale.edu

Yvan Saeys
Department of Plant Systems Biology
Flanders Interuniversity Institute for Biotechnology
Ghent University
K.L. Ledeganckstraat 35
B-9000 Gent
Belgium
yvan.saeys@gengenp.rug.ac.be

Manfred Schartl
Physiologische Chemie I
Biozentrum
University of Würzburg
Am Hubland
D-97074 Würzburg
Germany
phch1@biozentrum.uni-wuerzburg.de

Arend Sidow
Department of Pathology
Stanford University Medical Center
Room 248B
300 Pasteur Drive
Stanford, CA 94305-5324
USA
arend@stanford.edu

Cedric Simillion
Department of Plant Systems Biology
Flanders Interuniversity Institute for Biotechnology
Ghent University
K.L. Ledeganckstraat 35
B-9000 Gent
Belgium
cedric.simillion@gengenp.rug.ac.be

Jürg Spring
Institute of Zoology
University of Basel
Biocenter/Pharmacenter
Klingelbergstrasse 50
CH-4056 Basel
Switzerland
j.spring@unibas.ch

Kenta Sumiyama
Department of Molecular, Cellular, and Developmental Biology
Yale University
PO Box 208103
New Haven, CT 06520
USA
kenta.sumiyama@yale.edu

John S. Taylor
Department of Biology
University of Victoria
PO Box 3020
Victoria V8W 3N5
Canada
taylorjs@uvic.ca

Yves Van De Peer
Department of Plant Systems Biology
Flanders Interuniversity Institute for Biotechnology
Ghent University
K.L. Ledeganckstraat 35
B-9000 Gent
Belgium
yves.vandepeer@gengenp.rug.ac.be

Klaas Vandepoele
Department of Plant Systems Biology
Flanders Interuniversity Institute for Biotechnology
Ghent University
K.L. Ledeganckstraat 35
B-9000 Gent
Belgium
klaas.vandepoele@gengenp.rug.ac.be

Philippe Vernier
D.E.P.S.N., UPR2197
Institut de Neurobiologie A. Fessard
C.N.R.S.
Avenue de la Terrasse
F-91198 Gif-sur-Yvette cedex
France
vernier@iaf.cnrs-gif.fr

Jean-Nicolas Volff
Biofuture Research Group 'Evolutionary Fish Genomics'
Biozentrum
University of Würzburg
Am Hubland
D-97074 Würzburg
Germany
volff@biozentrum.uni-wuerzburg.de

Wayne C.H. Wang
Department of Dairy and Animal Science
The Pennsylvania State University
324 Henning Building
University Park, PA 16802-3503
U.S.A.
wcw116@psu.edu

Kenneth H. Wolfe
Department of Genetics
Smurfit Institute
University of Dublin
Trinity College
Dublin 2
Ireland
khwolfe@tcd.ie

Ziheng Yang
Department of Biology
Galton Laboratory
University College London
Darwin Building
Gower Street
London WCIE 6BT
United Kingdom
z.yang@ucl.ac.uk

A. Meyer, Y. Van de Peer (eds.), Genome Evolution, 1-17.

Gene duplication and other evolutionary strategies: from the RNA world to the future

Jürgen Brosius
Institute of Experimental Pathology, Center for Molecular Biology of Inflammation, University of Münster, Von-Esmarch-Str. 56, D–48149 Münster, Germany
E-mail: RNA.world@uni-muenster.de; fax:+49 251 835 8512

Received 16.05.2002; Accepted in final form 29.08.2002

Key words: barriers of genetic exchange, Darwinian principles, evolutionary transitions, exaptation, group selection, Lamarckism, memes, nuon duplication, nuons, primordial sex, retroposition, RNA world, RNP world

Abstract

Beginning with a hypothetical RNA world, it is apparent that many evolutionary transitions led to the complexity of extant species. The duplication of genetic material is rooted in the RNA world. One of two major routes of gene amplification, retroposition, originated from mechanisms that facilitated the transition to DNA as hereditary material. Even in modern genomes the process of retroposition leads to genetic novelties including the duplication of protein and RNA coding genes, as well as regulatory elements and their juxtapositon. We examine whether and to what extent known evolutionary principles can be applied to an RNA-based world. We conclude that the major basic Neo-Darwinian principles that include amplification, variation and selection already governed evolution in the RNA and RNP worlds. In this hypothetical RNA world there were few restrictions on the exchange of genetic material and principles that acted as borders at later stages, such as Weismann's Barrier, the Central Dogma of Molecular Biology, or the Darwinian Threshold were absent or rudimentary. RNA was more than a gene: it had a dual role harboring, genotypic and phenotypic capabilities, often in the same molecule. Nuons, any discrete nucleic acid sequences, were selected on an individual basis as well as in groups. The performance and success of an individual nuon was markedly dependent on the type of other nuons in a given cell. In the RNA world the transition may already have begun towards the linkage of nuons to yield a composite linear RNA genome, an arrangement necessitating the origin of RNA processing. A concatenated genome may have curbed unlimited exchange of genetic material; concomitantly, selfish nuons were more difficult to purge. A linked genome may also have constituted the beginning of the phenotype/genotype separation. This division of tasks was expanded when templated protein biosynthesis led to the RNP world, and more so when DNA took over as genetic material. The aforementioned barriers and thresholds increased and the significance and extent of horizontal gene transfer fluctuated over major evolutionary transitions. At the dawn of the most recent transformation, a fast evolutionary transition that we will be witnessing in our life times, a form of Lamarckism is raising its head.

Introduction

A very reasonable, albeit unproven scenario, maintains that early cellular life went through a stage in which RNA macromolecules constituted the major biopolymers (Woese, 1967; Crick, 1968; Orgel, 1968). Their functions in such primordial cells were not only the storage and replication of genetic information, but they were also involved in structural and enzymatic tasks as well. Earlier forms of life based on inorganic materials or other polymers related or unrelated to extant polynucleic acids (Orgel, 1998) are likely to have preceded RNA-based forms (Cairns-Smith and Davis, 1977; Cairns-Smith, 1982; Wächtershauser, 1992). At this point, it is still difficult to explain the origin of nucleotides (Ferris, 1994;

Orgel, 1998) and there is a vast void concerning initial steps towards the polymerization of nucleotides into oligo- and polynucleotides (Kiedrowski, 1986; Li and Nicolaou, 1994; James and Ellington, 1995; Szathmáry, 1997) and their self-organization into living entities (Jantsch, 1979; Kauffman, 1993). For more detailed discussions of these issues, the reader is referred to excellent reviews and books (multiple authors, 1987; Gesteland *et al.*, 1999; Orgel, 1998; Fry, 2000; Wills and Bada, 2000; Joyce, 2002).

Yet, in this seemingly despairing situation, one should remember that only two decades ago, an answer to the question whether protein biopolymers preceded nucleic acid biopolymers or *vice versa* seemed equally elusive. The discovery of catalytic RNA resolved this dilemma virtually overnight (Kruger *et al.*, 1982; Guerrier-Takada *et al.*, 1983; Gilbert, 1986; Westheimer, 1986; Gesteland *et al.*, 1999). Despite some lingering uncertainties we can consider the RNA world as a reasonable theoretical outpost from which we can make rather effortless associations to extant cells as they still contain RNAs that function in various compartments and diverse biochemical tasks. By hypothesizing the nature of this RNA world, we may also eventually be able to venture back to examining earlier, even simpler stages of life. Thus, we can begin to integrate a hypothetical RNA world into an attempt to understand biological and evolutionary principles spanning 3–4 billion years of macromolecular, cellular and organismal transitions. Importantly, the RNA world is not yet behind us; particularly in eukaryotic cells, RNA still plays a pervasive role in numerous biochemical and regulatory pathways (Brosius, 1999d; Herbert and Rich, 1999; Filipowicz, 2000; Eddy, 2001; Erdmann *et al.*, 2001a, 2001b; Lagos-Quintana *et al.*, 2001; Lau *et al.*, 2001; Lee and Ambros, 2001; Mattick, 2001; Hüttenhofer and Brosius, 2002; Storz, 2002).

In this chapter, we will first discuss the impact of RNA in extant cells with an emphasis on its continuous conversion to DNA. This section will remain brief as much of it has been dealt with in earlier articles (Brosius, 1991, 1999a, 1999b, 1999c, 1999d; Brosius and Gould, 1992; Brosius and Tiedge, 1995b). The rest of the chapter will consider what effect envisioning an RNA world built on Neo-Darwinian principles has on accepted evolutionary views – always with the caveat of our limited understanding of such a distant era.

Duplication of genes and regulatory elements by retroposition

One of the major transitions in life occurred when the RNA world gradually evolved into the RNP world (Maynard Smith and Szathmáry, 1995; Szathmáry and Smith, 1995). At that stage an extremely versatile biopolymer, protein, joined RNA in structural and catalytic tasks. Simple, untemplated peptides were probably already present in the RNA world. What set the RNP world apart was the advent of translation; the templated synthesis of polypeptides at proto-ribosomes, that became increasingly sophisticated over time (Woese, 1980, 2001; Maizels and Weiner, 1987; Szathmáry, 1999b; Brosius, 2001). The advent of versatile protein biopolymers was an important prerequisite for the next major transition, when DNA gradually replaced RNA as genetic material (Brosius, 1999d). The RNA to DNA transition probably occurred via a process of retroposition with the aid of an early form of the enzyme reverse transcriptase and other enzymes presumably as well. Hence, retroposition led to the establishment and growth of a DNA genome at the onset of the DNA world. When such retroposition events occurred in the 'germ line' some became fixed in the genomes of populations. The process of retroposition apparently did not vanish when all genomic RNA was converted into DNA but continues in many lineages even to this day (Jurka, 1998; Brosius, 1999a, 1999d). Despite the fact that retroposition is ancient it is still frequent today. We observe that at least 42% of the human genome is still discernable as having been contributed by retronuons (Lander *et al.*, 2001; Venter *et al.*, 2001). Even this is likely an underestimate, as retropositions that occurred more than 150–200 million years ago, are no longer identifiable due to the continuous onslaught of mutations. In fact, one might assume that all founding DNA sequences were generated by retroposition. Once the DNA genome became the template of enzymatic replication, additional duplication events based mainly on DNA replication errors expanded the genome. Such mechanisms, also occurring in modern cells, included simple stuttering at expanding homopolymers as well as oligonucleotide repeats involving a few base-pairs at a time or segmental duplications involving megabases.

While only a little more than a decade ago it was not trivial to convey the message that retroposition has the potential to generate functional genes (Brosius, 1991) and not just junk in the form of retropseudogenes, it is now apparent that many intronless genes were duplicated by retroposition via processed mRNA intermediates. One of the difficulties was the fact that retroposition had an apparent disadvantage compared to segmental duplication or whole genome duplication; mechanisms that were once thought to be solely responsible for generating novel genes (Bridges, 1936b; Lewis, 1951; Stephens, 1951; Nei, 1969; Ohno, 1970; and other chapters in this volume). Usually the mRNA-derived gene copy transposes without its associated promoters and other regulatory elements and hence, is most likely to be inactive. On the other hand, a retrogene that chances to integrate near resident promoter elements might lead to a selective advantage by immediately exhibiting a different temporal and or spatial expression pattern in comparison to the source gene. In contrast, a segmentally duplicated gene is likely to be expressed in the same way as the source gene. There is now a steadily growing list of genes generated by retroposition that often exhibit differential expression patterns in comparison to their source genes and were recruited or exapted (Gould and Vrba, 1982) into variant or novel functions (see for example http://exppc01.uni-muenster.de/expath/alltables.htm #table3). In fact, the majority of genes that lack introns may have been generated by retroposition (Brosius, 1999a). According to J.J. Emerson and M. Long, ~15–25% of all genes in the human genome lack introns in their coding regions (personal communication).

Retroposition also generates novel genes encoding non-messenger RNAs. A decade ago, the idea that cells might rely on novel RNA molecules rather than proteins for potential functional innovations was peculiar to scientists. However, a novel, neuron-specific, small non-messenger RNA (BC1 RNA, age ~60–110 MY) has been observed in rodents and another analogous neuron-specific RNA (BC200 RNA, age ~35–55 MY) has been discovered in anthropoid primates (Sutcliffe *et al.*, 1982; DeChiara and Brosius, 1987; Watson and Sutcliffe, 1987; Martignetti and Brosius, 1993a, 1993b; Brosius and Tiedge, 1995a, 2001; Tiedge *et al.*, 1991, 1993). When we are finished analyzing eukaryotic genomes, we believe we will not only have discovered a plethora of ancient RNA genes but also a number of RNA genes that arose well past the stages of the RNA/RNP worlds (Hüttenhofer *et al.*, 2001; Hüttenhofer and Brosius, 2002); as recently as a few million years ago (Wang *et al.*, 2002).

The continued use of RNAs in diverse cellular tasks and the recruitment of novel RNAs into biological systems is further testimony to the fact that the RNP world– at least in Eukarya – is not yet history. Despite their antiquity, RNA molecules are still superior to proteins in certain aspects. For example, RNAs are able to interact with other nucleic acids at higher levels of specificity and RNA molecules are capable of bringing together protein molecules from different cellular compartments (e.g., via RNA binding sites) that otherwise would never interact with each other.

Demonstrably, retroposition has the potential to generate novel genes that may encounter different regulatory elements at the respective loci of integration. The process also facilitates the opposite scenario: retronuons can insert at more or less random positions in the genome and potentially modulate the expression of targeted genes. Retronuons of the SINE, LINE and LTR subfamilies frequently contribute promoter elements, enhancers, silencers, polyadenylation signals, and splice sites that act on the targeted genes. LINE retronuons have been implicated in the inactivation of X-chromosomes (Lyon, 1998, 2000; Bailey *et al.*, 2000). For further details the reader is referred to a number of reviews on the subject (Brosius and Gould, 1992; Shapiro, 1992; McDonald, 1993, 1995; Britten, 1996, 1997; Kidwell and Lisch, 1997; Brosius, 1999a, 1999b, 1999c, 2003b; Makalowski, 2000; Sverdlov, 2000; Mattick, 2001) or to our www page (http://exppc01.uni-muenster.de/expath/alltables. htm#table1 and http://exppc01.uni-muenster.de/ expath/alltables.htm#table2).

In summary, retropositional duplication of protein encoding genes can lead to their admixture with resident regulatory regions at the locus of retronuon integration and conversely, novel regulatory regions can combine with resident genes through retronuon dispersal. Hence, retroposition is a process that constantly and pervasively modulates genomes.

What can the RNA world reveal about the validity of evolutionary principles?

Like a rock in the surf, the RNA world can be viewed as a vantage point in a sea of uncertainty. From there, we can almost reach on dry feet (and better understand) the shores of modern cells. We may also have a chance to venture out into the stormy and uncharted waters of the pre-RNA world life.

In addition to imagining molecular events at the beginning of cellular life, the RNA world may, in the not so distant future, be reconstructed in an experimental setting offering the chance to physically reverse engineer primitive cells based on RNA as enzymes, genetic material and structural components. There are efforts under way to generate selected functional RNA nuons in the test tube via selection/amplification schemes (Famulok and Jenne, 1999; Wilson and Szostak, 1999) that may eventually contribute to man-made life forms capable of reenacting life in a primitive cell reminiscent of the RNA world (Bartel and Unrau, 1999; Szostak *et al.*, 2001). It is obvious that such a reconstruction would not accurately retrace the path of evolution, nevertheless, the resulting RNA-based cells could serve as models for life in the RNA world. One can expect this feat to have been achieved within a few decades.

By extrapolating from such hypothetical but reasonable scenarios, aspects of which may be testable in the future, we may be able to address important questions in evolutionary biology at their roots. Also it should be remembered that modern cells have still not completely left the RNA world. It is apparent that extant cells harbor many more RNA species than anticipated (Hüttenhofer *et al.*, 2001). These RNAs still function in a wide spectrum of cellular regulation (Eddy, 1999, 2001; Hüttenhofer and Brosius, 2002; Storz, 2002). Even RNAs that arose quite recently (i.e., after the mammalian radiation) have been exapted into novel functions (DeChiara and Brosius, 1987; Brosius and Gould, 1992; Martignetti and Brosius, 1993a, 1993b; Brosius, 1999c; Cavaillé *et al.*, 2000). The vigorous and diverse 'RNA' life in contemporary cells is an important link to primordial RNA/RNP worlds (Jeffares *et al.*, 1998; Poole *et al.*, 1998; Brosius, 1999d; Herbert and Rich, 1999; Mattick, 2001).

If evolutionary theories are correct, they should apply to all stages of evolution, including primal ones. As we will see, the core predications including the evolutionary importance of unit duplications (genes, nuons) and the Darwinian principles can remain valid in the RNA or RNP worlds, while the roles that other processes, fundamental to evolution, might have fluctuated quantitatively or require modification at different stages of cellular or organismal organization and complexity.

Scenes from an imaginary RNA world

In the following we will demonstrate that, in the RNA world gene (nuon) duplication must have played a decisive role in evolution, selection acted upon individual RNA nuons as well as on groups of nuons, and primordial sex (exchange of RNA nuons) was promiscuous. Due to the latter the evolution of distinct species in the RNA world was virtually impossible. As RNA nuons determined genotype *and* phenotype of a cell, we can conceive of a world with Neo-Lamarckian elements defined as the inheritance of acquired characteristics barring directed modification of genetic material. Importantly, the significance of amplification of genetic material, its random variation (including recombination) and selection, remains a pillar of evolutionary theory even in the RNA/RNP worlds. Several other principles did not apply. For example, Weismann's Barrier and the Central Dogma of Molecular Biology were virtually absent or rudimentary.

Modest beginnings of an RNA-based world

As stated above, it is not yet clear, how ribonucleotide phosphates, the building blocks of RNA, were generated from inorganic matter and what drove their polymerization. In the beginning, a self-replicating biopolymer could have arisen by chance. However, if more than one replicator was necessary their containment was vital (Szostak *et al.*, 2001). Thus, unless these replicators were present at high concentrations in a given environment, co-optation of a semi-permeable (lipid) membrane should have occurred at an early stage (Maynard Smith and Szathmáry, 1995; Szostak *et al.*, 2001). Cycles of replication and cellular growth (Figure 1A) with spontaneous 'cell' division could have constituted an early form of RNA-based life (Figure 1B). If a larger undivided 'cell' (Figure 1A) had a selective advantage over a 'cell' with less nuons, we can perhaps equate such a scenario with the first 'genome' ('nuonome') duplication in the RNA world. Some of the replication-incompetent nuon variants (mutants, shown by thinner lines in Figure 1C -) were possibly the first 'genomic parasites'. Through gradual alteration of sequence/structure and subsequent exaptation of the variant biopolymers into novel functions, some of these may have conveyed a selective advantage to the cell (Figure 1C +) by facilitating recruitment of or by enzymatically aiding the synthesis of precursor mole-

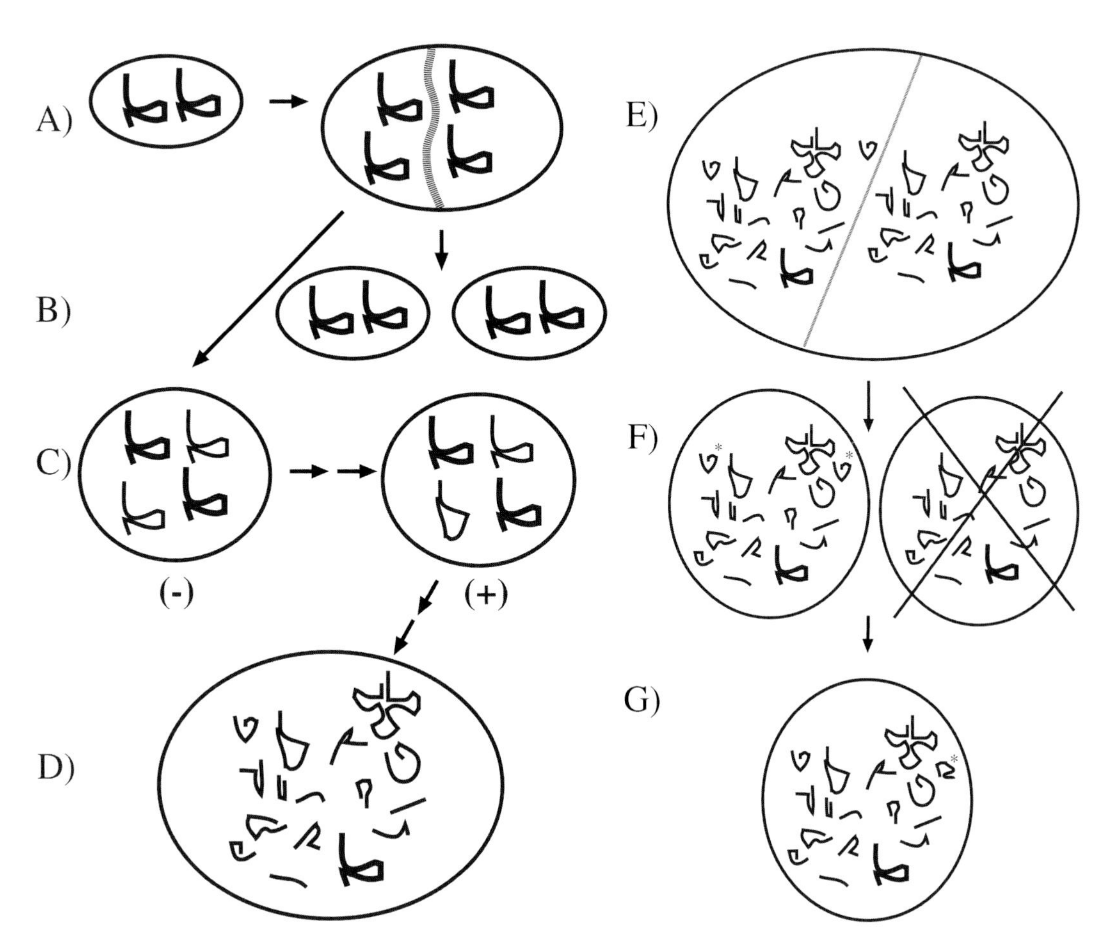

Figure 1. An RNA world can be imagined that already encompassed 'gene' duplication, generating new nuons by amplification/mutation. Two RNA biopolymers capable of replication enclosed by a primordial membrane multiplied and the primordial cell grew in size (A). The normal route was eventual division by physical forces resulting in two daughter cells containing similar sets of RNA nuons (B). Alternatively, some of the RNA nuons lost their ability to replicate (thin lines) but were propagated by the replicators (bold lines). Carrying a functionless nuon as ballast must have translated into a high cost and thus, cells containing 'parasitic' nuons were negatively selected (C, -). Like all nuons however, the superfluous nuon may have mutated and on rare occasions, altered forms may have accidentally added a function for the benefit of the cell. Such a cell was positively selected (C, +). Continued recruitment of duplicated and altered nuons established more complex cells with different functional nuons and a replicator (D). Duplication (E) and unequal distribution of nuons to daughter cells (F) resulted in one that possessed an extra copy of a given nuon type (asterisk) and another (on the right) with none. This led to a reduced or loss of viability of the latter (crossed out). The daughter cell with duplicate nuons might have had a further advantage over daughter cells resulting from an equal distribution of nuons if the duplicated and subsequently altered (asterisk) nuon was exapted into a variant or novel function (G).

cules. If the depicted scenario or a similar one unfolded, early symbiosis and group selection emerged. In the same vein, an 'altruistic' replicator had a selective advantage over a 'selfish' replicator as it facilitated provision of 'raw material' for its own success: the altered nuon, itself replication incompetent, might have provided a different auxiliary metabolic function and thus would have been among the first exaptations (Gould and Vrba, 1982; Brosius and Gould, 1992).

Various RNA nuons could have been generated in a similar fashion (Figure 1D) leading to more complicated cells with enhanced metabolic and reproductive capabilities. Thus, the principle of nuon (gene) duplication (Bridges 1936a; Lewis 1951; Muller *et al.* 1936; Nei 1969; Ohno 1970; Stephens 1951; Sturtevant 1925), as well as other principles of Neo-Darwinian evolution, namely random variation and selection, would have played major roles in an evolving RNA world.

Evolution by nuon duplication and exchange

As described above on a minimal scale, replication of RNA molecules would have led to their amplification and, after growth of cell content and incorporation of external or internally synthesized lipids into the membrane, there could have been spontaneous or triggered division (Figure 1E,F). For simplification only a duplication of each nuon type ($n = 2$) rather than multiplication ($n > 2$) is shown; of course, multiple copies of each nuon ($n > 2$) were most likely generated. In the absence of sophisticated mechanisms that would have enabled equal distribution of each nuon type to the daughter cells, multiple copies would be vital. According to the formula $P = (1 - 1/2^n)^N$ where P is the probability, N, the number of different nuons per cell and n, the number of copies of each component, a genome of 10, 100, or 1000 different nuons would require a copy number of 10, 14, or 17, respectively, to have a 99% chance that each nuon type be found in both daughter cells. It is not clear how a daughter cell with a low copy number of a given nuon could later re-adjust its copy number. Therefore, above a certain genome size, linkage of different types of RNA nuons into heteromeric concatemers ('chromosomes') might have been advantageous.

The unequal distribution of individual nuons would have resulted in an early form of nuon (gene) duplication. While the absence or underrepresentation of one nuon type conceivably could have resulted in a selective disadvantage or non-viability of the affected daughter cell (Figure 1F, right), the presence of an extra nuon copy or overrepresentation of one nuon type (Figure 1F, left), especially at a high error rate of RNA replication, might have altered, modulated and/or changed the function of such nuons. Such a variant nuon may have provided a selective advantage (Figure 1G).

The exchange of genetic material (primordial sex) was probably rampant among RNA-based cells. Just as cells could have spontaneously divided to yield two daughter cells, two cells with different sets of RNA nuons (genetic material) may have fused. As a result, nuons would have been shuffled and re-distributed to daughter cells yielding sets of nuons ('nuonomes') that were different from those of the initial cells. Depending on the distribution, this may have been advantageous (e.g., in a different environment) or disadvantageous (if important nuons were underperforming in the different 'nuonomic' environment) to one or both cells (Figure 2).

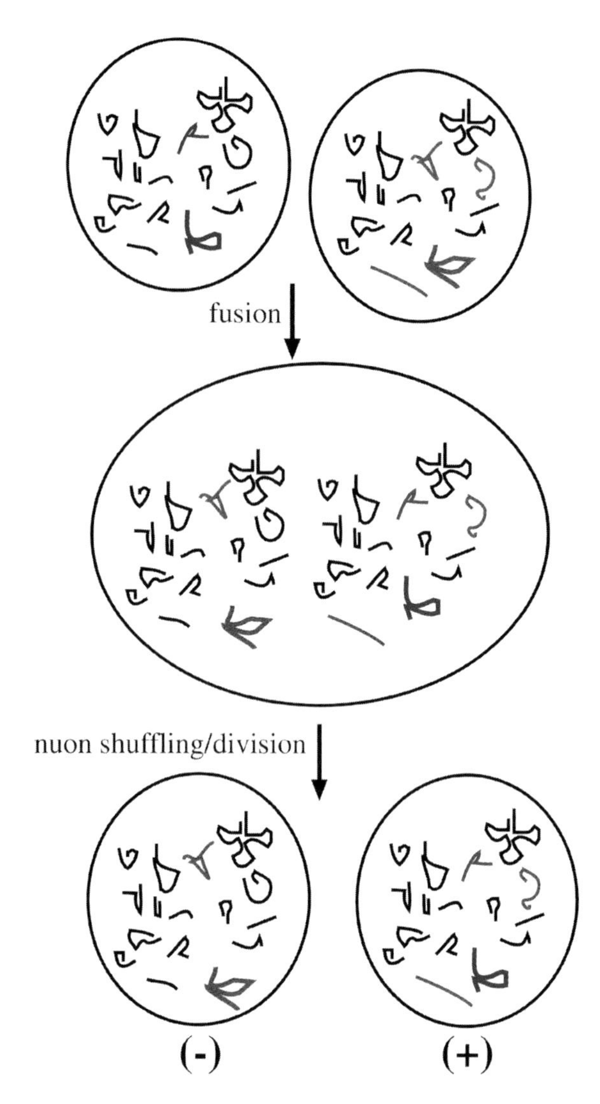

Figure 2. Sex in an imaginary RNA world. Two cells with both different (blue and red) and identical (black) sets of nuon types (top) fused and after diffusion and amplification (not shown) of some of the nuons (center), divided again (bottom). Some nuon combinations in a cell had a selective advantage (+) over others (-). Replicators are in bold. The bottom left cell may have possessed a better replicator but the replicator was better only in association with different accessory nuons ('genetic' background).

Phenotype was genotype

The following example is included to demonstrate that in the RNA world phenotype and genotype were inseparable in a given nuon. The scenario is highly speculative as it implies that homo- or hetero-multimeric RNA biopolymers associated to form functional units for such things as motility (flagelloids), for (chemo)sensation or signal transduction. Nevertheless, it illustrates a point that is also valid for less complicated and not so speculative functions of RNA biopolymers.

Even though there is no evidence whether motility evolved in RNA-based cells, movements surely would have been advantageous to a part of a population that could then move away from nutrient depleted areas or otherwise unfavorable conditions. The evolution of sensory systems and signaling pathways involving RNA biopolymers and other small molecules would have provided a significant selective advantage in the search for favorable and avoidance of unfavorable conditions (Figure 3). Despite its speculative nature, the example illustrates how in the RNA world genotype and phenotype were a single entity. A fortuitous mutation in one of the RNA components of the sensory cluster may have improved its performance. From there it is not hard to imagine that this RNA nuon spread in cells and populations (through primordial sex) over time. Such a mechanism is also conceivable for less hypothetical RNA nuons if a newly arisen mutation was under positive selection.

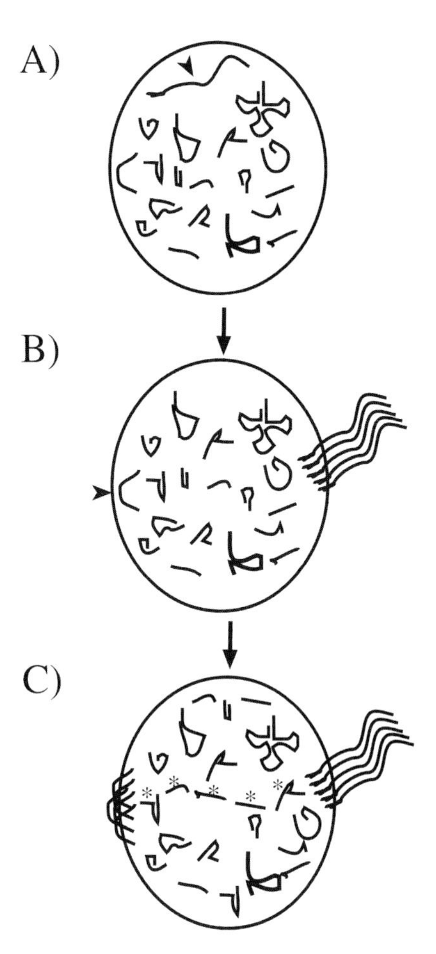

Figure 3. In an imaginary RNA world phenotype equaled genotype. In a highly speculative scenario, an RNA nuon (arrowhead) generated by duplication and modification (perhaps in association with duplicates as a homopolymer) acquired a function in cell motility (B). This was advantageous. Likewise, other nuons (arrowhead, B) arose by amplification/modification and generated a complex of molecules that was able to sense cues from the environment (photons, pH, temperature, nutrients) (C). A hypothetical chain of signal transducing molecules (asterisks) may have linked the two systems (C).

Chaining nuons – an early evolutionary compromise

For reasons discussed above, linkage of several nuons into an RNA genome might have been advantageous. In addition to the joined genomic form (the 'germ line') there would still have been free RNA nuons (somatic nuons) in the 'cells'. Occasional replication of these individual nuons could still have taken place (Figure 4A,B). After replication, chiefly of the joined genetic form, the cell divided (Figure 4B,C) and a portion of the joined forms was processed into individual functional nuons (Figure 4C). A major problem of this scenario is posed by the necessity of maintaining some of the genetic chains intact while producing enough of the functional free nuons. However, RNA processing is still an essential mechanism during RNA biogenesis in all extant cells (Herbert and Rich, 1999). Despite the absence of any direct proof, it is attractive to contemplate that many forms of RNA processing had their roots in the RNA world (Brosius, 1999d).

An interesting aspect is that the appearance of chiefly genetic (versus somatogenetic) RNA already began to establish a barrier. It was not impenetrable, as a randomly altered 'somatic' nuon could occasionally be added to the linked 'germ line' RNA chain, or an altered nuon could replace the corresponding nuon in the 'genome' by some sort of recombination (Figure 4D,E). While establishing a barrier in one sense, recombination of nuons with the genome generated a 'safer' environment for the survival of selfish nuons: interconnected with proto-chromosomes they would have been more refractory to purging. This situation is reminiscent of modern genomes, in which disadvantageous nuons can hitch a ride by link-

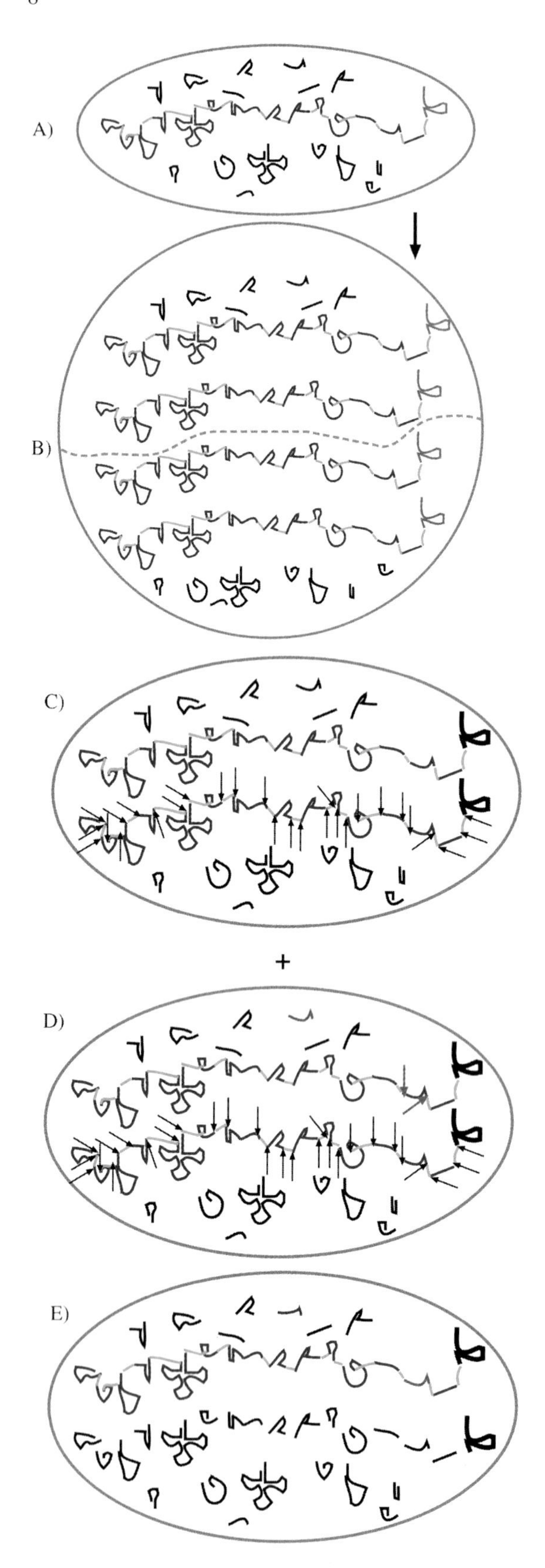

age to positive traits (Brosius and Tiedge, 1995b; Brosius, 1999a).

Discussion and Conclusions

Despite, or perhaps because of, our gaps in knowledge about early life forms, it was not difficult to construct a hypothetical RNA world that is solidly anchored within the framework of Neo-Darwinian principles. Akin to established *in vitro* amplification/selection experiments (Szathmáry, 1990; Ellington and Szostak, 1990; Robertson and Joyce, 1990; Tuerk and Gold, 1990; Wilson and Szostak, 1999; Szostak *et al.*, 2001) evolutionary change is always mediated by amplification of nuons and the selection of ramdom alterations. A minor differences between this type of *in vitro* evolution and natural selection is that, *in vitro*, usually only the best performers are selected for subsequent rounds of amplification while the latter usually removes only individuals of lower fitness from the population. This ensures maintenance of a large genetic diversity in populations (Mayr, 2001). In natural settings, there are additional principles at work such as the conflict between integration of and insulation from foreign genetic material (see below).

Group selection, group sex and chains of nuons

As outlined above and previously noted by Szathmáry (1999a), in the RNA/RNP worlds the selection of RNA nuons enclosed by a membrane can only be interpreted as group selection. In an RNA world with limited numbers of nuons, a selfish parasitic RNA nuon could pose a serious disadvantage to the cell. This situation could only have a positive outcome if the cell lost the selfish nuon or by fortuitous alteration

Figure 4. Linkage of nuons and the origin of RNA processing in an imaginary RNA world. In addition to different individual nuons (black) a more advanced RNA-based cell could also have featured the same nuons in a linked form (blue) giving rise to one or several proto-chromosomes that might have included a terminal replicator as a tag shown in red (Maizels and Weiner, 1987). The joined nuons (A) were interspersed by short linkers (green). RNA chromosomes may have facilitated replication (B) and equal nuon distribution to daughter cells (C, D). The 'intervening' linker sequences may have been necessary for efficient processing (arrows) of individual nuons from the joined proto-chromosome (C, D, E). A free variant nuon (magenta) may have recombined with the linked homologue by the appropriate cuts (magenta arrows) in the proto-chromosome (D) followed by replacement and re-ligation (E).

and adaptation/exaptation of the formerly selfish nuon into a functional role. Already at the level of the RNA world, the types and variants of other nuons in the cell, its 'genetic background', could have played a decisive role in selecting for or against a given nuon. Clearly, a nuon's performance greatly depended on its 'co-players' and a given environment.

The reshuffling of nuons creating ever – changing genetic backgrounds was favored by the unlimited exchange of genetic material. Free sex was gradually curbed by linkage of different nuons. This would have facilitated equal distribution of genetic material during cell division and begun to establish more specialized genomes; gradually separating genotyope and phenotype. This eventually led to the complex DNA-based genomes found in extant organisms. Hence, evolutionary novelties would have been selected more directly in the RNA world and more indirectly in increasingly complex systems at the molecular (chromosomes, DNA), subcelluar (nuclear compartmentalization) and organismal (compartmentalization into the germ line versus somatic cells) levels.

Evolutionary transitions

It is dangerous to categorize life into lower or higher, primitive or advanced forms (Dawkins, 1992; Gould, 1996). All too often, progress is viewed from an

bxanthropocentric pedestal. One forgets that even reduction in complexity, such as reduction in genome size, or reduction of gene numbers can be evolutionary progress, dependening on the 'demands' of changing environments. Nevertheless, when considering extinct and extant forms of life, common sense dictates that there must have been an increase of complexity over time. Such an increase is apparent when comparing the limited number of different nuons in a fledging RNA world to the thousands or tens of thousands of genes in extant organisms (see also Adami *et al.*, 2000). Over time, one observes the evolution and increasing use of novel biopolymers. Furthermore, mergers of cells occurred, in which one partner became an organelle (Margulis, 1970) or, on more subtle terms, in which individual cells associated to form multicellular and differentiated organisms. Transitions to levels of increasing complexity have been characterized by, to name a few, the origin of translation, chromosomes, eukaryotes, sex, multicellular organisms and social groups (Maynard Smith and Szathmáry, 1995). Numerous other transitions such as the rise of epigenetic phenomina, (e.g., imprinting of

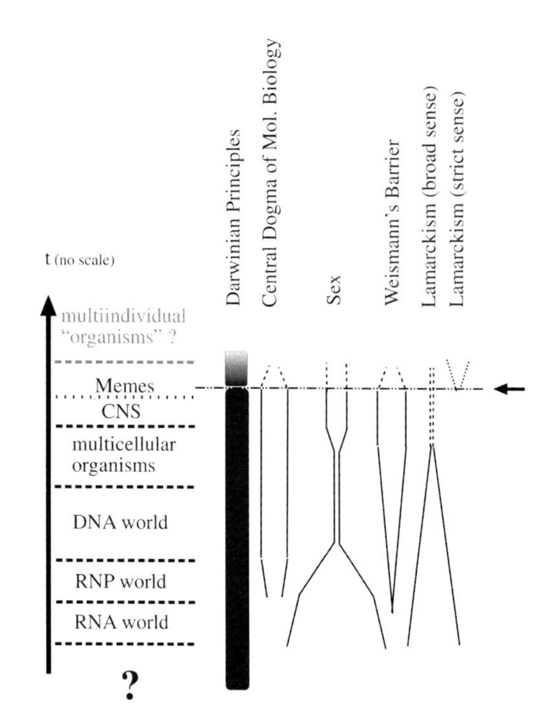

Figure 5. Stages of evolutionary transitions versus variance of evolutionary dogmas. The left panel shows selected major transitions in evolution resulting in different stages over time (a time scale was not attempted). The nature of such stages and transitions prior to the RNA world are not yet perceived (question mark). Major principles (amplification, variation, and selection of units) remained constant over the various transitions, others shifted in presence or scope. Broken lines indicate enhanced uncertainty. The stippled horizontal line indicated by the arrow on the right denotes present time, beyond which we can only speculate.

genes that arose in placental mammals) are not considered here.

In Figure 5, Maynard Smith's and Szathmáry's concept of major transitions has been modified slightly (Maynard Smith and Szathmáry, 1995). Only a selection from numerous transitional stages is included. Starting from a likely cellular progenitor in the RNA world (see above), the first major transition considered was the advent of templated protein biosynthesis, transferring the catalytic and structural tasks of RNA macromolecules, to a totally different class of macromolecules. The next major transition considered was the transfer of genetic information from RNA to DNA. A prerequisite of which was a major transition itself, namely the concatenation of RNA nuons into 'RNA chromosomes', perhaps allowing improved replication and/or more equal distribution of RNA nuons to daughter cells. The transition from RNA to DNA as genetic material was a gradual process that obviously continues in many modern genomes in the form of retroposition (Brosius, 1999a). The next evolutionary stage was reached when cells associated and gradually formed multicellular and eventually increasingly differentiated organisms some of which subsequently sequestered germ line cells from somatic cells.

The evolution of a nervous system heralded the beginning of another major transition. It provided a notable selective advantage as information about the environment could be processed rapidly and transformed the organism into a better 'survival machine' for its genes (Dawkins, 1976). There are precedents of non-genetic transfer of information in social insects, such as the locations of food sources in bees, or transmission of 'culture' in mammals, as exemplified by songs of whales or tool utilisation of chimpanzees (Bonner, 1980). In one lineage, however, a highly developed central nervous system (CNS) was exapted into an instrument that permitted the direct, non-genetic transfer of information at an unprecedented magnitude (Gould and Vrba, 1982). This transfer could not only flow horizontally, but also from one generation to the next with equal efficiency. The vehicle of transfer is spoken, written or digitized information, memes in a broad sense (Dawkins, 1976; Blackmore, 1999).

At this point, it should be stressed that at least the upper levels of biological complexity depicted in Figure 5 are biased, to all intents and purposes, towards the evolutionary fate of our own species. In a correct and all-encompassing representation this scheme would include all lineages of extant and extinct life (Gould, 1996, 2002). A large proportion of species (presumably all Prokaryotes) have apparently not made major transitions beyond the one leading to the use of DNA as genetic material. The scheme would even encompass organisms, such as viruses, endosymbionts, or parasites, which made the transition to a new 'stage' via simplification of complexity. It is noteworthy that the evolutionary stages depicted in Figure 5 seem to imply, although there was no attempt to introduce a time scale, that the more recent transitions occurred in shorter successions, as if evolutionary complexity accelerated when higher levels of complexity were reached.

We are possibly at the threshold of a novel stage representing analogues of multicellular organisms, namely super-organisms whose units are not cells but individual organisms, even social entities. Again, this is not the first time in evolution that such a transition occurred, there are already notable examples in insects (Giraud *et al.*, 2002).

The 'softer' evolutionary principles are neither present nor constant over the entire transitional spectrum

The aforementioned evolution of a linear 'genome' also illustrates how, unlike the constant Neo-Darwinian tenets, other evolutionary principles were subject to modification during major evolutionary transitions. Weismann's Barrier could be considered a continuation of the demarcation of the initial 'germ line'. While, in a strict sense, Weismanns's barrier applies only to organisms with a germ line that is separate from somatic cells (Pollard, 1984), one may postulate that barriers existed to varying degrees at several evolutionary stages. Initially, in the RNA world, there was no barrier whatsoever, apart from a presumed membrane. Primordial sex which experimented with new combinations of different nuons, was virtually unlimited (Figure 3). When nuons merged into proto-chromosomes (Figure 4), a distinct novel cellular RNA nuon, generated either by amplification and subsequent mutation of a free endogenous nuon or by exchange with another cell harbouring different nuons, would encounter a barrier. Such a barrier was evident simply by the reduced ease at which the novel nuon could be integrated into the 'germ line'. This contrasts with the simple uptake in an unlinked genome of separate nuons, a frequent event that would have been sufficient to ensure propagation into daughter cells. Conceivably, integration into the proto-chromosome was a means to overcome the barrier (as depicted in Figure 4D,E). As different levels of complexity were reached, the barrier became temporarily less penetrable when, for example, DNA became the genetic material. Nevertheless, penetration of the barrier may have varied over evolutionary transitions, for example, by cells evolving various recombinatorial mechanisms. With the establishment of additional barriers against the indiscriminate exchange of genetic material, but not necessarily concommitant with these events as additional parameters might also have varied, unlimited sex (horizontal gene transfer) was curbed.

One of the consequences of this primordial puritanism was the gradual separation of phenotype and genotype. It might have constituted one of the first evolutionary barriers. The Darwinian Threshold (Woese, 2002) triggered by templated translation leading to more complex and interconnected cell designs is yet another important barrier; according to Woese (2002) the one that 'truly represents the origin of species'. I myself tend to consider the origin of species as a more prolongated process that might have included several critical transition points requiring several barriers, such as (i) the aforementioned linkage of individually replicated RNA nuons into a concatenated RNA genome, (ii) the Darwinian Threshold (Woese, 2002) and, in particular, (iii) the action of reverse transcriptase, the enzyme responsible for converting RNA into DNA genomes (Brosius, 2003a).

Evolution is conflict – conflict is evolution

Perhaps a further evolutionary force, in addition to amplification, variation, and selection, is conflict – an erratic, yet everpresent dynamism. Over evolutionary time, there must have been a continuous oscillation between the optimal value of accuracy in replication of genetic information and the availability of sufficient variation for Darwinian natural selection. Similarly, we observe the expansion and contraction of 'superfluous' genetic information in intergenic sequences and introns as a constant fluctuation. To a significant extent, retroposition, itself a product of an ancient but still ongoing conflict of reverse transcriptase encoding retronuons with their host genomes, is responsible for major genome expansions. Sex poses yet another riddle: An organism devotes a large proportion of its energy to protecting and promoting every single nucleotide in its genome just to squander 50% of them in a single sweep in the act of propagation. In this context, Ghiselin's suggestion that a species should be considered an invididual deserves additional scrutiny (Ghiselin 1974; Ghiselin 1989; Ghiselin 1997). The need to co-opt genetic material from other cells, organisms and even species is in apparent conflict with the simultaneous establishment of barriers against invasion by foreign nucleic acids. Similar conflicts involving insulation against and integration of evolutionary units not only apply to nuons, genes and genomes but also extend to organellar symbionts, cells, individuals and societies. Often, one gets the impression of a delicate tug of war on a ridge – if one team is too strong and wins, it pulls both into the abyss, what we would call extinction or an evolutionary dead-end street. Perhaps, co-operational strategies evolved, in part, to alleviate evolutionary conflicts.

Lamarckism raises its head

At the dawn of a major transition in our own lineage we witness major shifts in the evolutionary impact of Weismann's barrier, sex, and the Darwinian Threshold (Woese, 2002) without abandoning the basic Neo-Darwinian principles. In the RNA/RNP worlds (Figure 5) there was little separation between molecules acting as functional units and those acting as genetic material. Nevertheless, variations in RNA nuons and their immediate selection cannot be considered Lamarckism in the strict sense as the mutations were random and not geared towards a specific nuon in 'need' of modification. Only at the stage of memes are we able to observe Neo-Lamarckism: a learned behavior can be transmitted from individual to individual not only in a vertical but also in a horizontal transfer (Jablonka and Lamb, 1995; Jablonka *et al.*, 1998). Of course, this knowledge or meme could not direct the alteration of genetically based behavior – up to now.

There is no doubt that somatically rearranged genes could make their way to the germ line by viral transfer (Steele *et al.*, 1998). Such events, however, are few and far between and they are certainly, like retroposition, random events with no directionality. Almost timidly however, the same authors allude to future possibilities: 'we may very well discover and harness and oversee our genetic destiny' (Steele *et al.*, 1998). It cannot be denied that a novel form of Lamarckian evolution is on the rise and it might remain in effect for some time. The same species that obviously uses memes in a most efficient manner is not only able to conceptualize the genetic mechanisms of evolution, but has acquired the possibility to free itself from the dictatorship of the genes by engineering the germ line of virtually all living species including its own. It can, for example, strive to correct genetic disease, introduce desired traits, design genes from scratch, and introduce additional chromosomes (Silver, 1997). Weismann's Barrier and the Darwinian Threshold will recede, sex may disappear or experience transformation akin to the one from primordial sex to meiotic sex and Lamarckian mechanisms will persist and expand. Traits that have been acquired and back translated into the genetic code during a lifetime will be passed genetically to the next generation. The Central Dogma of Molecular Biology that sought to extend Weismann's Barrier to the molecular level will receive yet another, perhaps its most serious, blow. Is this the first time in 3.5 billion years that something akin to Lamarckian evolution arose or has it been tried before and proven to be inferior (e.g., too costly) compared to Darwinian selection (Hayes, 1999)?

Peaceful coexistence of Darwinism and Lamarckism?

The Darwinian principles of amplification, variation and selection appear to have permeated all major transitions in all forms of life like a monolith (Figure 5). Despite the rise of Lamarckism they will remain the cornerstones of evolution in the distant future, if our particular lineage has one.

Acknowledgements

Thanks to Michael Famuluk, Michael Ghiselin, Maryuan Long, Ernst Mayer, Günter Theißen, Leigh Van Valen and Carl Woese for references or discussions. I would like to dedicate this paper to Wally Gilbert on occasion of his seventieth birthday. The author is supported by the German Human Genome Project through the BMBF (#01KW9966).

References

Adami, C., Ofria, C. and Collier, T.C. (2000) Evolution of biological complexity. *Proc. Natl. Acad. Sci. USA*, **97**, 4463–4468.

Bailey, J.A., Carrel, L., Chakravarti, A. and Eichler, E.E. (2000) Molecular evidence for a relationship between LINE-1 elements and X chromosome inactivation: the Lyon repeat hypothesis. *Proc. Natl. Acad. Sci. USA*, **97**, 6634–6639.

Baltimore, D. (1970) Viral RNA-dependent DNA polymerase. *Nature*, **226**, 1209–1211.

Bartel, D.P. and Unrau, P.J. (1999) Constructing an RNA world. *Trends Cell Biol.*, **9**, M9-M13.

Blackmore, S. (1999) *The Meme Machine*. Oxford University Press, Oxford, UK.

Bonner, J.T. (1980) *The Evolution of Culture in Animals*. Princeton University Press, Princeton, NJ.

Bowler, P.J. (1992) Lamarckism. In *Keywords in Evolutionary Biology* (Keller, E.F. and Lloyd, E.A.), Harvard University Press, Cambridge, MA, pp. 188–193.

Bridges, C. (1936a) The Bar 'gene': a duplicaton. *Science*, **83**, 210–211.

Bridges, C.B. (1936b) Genes and chromosomes. *Teaching Biol.*, 17–23.

Britten, R.J. (1996) DNA sequence insertion and evolutionary variation in gene regulation. *Proc. Natl. Acad. Sci. USA*, **93**, 9374–9377.

Britten, R.J. (1997) Mobile elements inserted in the distant past have taken on important functions. *Gene*, **205**, 177–182.

Brosius, J. (1991) Retroposons--seeds of evolution. *Science*, **251**, 753.

Brosius, J. (1999a) Genomes were forged by massive bombardments with retroelements and retrosequences. *Genetica*, **107**, 209–238.

Brosius, J. (1999b) Many G-protein-coupled receptors are encoded by retrogenes. *Trends Genet.*, **15**, 304–305.

Brosius, J. (1999c) RNAs from all categories generate retrosequences that may be exapted as novel genes or regulatory elements. *Gene*, **238**, 115–134.

Brosius, J. (1999d) Transmutation of tRNA over time. *Nat. Genet.*, **22**, 8–9.

Brosius, J. (2001) tRNAs in the spotlight during protein biosynthesis. *Trends Biochem. Sci.*, **26**, 653–656.

Brosius, J. (2003a) The contribution of RNAs and retroposition to evolutionary novelties. *Genetica*, in press.

Brosius, J. (2003b) Echoes from the past – are we still in an RNP world? In preparation.

Brosius, J. and Gould, S.J. (1992) On 'genomenclature': a comprehensive (and respectful) taxonomy for pseudogenes and other 'junk DNA'. *Proc. Natl. Acad. Sci. USA*, **89**, 10706–10710.

Brosius, J. and Gould, S.J. (1993) Molecular constructivity. *Nature*, **365**, 102.

Brosius, J. and Tiedge, H. (1995a) Neural BC1 RNA: Dendritic localization and transport. In Lipshitz, H.D. (ed.) *Localized RNAs*. R.G. Landes, Austin, TX, pp. 289–330.

Brosius, J. and Tiedge, H. (1995b) Reverse transcriptase: mediator of genomic plasticity. *Virus Genes*, **11**, 163–179.

Brosius, J. and Tiedge, H. (2001) Dendritic BC1 RNA: Intracellular transport and activity-dependent modulation. In Richter, D. (ed.) *Cell polarity and subcellular RNA localization*. Springer Verlag, Berlin, pp. 129–138.

Cairns-Smith, A.G. (1982) *Genetic Takeover and the Mineral Origins of Life*. Cambridge University Press, Cambridge, UK.

Cairns-Smith, A.G. and Davis, C.J. (1977) In Duncan, R. and Weston-Smith, M. (eds.), *Encyclopaedia of Ignorance*. Pergamon Press, p. 397–403.

Cavaillé, J., Buiting, K., Kiefmann, M., Lalande, M., Brannan, C.I., Horsthemke, B., Bachellerie, J.P., Brosius, J. and Huttenhofer, A. (2000) Identification of brain-specific and imprinted small nucleolar RNA genes exhibiting an unusual genomic organization. *Proc. Natl. Acad. Sci. USA*, **97**, 14311–14316.

Crick, F.H.C. (1958) On protein synthesis. *Symp. Soc. Exp. Biol.*, **12**, 138–183.

Crick, F.H.C. (1968) The origin of the genetic code. *J Mol Biol*, **38**, 367–379.

Darwin, C. (1872) *The origin of species by means of natural selection or the preservation of favoured races in the struggle for life*. John Murray, London.

Dawkins, R. (1976) *The selfish gene*. Oxford University Press, Oxford, UK.

Dawkins, R. (1982) *The extended phenotype*. Freeman, San Francisco.

Dawkins, R. (1992) Progress. In Keller, E.F. and Lloyd, E.A. (eds.), *Keywords in evolutionary biology*. Harvard University Press, Cambridge, MA, pp. 263–272.

DeChiara, T.M. and Brosius, J. (1987) Neural BC1 RNA: cDNA clones reveal nonrepetitive sequence content. *Proc. Natl. Acad. Sci. USA*, **84**, 2624–2628.

Eddy, S.R. (1999) Noncoding RNA genes. *Curr. Opin. Genet. Dev.*, **9**, 695–699.

Eddy, S.R. (2001) Non-coding RNA genes and the modern RNA world. *Nat. Rev. Genet.*, **2**, 919–929.

Ellington, A.D. and Szostak, J.W. (1990) In vitro selection of RNA molecules that bind specific ligands. *Nature*, **346**, 818–822.

Erdmann, V.A., Barciszewska, M.Z., Hochberg, A., de Groot, N. and Barciszewski, J. (2001a) Regulatory RNAs. *Cell. Mol. Life Sci.*, **58**, 960–977.

Erdmann, V.A., Barciszewska, M.Z., Szymanski, M., Hochberg, A., de Groot, N. and Barciszewski, J. (2001b) The non-coding RNAs as riboregulators. *Nucleic Acids Res.*, **29**, 189–193.

Famulok, M. and Jenne, A. (1999) Catalysis based on nucleic acid structures. *Top. Curr. Chem.*, **202**, 101–131.

Ferris, J. (1994) Origins of life - chemical replication. *Nature*, **369**, 184–185.

Filipowicz, W. (2000) Imprinted expression of small nucleolar RNAs in brain: time for RNomics. *Proc. Natl. Acad. Sci. USA*, **97**, 14035–14037.

Fry, I. (2000) *The Emergence of Life on Earth: A Historical Overview*. Rutgers University Press, New Brunswick, NJ.

Gesteland, R.F., Cech, T.R. and Atkins, J.F. (1999) *The RNA World*. Cold Spring Harbor Laboratory Press, Cold Spring Harbor, NY.

Ghiselin, M.T. (1974) A radical solution to the species problem. *Systematic Zoology*, **23**, 536–544.

Ghiselin, M.T. (1989) *Intellectual Compromise. The Bottom Line*. Paragon House, New York.

Ghiselin, M.T. (1994) The imaginary Lamarck: a look at bogus 'history' in schoolbooks. *The Textbook Letter*. Sept-Oct.

Ghiselin, M.T. (1997) *Metaphisics and the Origin of Species*. State University of New York Press, Albany, NY. [ISBN 0-7914-3467-2].

Gilbert, W. (1986) The RNA world. *Nature*, **319**, 618.

Giraud, T., Pedersen, J.S. and Keller, L. (2002) Evolution of supercolonies: The Argentine ants of southern Europe. *Proc. Natl. Acad. Sci. USA*, **99**, 6075–6079.

Gould, S.J. and Vrba, E.S. (1982) Exaptation – a missing term in the science of form. *Paleobiology*, **8**, 4–15.

Gould, S.J. (1996) *Full House. The Spread of Excellence from Plato to Darwin*. Harmony Books, New York.

Gould, S.J. (2002) *The Structure of Evolutionary Theory*. The Belknap Press of Harvard University Press, Cambridge, MA.

Guerrier-Takada, C., Gardiner, K., Marsh, T., Pace, N. and Altman, S. (1983) The RNA moiety of ribonuclease P is the catalytic subunit of the enzyme. *Cell*, **35**, 849–857.

Haruna, I. and Spiegelman, S. (1965) Autocatalytic synthesis of a viral RNA in vitro. *Science*, **150**, 884–886.

Hayes, B. (1999) Experimental Lamarckism. *Am. Sci.*, **87**, 494–498.

Herbert, A. and Rich, A. (1999) RNA processing and the evolution of eukaryotes. *Nat. Genet.*, **21**, 265–269.

Hüttenhofer, A. and Brosius, J. (2002) Experimental RNomics. In *Functional Genomics* (Eds. Galperin, M. and Koonin, E.V.). Horizon Scientific Press, New York.

Hüttenhofer, A., Kiefmann, M., Meier-Ewert, S., O'Brien, J., Lehrach, H., Bachellerie, J.-P. and Brosius, J. (2001) RNomics: an experimental approach that identifies 201 candidates for novel, small, non-messenger RNAs in mouse. *EMBO J.*, **20**, 2943–2953.

Jablonka, E. and Lamb, M.J. (1995) *Epigenetic Inheritance and Evolution in the Lamarckian Dimension*. Oxford University Press, Oxford, UK.

Jablonka, E., Lamb, M.J. and Avital, E. (1998) Lamarckian mechanisms in Darwinian evolution. *Trends Ecol. Evol.*, **13**, 206–210.

James, K.D. and Ellington, A.D. (1995) A search for missing links between self-replicating nucleic acids and the RNA world. *Origins Life Evol Biosphere*, **25**, 515–530.

Jantsch, E. (1979) *Die Selbstorganisation des Universums: Vom Urknall zum Menschlichen Geist*. Carl Hanser Verlag, München, Germany.

Jeffares, D.C., Poole, A.M. and Penny, D. (1998) Relics from the RNA world. *J. Mol. Evol.*, **46**, 18–36.

Joyce, G.F. (2002) The antiquity of RNA-based evolution. Nature **418**, 241–221.

Jurka, J. (1998) Repeats in genomic DNA: mining and meaning. *Curr. Opin. Struct. Biol.*, **8**, 333–337.

Kauffman, S.A. (1993) *The Origins of Order. Self-Organization and Selection in Evolution*. Oxford University Press, Oxford, UK.

Kidwell, M.G. and Lisch, D. (1997) Transposable elements as sources of variation in animals and plants. *Proc. Natl. Acad. Sci. USA*, **94**, 7704–7711.

Kiedrowski, G.v. (1986) A self-replicating hexadeoxynucleotide. *Angew. Chem. Int. Ed. Engl.*, **25**, 932–935.

Kruger, K., Grabowski, P.J., Zaug, A.J., Sands, J., Gottschling, D.E. and Cech, T.R. (1982) Self-splicing RNA: autoexcision and autocyclization of the ribosomal RNA intervening sequence of Tetrahymena. *Cell*, **31**, 147–157.

Lagos-Quintana, M., Rauhut, R., Lendeckel, W. and Tuschl, T. (2001) Identification of novel genes coding for small expressed RNAs. *Science*, **294**, 853–858.

Lander, E.S., Linton, L.M., Birren, B., Nusbaum, C., Zody, M.C., Baldwin, J., Devon, K., Dewar, K., Doyle, M., FitzHugh, W., Funke, R., Gage, D., Harris, K., Heaford, A., Howland, J., Kann, L., Lehoczky, J., LeVine, R., McEwan, P., McKernan, K., Meldrim, J., Mesirov, J.P., Miranda, C., Morris, W., Naylor, J., Raymond, C., Rosetti, M., Santos, R., Sheridan, A., Sougnez, C., Stange-Thomann, N., Stojanovic, N., Subramanian, A., Wyman, D., Rogers, J., Sulston, J., Ainscough, R., Beck, S., Bentley, D., Burton, J., Clee, C., Carter, N., Coulson, A., Deadman, R., Deloukas, P., Dunham, A., Dunham, I., Durbin, R., French, L., Grafham, D., Gregory, S., Hubbard, T., Humphray, S., Hunt, A., Jones, M., Lloyd, C., McMurray, A., Matthews, L., Mercer, S., Milne, S., Mullikin, J.C., Mungall, A., Plumb, R., Ross, M., Shownkeen, R., Sims, S., Waterston, R.H., Wilson, R.K., Hillier, L.W., McPherson, J.D., Marra, M.A., Mardis, E.R., Fulton, L.A., Chinwalla, A.T., Pepin, K.H., Gish, W.R., Chissoe, S.L., Wendl, M.C., Delehaunty, K.D., Miner, T.L., Delehaunty, A., Kramer, J.B., Cook, L.L., Fulton, R.S., Johnson, D.L., Minx, P.J., Clifton, S.W., Hawkins, T., Branscomb, E., Predki, P., Richardson, P., Wenning, S., Slezak, T., Doggett, N., Cheng, J.F., Olsen, A., Lucas, S., Elkin, C., Uberbacher, E., Frazier, M., et al. (2001) Initial sequencing and analysis of the human genome. *Nature*, **409**, 860–921.

Lau, N.C., Lim, L.P., Weinstein, E.G. and Bartel, D.P. (2001) An abundant class of tiny RNAs with probable regulatory roles in *Caenorhabditis elegans*. *Science*, **294**, 858–862.

Lee, R.C. and Ambros, V. (2001) An extensive class of small RNAs in Caenorhabditis elegans. *Science*, **294**, 862–864.

Lewis, E.B. (1951) Pseudoallelism and gene evolution. *Cold Spring Harbor Symp. Quant. Biol.*, **16**, 159–174.

Li, T. and Nicolaou, K.C. (1994) Chemical self-replication of palindromic duplex DNA. *Nature*, **369**, 218–221.

Lyon, M.F. (1998) X-chromosome inactivation: a repeat hypothesis. *Cytogenet. Cell Genet.*, **80**, 133–137.

Lyon, M.F. (2000) LINE–1 elements and X chromosome inactivation: a function for 'junk' DNA? *Proc. Natl. Acad. Sci. USA*, **97**, 6248–6249.

Maizels, N. and Weiner, A.M. (1987) Peptide-specific ribosomes, genomic tags, and the origin of the genetic code. *Cold Spring Harbor Symp. Quant. Biol.*, **52**, 743–749.

Makalowski, W. (2000) Genomic scrap yard: how genomes utilize all that junk. *Gene*, **259**, 61–67.

Margulis, L. (1970) *Origin of Eukaryotic Cells*. Yale University Press, New Haven, CT.

Martignetti, J.A. and Brosius, J. (1993a) BC200 RNA: a neural RNA polymerase III product encoded by a monomeric Alu element. *Proc. Natl. Acad. Sci. USA*, **90**, 11563–11567.

Martignetti, J.A. and Brosius, J. (1993b) Neural BC1 RNA as an evolutionary marker: guinea pig remains a rodent. *Proc. Natl. Acad. Sci. USA*, **90**, 9698–9702.

Mattick, J.S. (2001) Non-coding RNAs: the architects of eukaryotic complexity. *EMBO Rep.*, **2**, 986–991.

Maynard Smith, J. and Szathmáry, E. (1995) *The Major Transitions in Evolution*. Oxford University Press, Oxford, UK.

Mayr, E. (1960) The emergence of evolutionary novelties. In Tax, S. (ed.) *Evolution after Darwin*, Vol. 1. The University of Chicago, Chicago, IL, pp. 349–380.

Mayr, E. (2001) *What Evolution Is*. Basic Books, New York.

McDonald, J.F. (1993) Evolution and consequences of transposable elements. *Curr. Opin. Genet. Dev.*, **3**, 855–864.

McDonald, J.F. (1995) Transposable elements: possible catalysts of organismic evolution. *Trends Ecol. Evol.*, **10**, 123–126.

Mendel, G. (1866) Versuche über Pflanzenhybriden. *Verhandl. naturforsch. Ver. Brünn*, **4**, 3–47.

Mendel, G. (1870) Über einige aus künstlicher Befruchtung gewonnene Hieraciumbastarde. *Verhandl. naturforsch. Ver. Brünn*, **8**, 26–31.

Muller, H.J., Prokofyeva-Belgovskaya, A.A. and Kossikov, K.V. (1936) Unequal crossing-over in the Bar mutant as a result of duplication of a minute chromosome section. *C.R. (Doklady) Acad. Sci. URSS*, **1**, 87–88.

multiple authors. (1987) *Evolution of Catalytic Function*, Cold Spring Harbor Laboratory Press, Cold Spring Harbor, NY.

Nei, M. (1969) Gene duplication and nucleotide substitution in evolution. *Nature*, **221**, 40–42.

Ohno, S. (1970) *Evolution by Gene Duplication*. Springer, New York, NY.

Orgel, L.E. (1968) Evolution of the genetic apparatus. *J. Mol. Biol.*, **38**, 381–393.

Orgel, L.E. (1998) The origin of life – a review of facts and speculations. *Trends Biochem. Sci.*, **23**, 491–495.

Pollard, J.W. (1984) Is Weissmann's barrier absolute?, pp. 291–314, In *Beyond Neo-Darwinism. An introduction to the new evolutionary paradigm*, edited by M.-W. Ho and P.T. Saunders. Academic Press, London.

Poole, A.M., Jeffares, D.C. and Penny, D. (1998) The path from the RNA world. *J. Mol. Evol.*, **46**, 1–17.

Robertson, D.L. and Joyce, G.F. (1990) Selection *in vitro* of an RNA enzyme that specifically cleaves singlestranded DNA. *Nature* **344**, 467–468.

Shapiro, J.A. (1992) Natural genetic engineering in evolution. Genetica **86**: 99–111.

Silver, L.M. (1997) *Remaking Eden. Cloning and Beyond in a Brave New World*. Avon, New York, NY.

Steele, E.J., Lindley, R.A. and Blanden, R.V. (1998) *Lamarck's Signature: How retrogenes are Changing Darwin's Natural Selection Paradigm*. Perseus, Cambridge, MA.

Stephens, S.G. (1951) Possible significance of duplication in evolution. *Adv. Genet.*, **4**, 247–265.

Storz, U. (2002) Counting all genes: how many other RNAs exist and what do they do? *Science*, **296**, 1260–1263.

Sturtevant, A.H. (1925) The effects of unequal crossing over at the *Bar* locus in *Drosophila*. *Genetics*, **10**, 117–147.

Sutcliffe, J.G., Milner, R.J., Bloom, F.E. and Lerner, R.A. (1982) Common 82-nucleotide sequence unique to brain RNA. *Proc. Natl. Acad. Sci. USA*, **79**, 4942–4946.

Sverdlov, E.D. (2000) Retroviruses and primate evolution. *BioEssays*, **22**, 161–171.

Szathmáry, E. (1990) Towards the evolution of ribozymes. *Nature*, **344**, 115.

Szathmáry, E. (1997) Origins of life. The first two billion years. *Nature*, **387**, 662–663.

Szathmáry, E. (1999a) The first replicators. In *Levels of Selection* (Ed. Keller, L.). Princeton University Press, Princeton, NJ, pp. 31–52.

Szathmáry, E. (1999b) The origin of the genetic code: amino acids as cofactors in an RNA world. *Trends Genet.*, **15**, 223–229.

Szathmáry, E. and Smith, J.M. (1995) The major evolutionary transitions. *Nature*, **374**, 227–232.

Szostak, J.W., Bartel, D.P. and Luisi, P.L. (2001) Synthesizing life. *Nature*, **409**, 387–390.

Temin, H.M. (1970) Viral RNA-depentent DNA polymerase. *Nature*, **226**, 1211–1213.

Tiedge, H., Chen, W. and Brosius, J. (1993) Primary structure, neural-specific expression, and dendritic location of human BC200 RNA. *J. Neurosci.*, **13**, 2382–2390.

Tiedge, H., Fremeau, R.T., Weinstock, P.H., Arancio, O. and Brosius, J. (1991) Dendritic location of neural BC1 RNA. *Proc. Natl. Acad. Sci. USA*, **88**, 2093–2097.

Tuerk, C. and Gold, L. (1990) Systematic evolution of ligands by exponential enrichment: RNA ligands to bacteriophage T4 DNA polymerase. *Science*, **249**, 505–510.

Venter, J.C., Adams, M.D., Myers, E.W., Li, P.W., Mural, R.J., Sutton, G.G., Smith, H.O., Yandell, M., Evans, C.A., Holt, R.A., Gocayne, J.D., Amanatides, P., Ballew, R.M., Huson, D.H., Wortman, J.R., Zhang, Q., Kodira, C.D., Zheng, X.H., Chen, L., Skupski, M., Subramanian, G., Thomas, P.D., Zhang, J., Gabor Miklos, G.L., Nelson, C., Broder, S., Clark, A.G., Nadeau, J., McKusick, V.A., Zinder, N., Levine, A.J., Roberts, R.J., Simon, M., Slayman, C., Hunkapiller, M., Bolanos, R., Delcher, A., Dew, I., Fasulo, D., Flanigan, M., Florea, L., Halpern, A., Hannenhalli, S., Kravitz, S., Levy, S., Mobarry, C., Reinert, K., Remington, K., Abu-Threideh, J., Beasley, E., Biddick, K., Bonazzi, V., Brandon, R., Cargill, M., Chandramouliswaran, I., Charlab, R., Chaturvedi, K., Deng, Z., Di Francesco, V., Dunn, P.,Eilbeck, K., Evangelista, C., Gabrielian, A.E., Gan, W., Ge, W., Gong, F., Gu, Z., Guan, P., Heiman, T.J., Higgins, M.E., Ji, R.R., Ke, Z., Ketchum, K.A., Lai, Z., Lei, Y., Li, Z., Li, J., Liang, Y., Lin, X., Lu, F., Merkulov, G.V., Milshina, N., Moore, H.M., Naik, A.K., Narayan, V.A., Neelam, B., Nusskern, D., Rusch, D.B., Salzberg, S., Shao, W., Shue, B., Sun, J., Wang, Z., Wang, A., Wang, X., Wang, J., Wei,

M., Wides, R., Xiao, C., Yan, C., et al. (2001) The sequence of the human genome. *Science*, **291**, 1304–1351.

Wächtershauser, G. (1992) Groundworks for an evolutionary biochemistry: the iron-sulphur world. *Prog. Biophys. Mol. Biol.*, **58**, 85–201.

Walbot, V. (1996) Sources and consequences of phenotypic and genotypic plasticity in flowering plants. *Trends Plant Sci.*, **1**, 27–32.

Wang, W., Brunet, F.G., Nevo, E. and Long, M. (2002) Origin of sphinx, a young chimeric RNA gene in Drosophilamelanogaster. *Proc. Natl. Acad. Sci. USA*, **99**, 4448–4453.

Watson, J.B. and Sutcliffe, J.G. (1987) Primate brain-specific cytoplasmic transcript of the Alu repeat family. *Mol. Cell. Biol.*, **7**, 3324–3327.

Weiner, A.M., Deininger, P.L. and Efstratiadis, A. (1986) Nonviral retroposons: genes, pseudogenes, and transposable elements generated by the reverse flow of genetic information. *Annu. Rev. Biochem.*, **55**, 631–661.

Weismann, A. (1892) *Das Keimplasma: Eine Theorie der Vererbung*. Gustav Fischer, Jena, Germany.

Weismann, A. (1893) *The Germ-plasm: a Theory of Heredity*. Walter Scott, London, UK.

Weismann, A. (1902) *Vorträge über Descendenztheorie gehalten an der Universität zu Freiburg im Breisgau*. Gustav Fischer, Jena, Germany.

Westheimer, F.H. (1986) Polyribonucleic acids as enzymes. *Nature*, **319**, 534–535.

Williams, G.C. (1966) *Adaptation and Natural Selection*. Princeton University Press, Princeton, NJ.

Wills, C. and Bada, J. (2000) *The Spark of Life: Darwin and the Primeval Soup*. Perseus Publishing, Cambridge, MA.

Wilson, D.S. and Szostak, J.W. (1999) In vitro selection of functional nucleic acids. *Annu. Rev. Biochem.*, **68**, 611–647.

Woese, C.R. (1967) *The Genetic Code: The Molecular Basis for Genetic Expression*. Harper and Row, New York, NY.

Woese, C.R. (1980) Just So Stories and Rube Goldberg machines: Speculations on the origin of the protein synthetic machinery. In *Ribosomes: Structure, Function, and Genetics* (Eds. Chambliss, G., Craven, G.R., Davies, J., Davis, K., Kahan, L. and Nomura, M.). University Park Press, Baltimore, MD, pp. 357–373.

Woese, C.R. (2001) Translation: in retrospect and prospect. *RNA*, **7**, 1055–1067.

Woese, C.R. (2002) On the evolution of cells. *Proc. Natl. Acad. Sci. USA*, **99**, 8742–8747.

Glossary

RNA world, a very early stage of cellular life, in which RNA biopolymers encoded genetic information and had catalytic and structural functions in the cell. Consequently there was no separation of phenotype and genotype.

RNP world, RNA/protein world, a subsequent stage of cellular life, in which genetically encoded polypeptides joined RNA as catalytically and structurally functional biopolymers.

DNA world, a further stage of cellular life, in which RNA as genetic material has gradually been exchanged and superseded by DNA, presumably by enzymatic conversion of RNA into DNA.

Retroposition, the conversion of RNA into DNA by reverse transcriptase. The resulting complementary DNA (cDNA) is more or less randomly integrated into the genome. Apparently, this proceeds rather intensely and unabated in many lineages to this day.

Exaptation, a nuon, gene, organ, or other fearures that now enhance fitness but were not built by natural selection for their current role (Gould and Vrba, 1982). Bird's feathers are exaptations as initially they served for temperature regulation and some were, at later stages, co-opted for flight. Another example is sex. Its major evolutionary advantage has been exchange of genetic material. It has also been exapted in many species for bonding between individuals, thus providing additional advantage to both the offspring and in some instances entire social groups. Charles Darwin already described the process of exaptation in his 'Origin of Species', page 398 (Darwin, 1872):

'Again, an organ may become rudimentary for its proper purpose, and be used for a distinct one: in certain fishes the swim-bladder seems to be rudimentary for its proper function of giving buoyancy, but has become converted into a nascent breathing organ or lung. Many similar instances could be given.'

Gene, originally a unit of heredity. The term underwent numerous modifications as knowledge in Molecular Biology accumulated. For example, its regulatory regions are sometimes included, sometimes it refers only to the DNA that is transcribed into RNA. Today, the gene concept is fuzzy.

Nuon, any discrete segment of nucleic acid, RNA or DNA (Brosius and Gould, 1992, 1993). A nuon can be defined by sequence similarity (repetitive elements) or function (exon, intron, promoter, enhancer, splice site), by its biogenesis (see retronuon), evolutionary fate or effect (e.g. naptonuon to indicate current non-aptation or xaptonuon for an exapted nuon). An advantage of nuon is that it not only denotes a purely genotypic nucleic acid but, also a biopolymer that unites both phenotype and genotype. This is not possible when using the term gene and hence very important for discussing the RNA and RNP worlds. For further information see: http://exppc01.uni-muenster.de/expath/retronuons.htm.

Retronuon, any nuon generated by reverse transcription of any RNA, viral RNA, messenger RNA or untranslated RNA yielding short interspersed repetitive elements (SINEs), long interspersed repetitive elements (LINEs) etc.

snmRNA, small (~50–500 nucleotides in length) non-messenger RNA. Some snmRNA species can be highly efficient templates for reverse transcriptase and are the founders for SINEs.

Gene Duplication, Susumu Ohno described in his book (Ohno, 1970) the importance of gene duplication for the evolution of genes, genomes and organisms. Although, by some, only the gene is considered the unit of selection (Williams, 1966; Dawkins, 1982), the concept might be expanded to signify 'unit' duplication if applied to nuons, cells, organs, organisms, social groups and states. As Ernst Mayr pointed out in 1960 (Mayr, 1960), Darwin already recognized the principle of duplication on page 147 (Darwin, 1872):

'Again, two distinct organs, or the same organ under two very different forms, may simultaneously perform in the same individual the same function, and this is an extremely important means of transition...'.

Amplification of units constitutes, together with random variation and selection, the basis of Neo-Darwinian evolution.

Darwinism, species have evolved from simpler ancestral types by the process of natural selection acting on the variability found within a population.

Neo-Darwinism, Darwinism merged with the Mendelian laws of genetics; makes use of the modern knowledge of chromosomes and genes to explain the source of the genetic variation upon which selection works. Darwin's (and Ohno's) concepts of unit duplication might be included to yield the evolutionary chain of amplification, variation (including recombination) and selection.

Lamarck, less well known for his theory of speciation through gradual change that had been published five decades before Darwin's, than for 'his' error that evolution took place through inheritance of modifications caused by the environment, and the effects of use and disuse of organs. In the strict sense such effects act directly and non-randomly on the gene(s) responsible for the respective phenotypes. For facts and fiction concerning Lamarck's ideas and contributions, see Ghiselin (1994).

Neo-Lamarckism, as understood today is more broadly defined as 'the evolutionary mechanism of the inheritance of acquired characteristics' (Bowler, 1992). Examples of Neo-Lamarckism emerge, wherever Weismann's Barrier recedes if one allows for its definition as the inheritance of acquired characteristics but barring directed modification of genetic material (see also Jablonka and Lamb, 1995).

Darwinian Threshold, a hitherto unrecognized phase transition in the evolutionary process (Woese, 2002). It corresponds to the emergence of a new level of order in terms of cell organization. Cellular machineries became more and more complex and interconnected. As a consequence, the interchangeability of genetic material became gradually less feasible and thus curbed the acquisition of novelties via horizontal gene transfer. Consequently short lived entities were replaced by more permanent ones - species. Hence, 'the Darwinian Threshold truly represents the origin of speciation as we know it' (Woese, 2002). Woese's concept might be misunderstood as an assault on the foundations of Darwinism. In contrast, it is solidly anchored on the three Neo-Darwinian pillars of amplification, modification and selection (see above). With his recent proposal, Woese has lifted the origin of species, always guided by Neo-Darwinian principles, out of the muddy layer of communal evolution, across the Darwinian Threshold onto a level where separatism allowed for the establishment of relatively stable 'species'. In molecular terms, Woese sees this transition occurring when templated translation produced ever more sophisticated protein molecules including the ones that are involved in the translational machinery itself.

Weismann's Barrier, recognized that the germ plasm remains unaffected by any changes affecting somatic cells during the life time of an organism Weismann's germ plasm theory (Weismann, 1892, 1893, 1902) based on Mendel's laws (Mendel, 1866, 1870) debunked Lamarckism. At the same time Weismann cemented the Neo-Darwinian doctrine. Yet, Weismann's Barrier applies only to a segment of all known forms of life, namely those that feature a germ line and do not permit genetic 'contributions' from somatic cells. Weismann's Barrier does not apply to unicellular organisms or multicellular organism that do not maintain a germ line or do so only temporarily by alternating between sexual and clonal reproduction, such as many plants (Polland, 1984; Walbot, 1996). An acquired mutation in a somatic cell that is beneficial for the respective plant may be passed on to future generations by somatic reproduction and subsequent switching to sexual reproduction.

Central Dogma of Molecular Biology, states that DNA makes RNA makes protein, in that order (Crick, 1958). This was amended twice; when S. Spiegelman showed that RNA could be replicated from RNA templates (Haruna and Spiegelman, 1965) and H. Temin and D. Baltimore discovered reverse transcriptase (Baltimore, 1970; Temin, 1970). If its often considered a molecular application of Weismann's Barrier.

A. Meyer, Y. Van de Peer (eds.), Genome Evolution, 19-25.

Major transitions in evolution by genome fusions: from prokaryotes to eukaryotes, metazoans, bilaterians and vertebrates

Jürg Spring
Institute of Zoology, University of Basel, Biocenter/Pharmacenter, Klingelbergstrasse 50, CH–4056 Basel, Switzerland
e-mail: j.spring@unibas.ch

Received 21.01.2002; accepted in final form 29.08.2002

Key words: allopolyploidy, gene duplication, genome evolution, hybridization

Abstract

The major transitions in human evolution from prokaryotes to eukaryotes, from protozoans to metazoans, from the first animals to bilaterians and finally from a primitive chordate to vertebrates were all accompanied by increases in genome complexity. Rare fusion of divergent genomes rather than continuous single gene duplications could explain these jumps in evolution. The origin of eukaryotes was proposed to be due to a symbiosis of Archaea and Bacteria. Symbiosis is clearly seen as the source for mitochondria. A fundamental difference of higher eukaryotes is the cycle from haploidy to diploidy, a well-regulated genome duplication. Of course, self-fertilization exists, but the potential of sex increases with the difference of the haploid stages, such as the sperm and the egg. What should be the advantage of having two identical copies of a gene? Still, genes duplicate all the time and even genomes duplicate rather often. In plants, polyploidy is well recognized, but seems to be abundant in fungi and even in animals, too. However, hybridization, rather than autopolyploidy, seems to be the potential mechanism for creating something new. The problem with chimaeric, symbiotic or reticulate evolution events is that they blur phylogenetic lineages. Unrecognized paralogous genes or random loss of one of the paralogs in different lineages can lead to false conclusions. Horizontal genome transfer, genome fusion or hybridization might be only truly innovative combined with rare geological transitions such as change to an oxygen atmosphere, snowball Earth events or the Cambrian explosion, but correlates well with the major transitions in evolution.

Introduction

Gene duplication is now well recognized as a source for evolutionary innovations. Since Ohno's book 'Evolution by Gene Duplication' (Ohno, 1970), many aspects of gene or genome duplication have been discussed. But even with the availability of complete genome sequences from all major lineages, it is not clear whether duplications of individual genes, chromosomal segments or entire genomes were more important. Gene duplications are abundant and new duplicates seem to appear and disappear at very high rates (Lynch and Conery, 2000), most likely by tandem duplication. Genome duplication would be a much more rare event, but could have a wider impact. Two rounds of genome duplication were discussed for vertebrate evolution (Lundin, 1993; Holland *et al.*, 1994; Spring, 1997; Pébusque *et al.*, 1998; Gibson and Spring, 2000), but except for the Hox clusters, a peculiarly conserved gene array of developmentally important transcription factors, this is not usually recognized.

One genome duplication is a likely source for duplicate genes in yeast (Wolfe and Shields, 1997), but comparing a wide variety of yeast species, consecutive single gene duplications appear to be sufficient to explain genome complexity (Llorente *et al.*, 2000). The complete genome of *Arabidopsis thaliana* clearly provides evidence for large-scale duplications (The Arabidopsis genome initiative, 2000), and in plants polyploidy is well accepted as a driving force for evolution. But even for plants, the fine-scale struc-

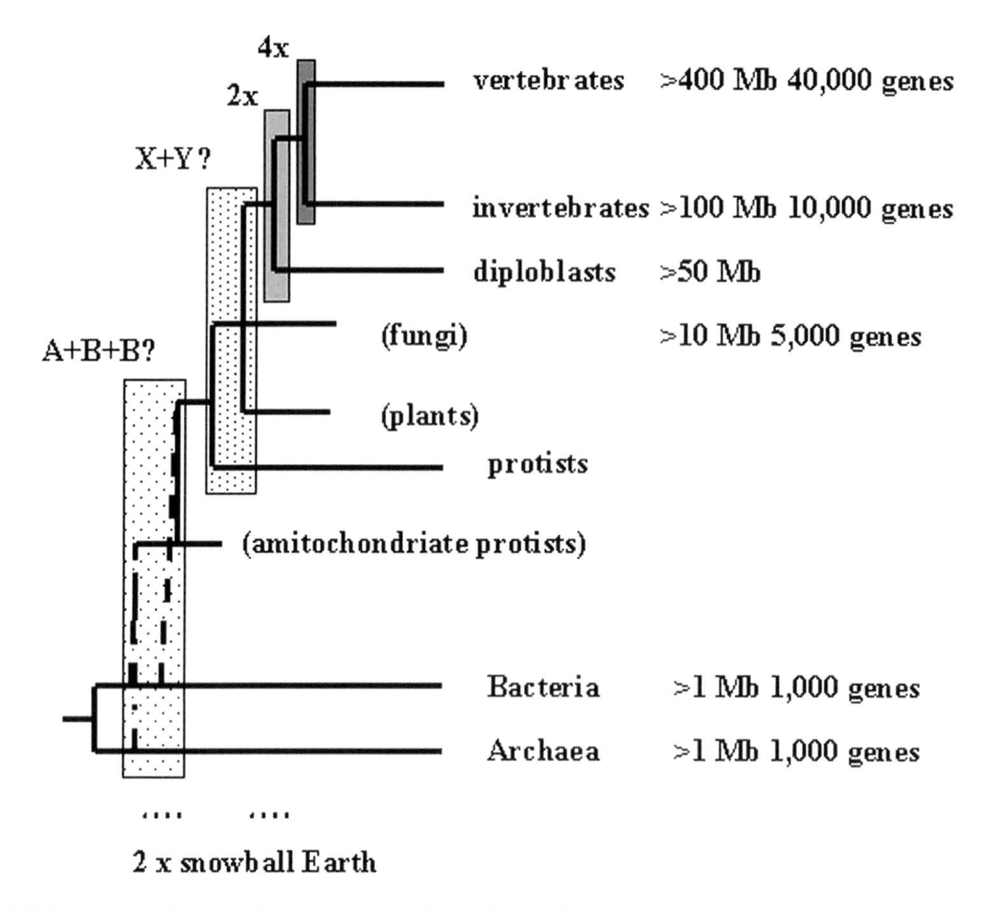

Figure 1. Multiple genome fusion events in vertebrate evolution. Major transitions in evolution, geological events and increase in genome complexity correlate, and suggest a few abrupt changes intermixed with continuous evolution.

ture of genomes seems to be more dynamic than assumed (Bancroft, 2001). Compared to yeast and plants, genome duplication is less obvious in animal evolution. Two vertebrate duplications could be complicated by additional duplications at the protozoan-metazoan and diploblast-triploblast transitions (Lundin, 1999). The presence of a ParaHox cluster next to the Hox cluster (Brooke *et al.*, 1998) might have just been the first indication of additional large scale duplications in animal evolution.

Gene duplication and subsequent divergence is of course the normal mechanism that has to be discussed for the continuous modulation of continuous lineages. But it is also well recognized that about four major transitions in the evolution from the first bacteria to complex animals occurred (Szathmary and Maynard Smith, 1995). These transitions, from prokaryotes to eukaryotes, from protozoans to metazoans, from diploblasts to triploblasts and from invertebrate chordates to vertebrates, were accompanied by increases in genome complexity, are not well documented in the fossil records and produced a lot of novelties in a short period of time, e.g. the so-called Cambrian explosion (Fig. 1). A symbiosis of Archaea and Bacteria as the origin of the first eukaryotes (Lake, 1988) would not follow the gradual duplicate and diverge pattern. Rather, divergence of separate lineages is followed by fusion, thus leading instantaneously to a potentially novel life form. While 'duplication and divergence' might be the normal mode of evolution for more stable environments, 'divergence and fusion' might be rarely successful except in extraordinary conditions, even in geological time. Symbiosis, chimaerism, hybridization, lateral genome transfer, reticulate evolution and other technical terms could all be combined as genome fusion. The presence of genome fusion in the history of a lineage has implications for its phylogenetic reconstruction, the timing of lineage separations and the nomenclature and functional assignments of paralogous genes.

From prokaryotes to eukaryotes

Archaea and Bacteria can be well distinguished, although horizontal gene transfer could confound individual gene trees (Eisen, 2000). A symbiosis of Archaea and Bacteria could have been at the beginning of eukaryotes (Lake, 1988; Horiike *et al.*, 2001). The origin of mitochondria (Margulis, 1970) and other organelles is clearly seen as an ancient endosymbiosis event. Endosymbiosis is actually not such a rare event, many protist lineages have species with additional prokaryotes or even other eukaryotes as endosymbionts. But widespread occurrence does not imply a high success rate in evolutionary time frames. The nucleus, the mitochondrion or the chloroplast might just be three out of numerous genome fusions that have left their traces, e.g. genes in the genome, while all other symbiosis events were only temporary. The big bang hypothesis necessary for the explanation of eukaryotic phylogeny (Philippe *et al.*, 2000) could reflect an unresolved number of genome fusion events.

From protozoans to metazoans

Even the best accepted phylogenies of single cell and multicellular organisms do not resolve the relation of the so-called protozoans and metazoans (Baldauf *et al.*, 2000; Philippe *et al.*, 2000). Multicellular animals usually group with fungi; only the little studied choanoflagellates might behave as expected from a single cell ancestor of metazoans. With genome sequences of other protozoans such as *Plasmodium falciparum* approaching, it will be important to confirm the absence of typical animal-specific genes evident from the available preliminary data. Where did collagens, tyrosine kinases or Hox clusters come from if they are not present in plants, fungi and protozoans, but are suddenly abundant in metazoans? Neither yeast, *Arabidopsis* nor *Plasmodium* has a gene complement that could be modified by a few single gene duplications and divergence to look like an animal genome. Genome fusion of a common ancestor of yeast and animals with a more specialized creature could have been responsible for a protozoan with an animal-specific gene set.

Multicellularity evolved several times in all lineages of life and is not a specialty of metazoans. Cell type-specific terminal differentiation into somatic neurons and muscle cells is a better description of mobile animals that can decide where they go. The innovations necessary for cell differentiation could have come from differently specialized ancestors by genome fusion. Even sponges have already two major collagen types, the fibrillar and the basement membrane collagens known from bilaterians. Also Hox and ParaHox clusters seem to be already present in cnidarians and other prebilaterians (Ferrier and Holland, 2001).

From prebilaterians to bilaterians

Sponges, placozoans, cnidarians and ctenophores are simpler animals than bilaterians and are probably best called prebilaterians as both diploblasts and Radiata are not accurate descriptions. Sponges might be even simpler and could be separated as parazoans from eumetazoans; but many animal-specific gene families are already present prominently in sponges and some of the simplicity of adult sponges could be secondary derivations from a more complex mobile ancestor due to the sedentary life-style. Such reductions are also possible for corals and sea anemones in cnidarians, and are obvious from comparisons with bilaterian groups such as bryozoans, echinoderms or ascidians. In all prebilaterian groups there are mobile larval stages which could be the connection with the first real multicellular animal. Initial studies in cnidarians suggest that probably all gene families are already present and are involved in processes comparable to those in bilaterians. Even mesoderm-like structures can be seen during medusa bud development of jellyfish and are correlated with the mesoderm specification factor Twist known from *Drosophila* to humans (Spring *et al.*, 2000).

The enigmatic placozoan *Trichoplax adhaerens* is the metazoan with the smallest known genome of only 40 Mb, about half the size of the smallest bilaterian genomes. Sponges and cnidarians are reported with small genomes of around 50 Mb but also with large genomes of the size of the human genome. A genome fusion of two prebilaterian could have let to the first bilaterian.

From invertebrate bilaterians to vertebrates

The model organisms *Drosophila* and *Caenorhabditis elegans* were long thought to be representatives of rather different animal groups. With the formation of the ecdysozoans, the moulting animals, our knowledge of animal diversity has changed (Aguinaldo *et al.*, 1997). Still, all bilaterians from flatworms to amphioxus seem to have a common core set of about 10,000 genes, called the 'core proteome' (Rubin *et al.*, 2000). Vertebrates have often 2, 3 or four paralogs of genes known from invertebrates on different chromosomes. In such comparisons it is important to distinguish the more closely related paralogs that could be due to genome duplications, the tetralogs, from all other duplicates. Additional copies could be due to tandem duplication, cis-paralogs, older genome duplications such as Hox and Parahox clusters or retrotransposition that can be distinguished in gene families that normally have conserved intron positions. Thus, protein or coding sequences alone are not sufficient to judge gene family evolution, but also chromosomal position and exon-intron structures should be considered.

With 'complete genomes' now available from *Drosophila*, *C. elegans* and human as well as yeast and *Arabidopsis* as useful outgroups, it should be now soon possible to decide whether whole genome duplications were involved in vertebrate evolution. Unfortunately, the first draft of the human genome is still poorly annotated and a simple one to four relationship is not visible but could also be obscured by massive gene loss and amplification of particular gene families (International Human Genome Sequencing Consortium, 2001; Venter *et al.*, 2001; Rubin, 2001). A further uncertainty stems from the lack of obvious homology between about half of the human and *Drosophila* and *C. elegans* genes. This could be due to fundamental differences; but the highly derived status of many *C. elegans* sequences and to a lesser extent also *Drosophila* sequences compared to other bilaterian invertebrates or even prebilaterians could suggest that we just do not see the homologies anymore. Typically, many secreted proteins such as growth factors are apparently vertebrate specific; alternatively, extracellular proteins or parts of proteins just change too fast to be recognized from *Drosophila* to humans. When vertebrate tetralogs are just barely recognizable with 25% of protein sequence identity, it is not surprising that *Drosophila* or *C. elegans* sequences are not recognized as homologs, especially if their exons are in addition wrongly assembled as indicated for about a third of the *C. elegans* genome derived protein predictions.

Another source of uncertainty is frequent gene loss. Collagens have very specific functions in animals and are already present in sponges as a fibrillar version and a basement membrane version, called type IV collagen in bilaterians. Genome wide comparison of human collagen genes suggests that the major fibrillar collagens, the minor fibrillar collagens, the interrupted fibrillar collagens and the basement membrane collagens duplicated along with Hox and ParaHox clusters or non-receptor tyrosine kinases and receptor tyrosine kinases (Fig. 2). Interestingly, collagen genes seem to be more resistant to gene loss than all other groups; even Hox genes can get lost. It is difficult to compare Hox clusters with 13 genes and ParaHox clusters with only 3 genes, but the association with major fibrillar and basement membrane collagens, respectively, suggests that two further classes of Hox-like genes were lost or are not recognized as cluster-like genes. But also collagen genes can get lost; *Drosophila* and *C. elegans* do not have real fibrillar collagen genes but retained highly conserved basement membrane collagens. This might be an ecdyosozoan specialty, other invertebrates such as sea urchins, annelids and even jellyfish or sponges still have fibrillar collagens. Nomenclature and phylogeny of tyrosine kinases are more complex than those of Hox or collagen genes. But receptor tyrosine kinases can be grouped in 16 ancient subfamilies and non-receptor tyrosine kinases in 8 subfamilies. Interestingly, the EGF receptor subfamily is most similar to the insulin receptor subfamily over the entire protein, but in the kinase domain is more similar to non-receptor tyrosine kinases such as Abl or Src. A further level of complexity can be illustrated with the EGF receptor like gene ERBB3 which is conserved from fish to humans as partner for heterodimerization with ERBB2, but is not an active kinase anymore (Kroiher *et al.*, 2001). It is not surprising then that the rate of accumulation of changes in an inactive kinase is different from its active paralogs. In duplicated gene arrays such as the Hox clusters and neighboring genes there are actually only a few cases with all four paralogs conserved from hundreds of genes with 1, 2 or 3 paralogs. The fact that these few fourfold gene families do not always show the same and a symmetric phylogenetic relationship could indicate that whole genome duplications were not the source of tetralogs (Skrabanek and Wolfe 1998; Hughes, 1999; Wang and Gu, 2000; Martin, 2001). But the unequal presence of the majority of gene families suggests that fourfold redundancy is not required and asymmetric evolution is the rule.

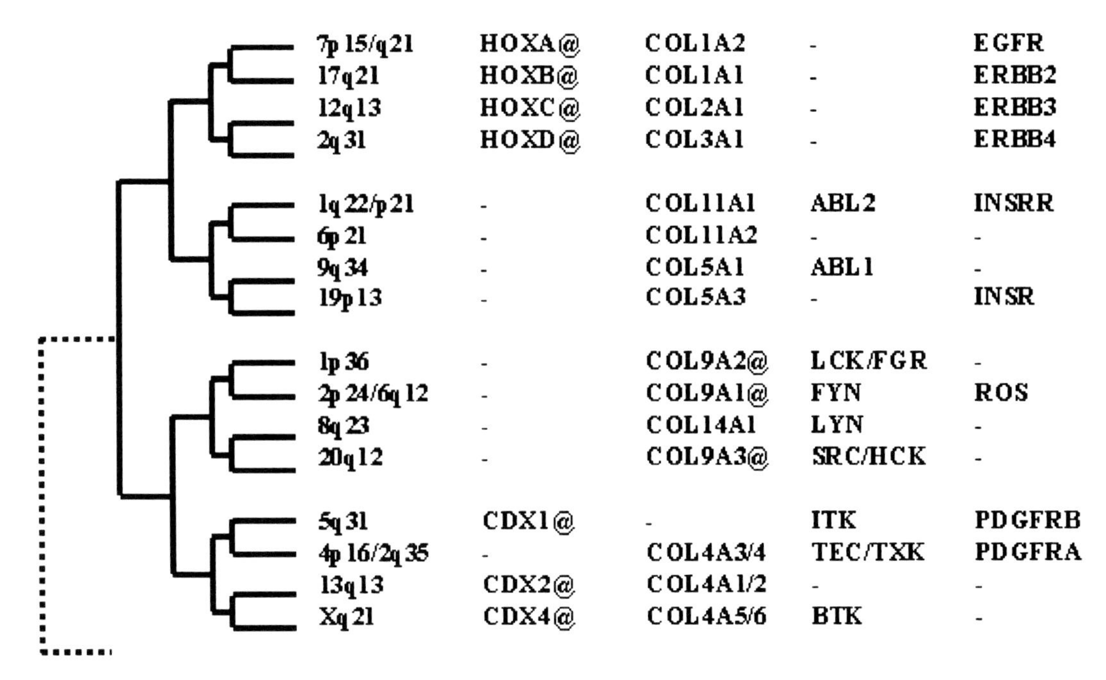

Figure 2. Comparison of Hox and ParaHox clusters, collagen and tyrosine kinase genes. Hox (HOX*@), ParaHox (CDX*@), collagen (COL*) and tyrosine kinase genes (ABL*-PDGFR*) were compared to available databases and grouped into subfamilies according to sequence similarities and chromosomal location (best viewed at http://www.ncbi.nlm.nih.gov/LocusLink/). Many additional conserved genes are found in vertebrate duplicates such as 7p15 and 17q21 next to Hox clusters of the major fibrillar collagen genes (COL1-3*) (Ruddle *et al.*, 1994) while the similarity between these four chromosomal regions and the four regions containing the minor fibrillar collagen genes (COL5*/COL11*) is much lower. Receptor (EGFR; etc.) and non-receptor tyrosine kinases ABL2, etc.) have a complicated nomenclature and relationship indicating that many additional mechanisms next to genome duplication were involved in the formation of the human genome.

A reference genome for vertebrates

Drosophila is the best available reference system for vertebrate genome comparisons, but is not closely related to the common ancestors of vertebrates. Data from amphioxus (Brooke *et al.*, 1998) would be ideal for comparisons to vertebrates, but cephalochordates are not easily accessible to genetic studies and have large genomes. The ascidian *Ciona intestinalis* seems to be better suited for genetic and genome studies (Simmen *et al.*, 1998), but might be more derived. Even without complete genomes, continuous DNA sequence, for example flanking the Hox and ParaHox clusters, could be useful to analyze the relationship of ascidians, amphioxus and the paralogous counterparts in hagfish, lamprey and higher vertebrates. The comparison of the human genome to the mouse genome, and eventually to those of rat, chimpanzee, zebrafish and pufferfish will help further to define vertebrate gene subfamilies in relation to invertebrate outgroups.

'Duplication and divergence' or 'divergence and fusion'

The most common type of gene duplicates, tandem duplicates, are initially identical and it is plausible that one copy retains the original function while the second copy can diverge and either gets lost or acquires a new function (Fig. 3). However, after separation into two lineages the same gene can also acquire two different subfunctions. These could be properties of the final protein product such as pH-tolerance or extended protein stability due to a few amino acid changes not affecting the original protein function, or even just changes in enhancer elements leading to tissue-specific expression of functionally identical proteins. Combination of such variants by a genome fusion event could lead immediately to a superior creature with elements of both parents. The principal of genome fusion is related to sexual reproduction. Recognition of hybridization, chimaerism, reticulate evolution or horizontal genome transfer as normal biological mechanisms of genome fusion

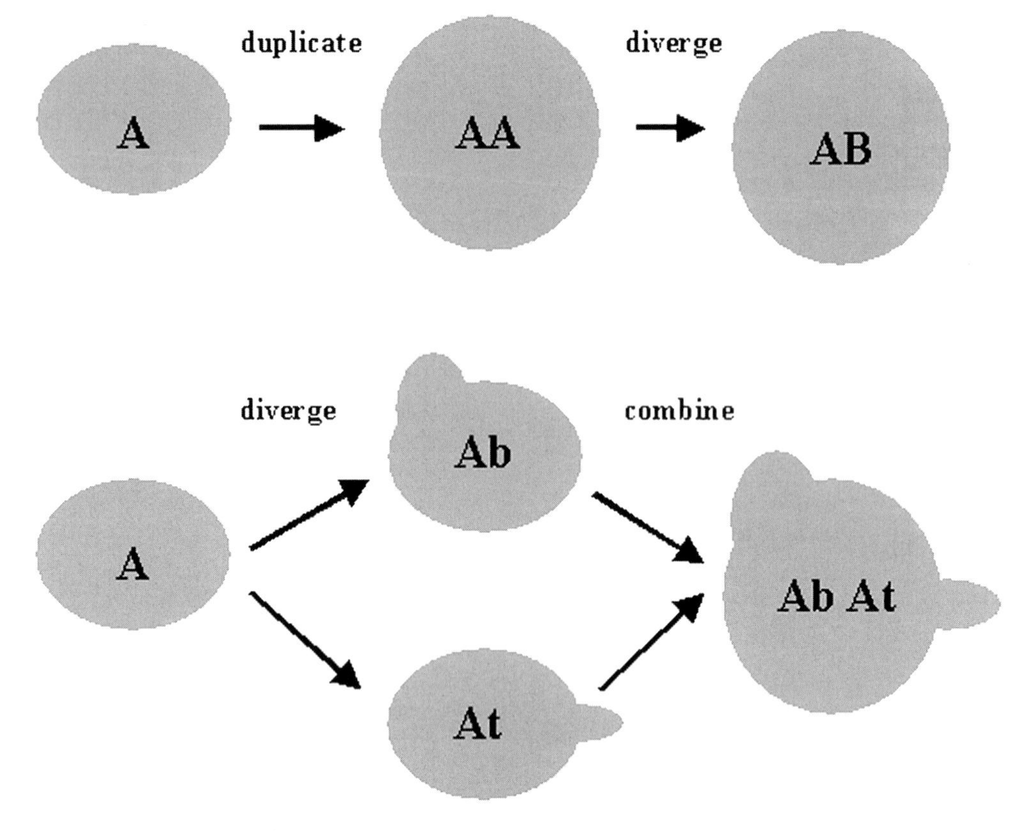

Figure 3. Duplication and divergence or divergence and fusion. Duplication and divergence is the common way of gene duplication and leads to continuous variation. Divergence and genome fusion are reminiscent of sexual reproduction. Successful fusion of genomes of established species should be rare; but when it happens under pressure of rare unusual geological events, it could lead immediately to evolutionary novelties.

could help to resolve some of the major problems in evolution, also called the major transitions in evolution. The fact that these transitions might correlate with major disturbances in geological history such as snowball Earth events (Hyde *et al.*, 2000), could be an indication that genome fusion is not rare, but it is rarely successful. From prokaryotes to humans four periods of genome fusions are sufficient to explain the stepwise increase of complexity.

Acknowledgments

I would like to thank all the people at the *Podocoryne* lab of Volker Schmid, the University of Basel, the Swiss National Science Foundation and the Treubel-Fonds for their support.

References

Aguinaldo, A.M., Turbeville, J.M., Linford, L.S., Rivera, M.C., Garey, J.R., Raff, R.A. and Lake, J.A. (1997) Evidence for a clade of nematodes, arthropods and other moulting animals. *Nature*, **387,** 489–493.

The Arabidopsis Genome Initiative. (2000) Analysis of the genome sequence of the flowering plant *Arabidopsis thaliana*. *Nature*, **408,** 796–815.

Baldauf, S.L., Roger, A.J., Wenk-Siefert, I. and Doolittle, W.F. (2000) A kingdom-level phylogeny of eukaryotes based on combined protein data. *Science*, **290,** 972–977.

Bancroft, I. (2001) Duplicate and diverge: the evolution of plant genome microstructure. *Trends Genet.*, **17,** 89–93.

Brooke, N.M., Garcia-Fernandez, J. and Holland, P.W. (1998) The ParaHox gene cluster is an evolutionary sister of the Hox gene cluster. *Nature*, **392,** 920–922.

Eisen, J.A. (2000) Assessing evolutionary relationships among microbes from whole-genome analysis. *Curr. Opin. Microbiol.*, **3,** 475–480.

Ferrier, D.E. and Holland, P.W. (2001) Ancient origin of the Hox gene cluster. *Nature Rev. Genet.*, **2,** 33–38.

Gibson, T.J. and Spring, J. (2000) Evidence in favour of ancient octaploidy in the vertebrate genome. *Biochem. Soc. Trans.*, **28,** 259–264.

Holland, P.W., Garcia-Fernandez, J., Williams, N.A. and Sidow, A. (1994) Gene duplications and the origins of vertebrate development. *Development*, **Suppl. 1994,** 125–133.

Horiike, T., Hamada, K., Kanaya, S. and Shinozawa, T. (2001) Origin of eukaryotic cell nuclei by symbiosis of Archaea in Bacteria is revealed by homology-hit analysis. *Nature Cell Biol.*, **3,** 210–214.

Hughes, A.L. (1999) Phylogenies of developmentally important proteins do not support the hypothesis of two rounds of genome duplication early in vertebrate history. *J. Mol. Evol.*, **48,** 565–576.

Hyde, W.T., Crowley, T.J., Baum, S.K. and Peltier, W.R. (2000) Neoproterozoic 'snowball Earth' simulations with a coupled climate/ice-sheet model. *Nature*, **405,** 425–429.

International Human Genome Sequencing Consortium. (2001) Initial sequencing and analysis of the human genome. *Nature*, **409,** 860–921.

Kroiher, M., Miller, M.A. and Steele, R.E. (2001) Deceiving appearances: signaling by 'dead' and 'fractured' receptor protein-tyrosine kinases. *BioEssays*, **23,** 69–76.

Lake, J.A. (1988) Origin of the eukaryotic nucleus determined by rate-invariant analysis of rRNA sequences. *Nature*, **331,** 184–186.

Llorente, B., *et al.* (2000) Genomic exploration of the hemiascomycetous yeasts: 20. Evolution of gene redundancy compared to *Saccharomyces cerevisiae*. *FEBS Lett.*, **487,** 122–133.

Lundin, L.G. (1993) Evolution of the vertebrate genome as reflected in paralogous chromosomal regions in man and the house mouse. *Genomics*, **16,** 1–19.

Lundin, L.G. (1999) Gene duplications in early metazoan evolution. *Semin. Cell Dev. Biol.*, **10,** 523–530.

Lynch, M. and Conery, J.S. (2000) The evolutionary fate and consequences of duplicate genes. *Science*, **290,** 1151–1155.

Margulis, L. (1970) *Origin of Eukaryotic Cells*, Yale University Press, New Haven, CT.

Martin, A. (2001) Is tetralogy true? Lack of support for the 'one-to-four rule'. *Mol. Biol. Evol.*, **18,** 89–93.

Ohno, S. (1970) *Evolution by Gene Duplication*, Springer-Verlag, Berlin, Germany.

Pebusque, M.J., Coulier, F., Birnbaum, D. and Pontarotti, P. (1998) Ancient large-scale genome duplications: phylogenetic and linkage analyses shed light on chordate genome evolution. *Mol. Biol. Evol.*, **15,** 1145–1159.

Philippe, H., Germot, A. and Moreira, D. (2000) The new phylogeny of eukaryotes. *Curr. Opin. Genet. Dev.*, **10,** 596–601.

Rubin, G.M., *et al.* (2000) Comparative genomics of the eukaryotes. *Science*, **287,** 2204–2215.

Rubin, G.M. (2001) The draft sequences. Comparing species. *Nature*, **409,** 820–821.

Ruddle, F.H., Bentley, K.L., Murtha, M.T. and Risch, N. (1994) Gene loss and gain in the evolution of the vertebrates. *Development*, **Suppl. 1994,** 155–161.

Simmen, M.W., Leitgeb, S., Clark, V.H., Jones, S.J. and Bird, A. (1998) Gene number in an invertebrate chordate, *Ciona intestinalis*. *Proc. Natl. Acad. Sci. USA*, **95,** 4437–4440.

Skrabanek, L. and Wolfe, K.H. (1998) Eukaryote genome duplication - where's the evidence? *Curr. Opin. Genet. Dev.*, **8,** 694–700.

Spring, J. (1997) Vertebrate evolution by interspecific hybridisation - are we polyploid? *FEBS Lett.*, **400,** 2–8.

Spring, J., Yanze, N., Middel, A.M., Stierwald, M., Groger, H. and Schmid, V. (2000) The mesoderm specification factor twist in the life cycle of jellyfish. *Dev. Biol.*, **228,** 363–375.

Szathmary, E. and Maynard Smith, J. (1995) The major evolutionary transitions. *Nature*, **374,** 227–232.

Venter, J.C., *et al.* (2001) The sequence of the human genome. *Science*, **291,** 1304–1351.

Wang, Y. and Gu, X. (2000) Evolutionary patterns of gene families generated in the early stage of vertebrates. *J. Mol. Evol.*, **51,** 88–96.

Wolfe, K.H., Shields, D.C. (1997) Molecular evidence for an ancient duplication of the entire yeast genome. *Nature*, **387,** 708–713.

Detection of gene duplications and block duplications in eukaryotic genomes

Wen-Hsiung Li[1*], Zhenglong Gu[1], Andre R.O. Cavalcanti[1,2] & Anton Nekrutenko[1]
[1]*Department of Ecology and Evolution, University of Chicago, 1101 East 57th Street, Chicago, IL 60637, USA;* [2]*Departamento de Quimica Fundamental, Universidade Federal de Pernambuco, Brazil;*
[*]*Author for correspondence: E-mail: whli@uchicago.edu*

Received 21.01.2002; accepted in final form 29.08.2002

Key words: codon usage bias, database cleaning, gene duplication rate, gene families

Abstract

Several eukaryotic genomes have been completely sequenced and this provides an opportunity to investigate the extent and characteristics (e.g., single gene duplication, block duplication, etc.) of gene duplication in a genome. Detecting duplicate genes in a genome, however, is not a simple problem because of several complications such as domain shuffling, the existence of isoforms derived from alternative splicing, and annotational errors in the databases. We describe a method for overcoming these difficulties and the extents of gene duplication in the genomes of *Drosophila melanogaster*, *Caenorhabditis elegans*, and yeast inferred from this method. We also describe a method for detecting block duplications in a genome. Application of this method showed that block duplication is a common phenomenon in both yeast and nematode. The patterns of block duplication in the two species are, however, markedly different. Yeast shows much more extensive block duplication than nematode, with some chromosomes having more than 40% of the duplications derived from block duplications. Moreover, in yeast the majority of block duplications occurred between chromosomes, while in nematode most block duplications occurred within chromosomes.

Introduction

Since Ohno (1970) gene duplication has been commonly thought to be the most important step for the origin of genetic novelties, because it creates gene copies whose functions can subsequently evolve in divergent directions. The importance of gene duplication in the evolution of a genome, however, cannot be fully appreciated without knowing the extent of gene duplication in a genome. This topic can now be fruitfully pursued, thanks to the availability of several completely sequenced eukaryotic genomes (e.g, human, yeast, *Drosophila*, *C. elegans*, *Arabidopsis*). However, evaluating the extent of gene duplication in a genome is a challenging task because there are many methodological problems and because the completed genomes are still not well annotated. Several authors have dealt with this issue (Lynch and Conery, 2000; Rubin *et al.*, 2000; Gu *et al.*, 2002). In this chapter, we shall review methods for handling these problems.

A closely related issue is the pattern of gene duplication in a genome. A duplication can involve (i) part of a gene, (ii) a single gene, (iii) a chromosome segment that contains more than one gene (a block duplication), (iv) an entire chromosome, or (v) the whole genome (see Li, 1997). According to Ohno (1970), whole genome duplications have been more important in evolution than regional duplications, because in regional duplications only parts of the regulatory system of structural genes may be duplicated, causing an imbalance that can disrupt the normal function of the duplicated regions. This view has not been well substantiated by data (see Wolfe, 2001). With the availability of complete genome sequences, it is now possible to study the frequency of each type of duplication in a genome and from such analyses one can evaluate the relative importances of different types of duplication. Here we address the question of how to detect block duplications in a genome, which has been a subject of several recent studies (Wolfe and Shields, 1997; Seoighe and Wolfe, 1999; Friedman and Hughes, 2001; Cavalcanti *et al.*, 2002).

To illustrate the problems and the methods, we shall discuss the application of the methods to the genomes of *Drosophila*, C. *elegans* and yeast (Gu *et al.*, 2001; Cavalcanti *et al.*, 2002), which have relatively good quality genomic sequence data.

Detection of duplicate genes in a genome

To determine whether two genes are derived from a duplication, that is, whether they are paralogous, one usually compares the sequences at the protein level because the divergence at the synonymous sites of the coding regions may have become saturated for relatively divergent duplicates. Of course, genomic sequence data can be of great help, especially the information on the exon-intron structure.

There are several difficulties in determining whether two proteins in a genome are paralogous. First, when the similarity between two proteins is in the so-called twilight zone (20–35%), it becomes very difficult to determine whether they are homologous (see Rost, 1999). This is particularly so for short proteins because two unrelated short peptide sequences can have a very high sequence similarity by chance. For example, even when 10 residues in an alignment of 16 amino acids are identical (>60% similarity), homology of the two sequences cannot be assured (see Rost, 1999). Second, domain shuffling is a common phenomenon in protein evolution (Doolittle, 1995; Ponting *et al.*, 2000) and two non-homologous proteins may share a domain and be mistaken as homologous. Third, the present protein databases are not well annotated and isoforms (from alternative splicing of a gene) in a genome may have been listed as independent proteins. Fourth, for the same reason (poor annotation), more than one name may have been given to the same gene and listed as independent genes or proteins in a database. Fifth, the present protein databases include many sequences that contain regions that were derived from repetitive elements (Brosius, 1999; Makalowski, 2000; Li *et al.*, 2001; Nekrutenko and Li, 2001); like domain sharing, two sequences sharing a repetitive element fragment may be mistaken as homologous. Finally, the huge amount of genomic data to be analyzed requires automated methods of analysis.

The first problem has been considered by many authors. In particular, Rost (1999) obtained an empirical formula in which the cut-off sequence identity (p) increases with decreasing length of the alignable regions; this is to avoid having a high identity by chance for a short alignable region. This method, however, cannot satisfactorily reduce the domain sharing effect (the second problem) and gave poor results when it was applied to the human protein database in November 2000 (Li *et al.*, 2001). This is also the problem for studies where the E score of BLASTP was used as the sole criterion to determine paralogous proteins (e.g., Rubin *et al.*, 2000). The current strategy to deal with the second problem is to add a second criterion that the total length of the alignable regions between the two sequences must exceed a certain proportion of the longer of the two sequences compared; this requirement may be called the minimum alignment coverage. In particular, to deal with the two first problems Gu *et al.* (2002) proposed to use the two criteria: (i) the similarity between the two sequences is $p = 30\%$ if $L \geqslant 150$ amino acids (a.a.) or

$$p = 0.01n + 4.8L^{-0.32(1+\exp(-L/1000))}$$

if $L < 150$ a.a. (Rost, 1999), where $n = 6$ and L is the length of the alignable region, and (ii) the length of the alignable region between the two sequences is $\geqslant 80\%$ of the longer protein.

To deal with all of the above problems Gu *et al.* (2002) proposed a two-step procedure as follows.

First round grouping

In the genome under study, every protein is used as the query to search against all other proteins in the same genome (database) using FASTA (E = 10). This fairly exhaustive search is to reduce the chance of missing potentially paralogous pairs. Note that for a potential pair to form a link or hit (i.e., to be in the same family) they must satisfy the above two criteria. The single linkage algorithm is used to group proteins into clusters (families), i.e., if protein A hits protein B and protein B hits protein C, then proteins A, B and C are put in the same cluster, regardless of whether protein A hits protein C or not.

Isoform cleaning

Based on gene and exon annotation, two genes are regarded as isoforms if their shared coding sequences are in total longer than 20% of the entire coding region of the shorter gene. Delete one of the two isoforms from the database as follows: Delete the shorter one if both are singletons; delete the one that is a singleton if the other belongs to a multigene family; delete the shorter one if the two isoforms form a two-member cluster; delete the shorter one if both are from the same gene family with more than 2 members and they have the same hits; delete the one with fewer hits if they belong to the same cluster but their hits are not all the same. Keep both proteins if they belong to different multigene families.

Cleaning of same genes with different names

Occasionally more than one name are assigned to the same gene and these names are presented as different genes in the database. Such a situation can be detected by comparing their sequence coordinates. In each of such cases only one copy is kept in the analysis.

Repetitive element (RE) cleaning

Each protein is used as the query to search against the repetitive element database for the same organism using FASTAX ($E = 10^{-5}$). Delete the whole protein sequence from the database if the part of the protein hit by an RE is longer than 80% of the protein itself. Delete only the part that is hit by an RE if it is shorter than 80% of the protein.

Second round grouping

Repeat the steps in the 'first round grouping' with the cleaned database. The new clusters are regarded as gene families.

Detection of block duplications in a genome

'Block duplication' is defined here as duplication of a chromosomal segment that includes more than one gene; it is also known as 'segmental duplication'. There has been considerable interest in block duplication (Wolfe and Shields, 1997; Friedman and Hughes, 2001), for it increases our understanding of the mechanism of gene duplication and the structure of a genome.

Friedman and Hughes (2001) proposed a method for detecting block duplications. They considered only proteins in the paranome (the set of all duplicate genes in the genome) and numbered them consecutively according to their positions on the chromosomes. They then divided the paranome into non-overlapping windows each consisting of 8 proteins and compared these windows. The statistical significance of the hits was decided by computer simulation (randomization tests); they found it statistically significant for two windows to share 4 paralogues. This method has several drawbacks. First, it relies solely on the E score of BLASTP to define paralogous proteins; as mentioned above, this criterion may lead to many false positives, even with a very stringent E value. Second, it uses non-overlapping windows, which may miss a block duplication if the block is broken into two consecutive windows. Third, it considers only the composition (content) of the window, but not whether the paralogues in the two windows appear in the same order (i.e., the gene order) and have the same orientation.

Cavalcanti *et al.* (2002) developed a method that may overcome these drawbacks. First, classify proteins in a genome into families using the method of Gu *et al.* (2002), which has been described above. Second, number the proteins in the paranome according to their positions on the chromosomes. Third, define windows in the paranome as follows. Each window of size n is defined as a chromosome segment containing n contiguous paranome members. For each position in the paranome, define a window containing the gene at that position and the following n−1 paranome members. This window is then compared with all other possible windows in the paranome, excluding those that overlap with it. This calculation is repeated for each position in the paranome. One first performs these calculations for windows of size two, and when two windows shared two paralogous proteins, one progressively increases the window size, adding one paranome member at each step, until the newly added member in one window is not paralogous to the newly added one in the other window. Fourth, one makes a choice of three criteria: (a) composition: two windows are counted as a hit if they have the same family composition, regardless of the order; (b) order: two windows are counted as a hit if they have the same family composition and the same gene order; and (c) orientation: two windows are counted as a hit if, besides fulfilling (a) and (b), the genes also have the same orientation. This method is completely automated, while part of Friedman and Hughes' method is visual.

After the window analysis, one excludes all redundant hits that superimpose. The main criterion for this filtering is: keep the larger block. For the blocks that share just the composition one filters superimposing blocks of the same size using only their position, and the first one to be detected is kept. For the blocks with the same order and/or orientation one filters the superimposing blocks of the same size according to the p_S values of the pairs of homologous genes, keeping the blocks that have the more uniform values of p_S, where p_S is the proportion of different nucleotides at synonymous sites. To evaluate the statistical significance of detected block duplications the above block detection procedure is applied to a large number (500 or more) of randomized paranomes, each of which is obtained by shuffling the real members in the paranome. That is, the randomization test is used.

Extent of gene duplication in yeast, *Drosophila* and *C. elegans*

Gu *et al.* (2002) applied the above method to the genomes of yeast, *Drosophila*, and *C. elegans* and the results are summarized below.

Databases used

Yeast: ftp://ncbi.nlm.nih.gov/genbank/genomes/S_cerevisiae/ The NCBI October 2000 version, which was part of the Reference Sequence (RefSeq) project, was used. The annotation for this version was based on the Saccharomyces Genome Database in the Stanford genomic resources (SGD, http://genome-www.stanford.edu/Saccharomyces/). A total of 6,297 protein sequences were in the database.

C. elegans: http://www.sanger.ac.uk/Projects/C_elegans/wormpep/ Wormpep release 40. The database contained 19,730 protein sequences; 48 sequences that did not have genomic position information and 22 sequences that did not have corresponding coding sequences were excluded.

Drosophila: ftp://ncbi.nlm.nih.gov/genbank/genomes/D_melanogaster/ Release 2, October 2000 from NCBI was used. A total of 14,335 protein sequences were in the database.

Table 1. Number of cases of a gene with different names, number of isoforms, and number of proteins hit by repetitive elements (REs) that were deleted from the databases.

	Family or group size	Yeast	*Drosophila*	*C. elegans*
Same gene				
	2	0	201	1
	> 2	0	215	1
	Total	0	416	2
Isoforms				
	1	53	197	166
	2	0	150	165
	> 2	2	108	116
	Total	55	455	447
REs				
	Proteins hit by RE ($E = 10^{-5}$)	110	116	506
	Hit length > 80% of protein itself	101	59	255

From Gu *et al.* (2002).

Database cleaning

Table 1 shows the number of cases of a gene with different names and the number of isoforms that were cleaned from each database used. In *Drosophila* database, more than 400 cases where a gene with different names were found, suggesting a relatively poor annotation. In yeast, isoforms were found mainly as singletons (i.e., they do not belong to any protein family), whereas in *Drosophila* and *C. elegans* the majority of isoforms were found in protein 'families' (Table 1). The number of proteins hit by repetitive elements is also listed in Table 1. In yeast and *C. elegans*, a large number of sequences deleted belong to the retrotranscriptase (RT) families. On the other hand, there are much fewer RTs in the current *Drosophila* protein database. It will be interesting to see whether the low number of RTs in *Drosophila* is real, or is due to incomplete genome sequencing. We note that in many cases the part of the protein derived from a repetitive element is less than 80% of the protein (9 in yeast, 57 in *Drosophila*, and 251 in *C. elegans*). Many of these proteins belong to non-RT protein families, although they include part(s) derived from a repetitive element. This observation suggests that repetitive elements play an important role in protein evolution (Brosius, 1999; Makalowski, 2000; Nekrutenko and Li, 2001).

Table 2. Number of gene pairs before and after database cleaning

K_S	Group size	Yeast Be-fore	Yeast After	*Drosophila* Be-fore	*Drosophila* After	*C. elegans* Be-fore	*C. elegans* After
< 0.01	< 6	34	32	660	2	301	76
	All	221	58	761	7	1,700	153
< 0.1	< 6	68	62	703	10	438	171
	All	930	172	822	26	2,113	379
< 0.25	< 6	95	88	729	22	535	254
	All	1,220	262	862	52	2,426	664

From Gu *et al.* (2002).
K_S: Number of substitutions per synonymous site betweem two genes.

Table 3. Distributions of singletons and gene families in yeast, *Drosophila* and *C. elegans*

Family size	Family number Yeast	*Drosophila*	*C. elegans*
1	4,768	10,786	12,858
2	415	404	665
3	56	113	188
4	23	46	93
5	9	21	71
6 ~ 10	19	52	104
11 ~ 20	8	26	57
21 ~ 50	0	11	33
50 ~ 80	0	0	5
> 80	0	1	3
No. of gene families	530	674	1,219
No. of unique gene types[a]	5,298	11,460	14,077

From Gu *et al.* (2002).
[a] One gene family is counted as one unique gene type

The dramatic effect of database cleaning is shown in Table 2. For example, for *Drosophila*, before the database was cleaned, there were 660 gene pairs with K_S < 0.01 within gene families having fewer than 6 members, but this number was reduced to 2 after database cleaning; 413 pairs were the same gene with different names and most of the rest (245 pairs) were isoforms.

Number of gene families

The numbers of protein families were estimated to be 530, 674 and 1,219 in yeast, *Drosophila* and *C. elegans*, respectively (Table 3). Our estimates are much smaller than those of Rubin *et al.* (2000) because we conducted a very detailed database cleaning and our criteria for homology are much more rigorous than theirs (they considered two proteins homologous if the BLASTP E value between them is = 10^{-6}, but did not require any minimal length of the alignable region). If we count one gene family as one unique gene type, the numbers of unique gene types are estimated to be 5,298, 11,460 and 14,077 in yeast, *Drosophila* and *C. elegans*, respectively. A striking point is that the number of gene families in *Drosophila* is only somewhat larger than that in yeast and much smaller than that in *C. elegans*, although the genome size and total protein number in the whole genome are similar in *Drosophila* and *C. elegans*, but much larger than those in yeast. The number of unique gene types in *Drosophila* is more similar to that in *C. elegans* than that in yeast.

Table 4. Top 5 multiple gene families in Yeast, *Drosophila* and *C. elegans*

Species	Size	Representative proteins
Yeast	20	Seripauperins
	19	Hexose transporters
	17	Amino acid permeases
	15	Putative helicase
	12	Heat shock proteins
Drosophila	111	Trypsins
	49	Cuticle proteins
	37	GTP binding proteins
	36	P450
	34	Cuticle proteins
C. elegans	242	Ofactory receptors
	181	Olfactory receptors
	154	Mostly hypoyhetical proteins
	76	Mostly hypoyhetical proteins
	73	Mostly hypothetical proteins, containing F-box domains

From Gu *et al.* (2002).

The top 5 gene families for each organism are listed in Table 4. In *C. elegans* it has been estimated that about 550 functional chemoreceptors (olfactory receptors) are scattered in the genome (Robertson, 1998). These proteins help nematode detect different kinds of chemicals. Chemoreceptors can be divided into different protein families, among which *srh*, *str*, *stl* and *srd* are large ones. The *srh* gene family was estimated to have 214 members (Robertson, 2000), most of which fall into the second largest family in *C. elegans* (Table 4), which has 181 members. Two closely related protein families, the *str* and *stl* (*str*-

like protein) protein families, constitute the largest gene family in Table 4, which has 242 members (the sum of proteins in these two families was estimated to be 240 by Robertson, 1998). The three other most common gene families in nematode (Table 4) are all hypothetical protein families; their function and phylogenetic relationship need to be investigated in the future. In yeast, among the 5 largest gene families there are two (seripauperins and putative helicases) located at the end of the chromosomes. Seripauperin genes encode serine-poor relatives of serine-rich proteins. Members in the putative helicase gene family are very similar to each other. Transcriptions of these genes are not detected under normal culture conditions (Viswanathan *et al.*, 1994; Yamada *et al.*, 1998). However, expression of putative helicases can be detected under some stress conditions (Yamada *et al.*, 1998). Another common gene family in yeast is also a stress response gene family: heat shock proteins. Most of the proteins within this family are cell stress chaperones. The remaining two of the 5 largest protein families are both membrane proteins. Although yeast has a gene duplication rate as high as that in *C. elegans* and much higher than that in *Drosophila*, there are no protein families in yeast that are as large as those in the other two organisms. In *Drosophila* cuticle proteins, like olfactory receptors in *C. elegans*, represent two out of the five largest protein families. The P450 proteins have been divided into two protein families (Tijet *et al.*, 2001), one of which is among the 5 largest protein families in *Drosophila* (Table 4) and the other is among the 10 most common protein families in flies (data not shown). The other two largest protein families in *Drosophila* are trypsin and GTP binding proteins, which are involved in metabolic and regulatory pathways.

Gene duplication rates

Table 5 lists the number of duplicate genes with $K_S < 0.01$. We compare K_S with the genetic distances in intron and flanking regions. Those pairs with both the genetic distance in intron and flanking regions (both up and down stream 150 bp) larger than 0.02 were excluded from the table. It is interesting to note that there are only 6 pairs of duplicate genes in *Drosophila* that have $K_S < 0.01$, while this number is 55 in yeast and 147 in *C. elegans* (35 in yeast if we exclude the putative helicase protein family). The recent gene duplication rates in the three organisms are shown in Table 5, under the assumption of similar synonymous mutation rates in these three organisms and using the estimated rate in *Drosophila* (15.6 substitutions per site per 10^9 years; Li, 1997). The recent gene duplication rate is estimated to be more than 10 fold lower in *Drosophila* than in yeast and *C. elegans*.

Table 5. Recent gene duplication rate in yeast, *Drosophila* and *C. elegans*

	Yeast	*Drosophila*	*C. elegans*
Original pair number with Ks < 0.01	58	7	153
Pair number after correction[a]	55[b] (32)[c]	6 (10)	147 (164)
Total protein used in the analysis	6,141	13,405	18,956
Gene duplicate rate (per gene per million years)[d]	0.028[e]	0.0014	0.024

From Gu *et al.* (2002).
[a] The gene pairs with the genetic distances in the intron and flanking regions (both up and down stream 150 bp) larger than 0.02 were excluded from the analysis.
[b] If the Y'-helicase (putative helicase, Table 4) protein family is excluded, this number is reduced from 55 to 35.
[c] The numbers in the parenthesis are from Lynch and Conery (2000).
[d] The estimated rate of silent-site substitution in *Drosophila* of 15.6 substitutions per site per 10^9 years was used for all three species.
[e] If the Y'-helicase protein family is excluded, this number is 0.018.

Table 6. Numbers of block duplications in worm (*C. elegans*) and yeast

Size	Worm			Yeast		
	Composition	Order	Orientation	Composition	Order	Orientation
2	590	693	475	83	85	79
3	179	168	61	15	15	15
4	80	62	42	12	12	13
5	11	8	5	7	6	5
6	8	6	3	6	6	6
7	3	2	1	1	1	1
8	3	3	2	1	1	1
9	0	0	0	0	0	0
10	2	0	0	0	0	0
11	1	0	0	0	0	0

From Cavalcanti *et al.* (2002)

Table 7. Mean number of blocks and standard deviation calculated from a set of 500 random genomes for worm and yeast

Species	Size of block	Composition No. of Blocks (S.D.)	Order No. of Blocks (S.D.)	Orientation No. of Blocks (S.D.)
Worm	2	791.06 (75.29)	798.78 (76.52)	402.83 (41.38)
	3	11.00 (4.17)	4.25 (2.21)	1.10 (1.09)
	4	0.18 (0.44)	0.02 (0.17)	0.00 (0.04)
Yeast	2	7.81 (2.81)	7.85 (2.82)	3.94 (1.97)
	3	0.05 (0.21)	0.01 (0.10)	0.00 (0.05)

From Cavalcanti *et al.* (2002).

Block duplications in yeast and *C. elegans*

Cavalcanti *et al.* (2002) applied the above new method to detect block duplications in the yeast and *C. elegans* genomes (Table 6); *Drosophila* was not included because the genomic sequences were not yet completely assembled in chromosomes. Table 7 gives the results of the analysis of the shuffled paranomes (randomization tests). For yeast all the observed numbers of block duplications are substantially higher than would be expected by chance. For worm, although the numbers of windows of size two are not significantly larger than expected by chance (at the 5% level), for larger block sizes there are significantly more hits than expected by chance. Because for worm the hits of size 2 are not statistically significant we excluded them from the analysis, and to use the procedure the same we also excluded the hits of size 2 for yeast.

For yeast, we (Cavalcanti *et al.*, 2002) detected fewer blocks than did Friedman and Hughes (2001). There are two reasons. First, our definition of protein families is more rigorous and has yielded fewer families. Second, we required the paralogous proteins detected in both windows to be contiguous; this is a conservative requirement in order to avoid spurious hits. Thus, our method tends to be more stringent than that of Friedman and Hughes. Yet surprisingly, in *C. elegans* we detected many more block duplications than did Friedman and Hughes, suggesting that dividing the paranome into a fixed number of windows in their method may overlook some block duplications. Another factor for the differences between the two studies is that in *C. elegans* there are several instances where more than one member of the same family is present in each block; in Friedman and Hughes (2001) these members would be counted just once, thus reducing the size of the hits.

Yeast shows much more extensive block duplication than worm, with some chromosomes, like chromosome XVI, having more than 40% of the duplications derived from block duplications of size larger than 2. But worm also shows evidence for block duplications, especially chromosome V, which has 12% of its duplicate genes derived from block duplications of size larger than 2.

The patterns of block duplication in the two species are markedly different. In worm the duplications are generally intrachromosomal, whereas in yeast interchromosomal. In worm some blocks were duplicated more than once, whereas in yeast all the detected blocks were duplicated just once. The scattering of the block duplications in the genome is also much stronger in yeast than in worm, which is consistent with Wolfe and Shields' (1997) hypothesis of massive gene loss following the duplication of the whole genome. What is surprising, on the other hand, is that yeast shows so little evidence of local block duplication, with just 4 intrachromosomal blocks found (sizes 7, 5, 3 and 3) and just one of these being in tandem (size 3), whereas local block duplication seems common in worm.

Concluding remarks

It should be noted that whether two genes are regarded as paralogous or not depends to some extent on the criteria used to detect them. For paralogous genes with high sequence similarity, they are likely to be detected for all criteria used, whereas whether or not two divergent paralogs are detected will depend much on the criteria used. Thus, the criteria used to define paralogs should be viewed as operational and the extent of gene duplication in a genome is not well defined unless the criteria used to detect paralogs are explicitly specified. Note further that when a criterion becomes more stringent, a gene family may split into two or more families, while when it becomes less stringent, two families may merge into one. Therefore, the number of gene families in a genome also depends on the criteria used. Clearly, how to find a set of best criteria is an important question and should be studied carefully.

Current methods for a large-scale analysis rely mainly on protein sequence data and use only limited information from genomic sequence data. Adding information from genomic sequence data such as the exon/intron structure of genes should increase the detecting power of a method. This direction of research should be pursued enthusiastically.

Acknowledgments

This work was supported was supported by NIH grants GM30998 and GM55759 and HD38287. A.R.O.C. was supported by CAPES – Brasilia.

References

Brosius, J. (1999) Genomes were forged by massive bombardments with retroelements and retrosequences. *Genetica*, **107**, 209–238.

Cavalcanti, A., Ferreira, R., Gu, Z. and Li W-H (2002) Patterns of gene duplication in yeast and *C. elegans*. *J. Mol. Evol.*, in press.

Doolittle, R.F. (1995) The multiplicity of domains in proteins. *Annu. Rev. Biochem.*, **64**, 287–314.

Friedman, R. and Hughes, A.L. (2001) Gene duplication and the structure of eukaryotic genomes. *Genome Res.*, **11**, 373–381.

Gu, Z., Cavalcanti, A., Chen, F.C., Bouman, P. and Li, W.-H. (2002) Extent of gene duplication in the genomes of Drosophila, nematode and yeast. *Mol. Biol. Evol.*, **19**, 250–262.

Gu, Z., Wang, H., Nekrutenko, A. and Li, W.H. (2000) Densities, length proportions, and other distributional features of repetitive sequences in the human genome estimated from 430 megabases of genomic sequence. *Gene*, **259**, 81–88.

Li, W.-H. (1997) *Molecular Evolution*. Sinauer Associates, Sunderland, MA.

Li, W.H., Gu, Z., Wang, H. and Nekrutenko, A. (2001) Evolutionary analyses of the human genome. *Nature*, **409**, 847–849.

Lynch, M. and Conery, J.S. (2000) The evolutionary fate and consequences of duplicate genes. *Science*, **290**, 1151–1155.

Makalowski, W. (2000) Genomic scrap yard: how genomes utilize all that junk. *Gene*, **259**, 61–67.

Nekrutenko, A. and Li, W.-H. (2001) Transposable elements are found in a large number of human protein-coding genes. *Trends Genet.*, **17**, 619–621

Ohno, S. (1970) *Evolution by Gene Duplication*, Springer-Verlag, Berlin, Germany.

Ponting, C.P., Schultz, J., Copley, R.R., Andrade, M.A. and Bork, P. (2000) Evolution of domain families. *Adv. Protein Chem.*, **54**, 185–244.

Robertson, H.M. (1998) Two large families of chemoreceptor genes in the nematodes *Caenorhabditis elegans* and *Caenorhabditis briggsae* reveal extensive gene duplication, diversification, movement, and intron loss. *Genome Res.*, **8**, 449–463.

Rost, B. (1999) Twilight zone of protein sequence alignments. *Protein Eng.*, **12**, 85–94.

Rubin, G.M., Yandell, M.D., Wortman, J.R., Gabor Miklos, G.L., Nelson, C.R., Hariharan, I.K., Fortini, M.E., Li, P.W., Apweiler, R., Fleischmann, W., Cherry, J.M., Henikoff, S., Skupski, M.P., Misra, S., Ashburner, M., Birney, E., Boguski, M.S., Brody, T., Brokstein, P., Celniker, S.E., Chervitz, S.A., Coates, D., Cravchik, A., Gabrielian, A., Galle, R.F., Gelbart, W.M., George, R.A., Goldstein, L.S., Gong, F., Guan, P., Harris, N.L., Hay, B.A., Hoskins, R.A., Li, J., Li, Z., Hynes, R.O., Jones, S.J., Kuehl, P.M., Lemaitre, B., Littleton, J.T., Morrison, D.K., Mungall, C., O'Farrell, P.H., Pickeral, O.K., Shue, C., Vosshall, L.B., Zhang, J., Zhao, Q., Zheng, X.H. and Lewis, S. (2000) Comparative genomics of the eukaryotes. *Science*, **287**, 2204–2215.

Seoighe, C. and Wolfe, K.H (1999) Updated map of duplicated regions in the yeast genome. *Gene*, **238**, 253–261.

Tijet, N., Helvig, C. and Feyereisen, R. (2001) The cytochrome P450 gene superfamily in *Drosophila melanogaster*: annotation, intron-exon organization and phylogeny. *Gene*, **262**, 189–198.

Viswanathan, M., Muthukumar, G., Cong, Y.S. and Lenard, J. (1994) Seripauperins of Saccharomyces cerevisiae: a new multigene family encoding serine-poor relatives of serine-rich proteins. *Gene*, **148**, 149–153.

Wolfe, K.H. and Shields, D.C. (1997) Molecular evidence for an ancient duplication of the entire yeast genome. *Nature*, **387**, 708–713.

Yamada, M., Hayatsu, N., Matsuura, A. and Ishikawa, F. (1998) Y'-Help1, a DNA helicase encoded by the yeast subtelomeric Y' element, is induced in survivors defective for telomerase. *J. Biol. Chem.*, **273**, 33360–33366.

A. Meyer, Y. Van de Peer (eds.), Genome Evolution, 35-44.

The evolutionary demography of duplicate genes

Michael Lynch[1*] & John S. Conery[2]

[1]*Dept. of Biology, Indiana University, Bloomington, Indiana 47405;* [2]*Dept. of Computer and Information Science, University of Oregon Eugene, Oregon 97403*

Received 21.05.2002; accepted in final form 29.08.2002

Key words: gene duplication, genome evolution, genome size

Abstract

Although gene duplication has generally been viewed as a necessary source of material for the origin of evolutionary novelties, the rates of origin, loss, and preservation of gene duplicates are not well understood. Applying steady-state demographic techniques to the age distributions of duplicate genes censused in seven completely sequenced genomes, we estimate the average rate of duplication of a eukaryotic gene to be on the order of 0.01/gene/million years, which is of the same order of magnitude as the mutation rate per nucleotide site. However, the average half-life of duplicate genes is relatively small, on the order of 4.0 million years. Significant interspecific variation in these rates appears to be responsible for differences in species-specific genome sizes that arise as a consequence of a quasi-equilibrium birth-death process. Most duplicated genes experience a brief period of relaxed selection early in their history and a minority exhibit the signature of directional selection, but those that survive more than a few million years eventually experience strong purifying selection. Thus, although most theoretical work on the gene-duplication process has focused on issues related to adaptive evolution, the origin of a new function appears to be a very rare fate for a duplicate gene. A more significant role of the duplication process may be the generation of microchromosomal rearrangements through reciprocal silencing of alternative copies, which can lead to the passive origin of post-zygotic reproductive barriers in descendant lineages of incipient species.

For practical reasons, much of the past focus on genome evolution has been on divergence at the nucleotide level in specific genes. But with the growing proliferation of whole-genome sequences, a more global view of genomic evolution is beginning to emerge. Just as nucleotide changes continuously arise within populations via mutation, accidents at the level of chromosomal regions regularly give rise to losses and duplications of entire genes. Such genomic turnover is ultimately responsible for interspecific divergence in gene content, which may be exploited for adaptive reasons, and for modifications of gene location, which may passively give rise to post-zygotic reproductive isolating barriers (for review, see Lynch, 2002). Thus, it is of some interest to determine the rate at which new genes arise via duplication events and the frequency and mechanisms by which they are preserved.

Because of the difficulties with quantifying low probability events at the molecular level, we are almost completely lacking in direct estimates of the rate of gene duplication, although rates as high as 10^{-6} to 10^{-4} per gene per generation have been reported for *Drosophila* (Shapira and Finnerty 1986). We recently obtained indirect estimates of the rates of birth and loss of new genes through censuses of the contents of the then largely sequenced nuclear genomes of several eukaryotes (Lynch and Conery 2000), and additional estimates using somewhat different criteria have been published by Gu *et al.* (2002). Since these studies were performed, nearly complete genomic sequences have emerged for several species and all of the pre-existing databases have been refined considerably. We, therefore, take this opportunity to update and expand our previous results.

Sources of data and methods of analysis

For each of the fully sequenced eukaryotic genomes, we downloaded all coding sequences and their corresponding amino-acid sequences from the most recently curated database (as of 1 April 2001), removing all suspected pseudogenes, transposable elements, and overlapping genes prior to subsequent analyses: *Schizosaccharomyces pombe* – The Sanger Centre (ftp://ftp.sanger.ac.uk/pub/yeast/sequences/pombe); *Saccharomyces cerevisiae* – National Center for Biotechnology Information ftp://ftp.ncbi.nih.gov/genbank/genomes/S_cerevisiae; *Arabidopsis thaliana* – The Institute for Genomic Research (ftp://ftp.tigr.org/pub/data/athaliana/ath1); *Caenorhabditis elegans* – WormBase (http://www.wormbase.org); *Drosophila melanogaster* – Berkeley *Drosophila* Genome Project (http://www.fruitfly.org/sequence/download.html); and *Homo sapiens* – The Ensembl Project (ftp://ftp.ensembl.org/current/data/).

To identify duplicate genes, we used BLAST (Altschul *et al.* 1997) to compare all pairs of amino-acid sequences within each genome, retaining only those pairs for which the alignment score was below 10^{-10}. To minimize the inclusion of members of large multigene families, we excluded all genes that identified more than five matching sequences. Using each protein alignment generated by BLAST as a guide, we aligned the nucleotide sequences, and then prior to sequence analysis, we used a gap-expansion algorithm to remove ambiguous portions of the alignments (Conery and Lynch 2001).

The numbers of nucleotide substitutions per silent and replacement sites (S and R, respectively) were then estimated for each pair by using the maximum-likelihood procedure in the PAML software package (version 2.0k) (Yang 1997). Estimated rates of nucleotide substitution are sensitive to the relative rates of occurrence of transitions and transversions, which cannot be estimated accurately when the amount of sequence divergence is high. Therefore, to obtain precise estimates of the transition/transversion bias among newly arisen mutations, prior to the analyses of sequence divergences for each species, we tallied the observed substitutions at all four-fold redundant sites in all pairs of duplicate sequences that were similar enough that multiple substitutions per site were unlikely (by confining these computations to loci for which the divergence at such sites was $\leq 15\%$, after verifying that the transition/transversion ratio is essentially constant below this point). Each species-specific estimate of the transition/transversion ratio was then treated as a constant in the maximum-likelihood analyses.

In genome-wide surveys, there is a need to distinguish the number of duplication events from the number of duplicate pairs. When neither member of a duplicate pair is homologous to another gene in the data set, then the pair represents a single duplication event. However, when three or more genes are mutually related, the number of duplication events is necessarily less than the number of pairs. For example, a trio of related genes revealed as three pairs must be the result of two duplication events. With a closed loop of three similar genes, the ancestral relationships are ambiguous, so we counted each pair as two-thirds of a duplication event. For cases in which four or more genes constituted a family, we constructed a graph with nodes corresponding to genes and edges connecting two nodes whenever the PAML estimate of S was less than 5.0 for the corresponding genes. For a particular family of such genes, all possible spanning trees (excluding closed loops) were constructed (Cormen *et al.* 1990, Shioura *et al.* 1997), and the weight for each gene pair was taken to be the fraction of times the edge connecting the pair was used in the total set of trees for the family.

The age distribution of duplicate genes

Assuming the number of silent substitutions per site increases approximately linearly with time, the relative age-distribution of gene duplicates within a genome can be inferred indirectly from the distribution of S (Figure 1). As in more traditional forms of demographic analysis, the forms of such distributions are diagnostic. For all species, the highest density of duplicates tends to be contained within the youngest age classes, with the density dropping off rapidly with increasing S (Figure 1). A smooth, nearly exponential decay with age is seen for both *H. sapiens* and *C. elegans*, for which the sample sizes of duplicate genes are very large. Similar distributions are seen for the two yeasts, although these are less smooth presumably because of the lower incidence of duplicate genes in these species. On the other hand, *A. thaliana* is exceptional in showing a pronounced secondary peak in the age distribution at $S \simeq 0.75$. This bulge

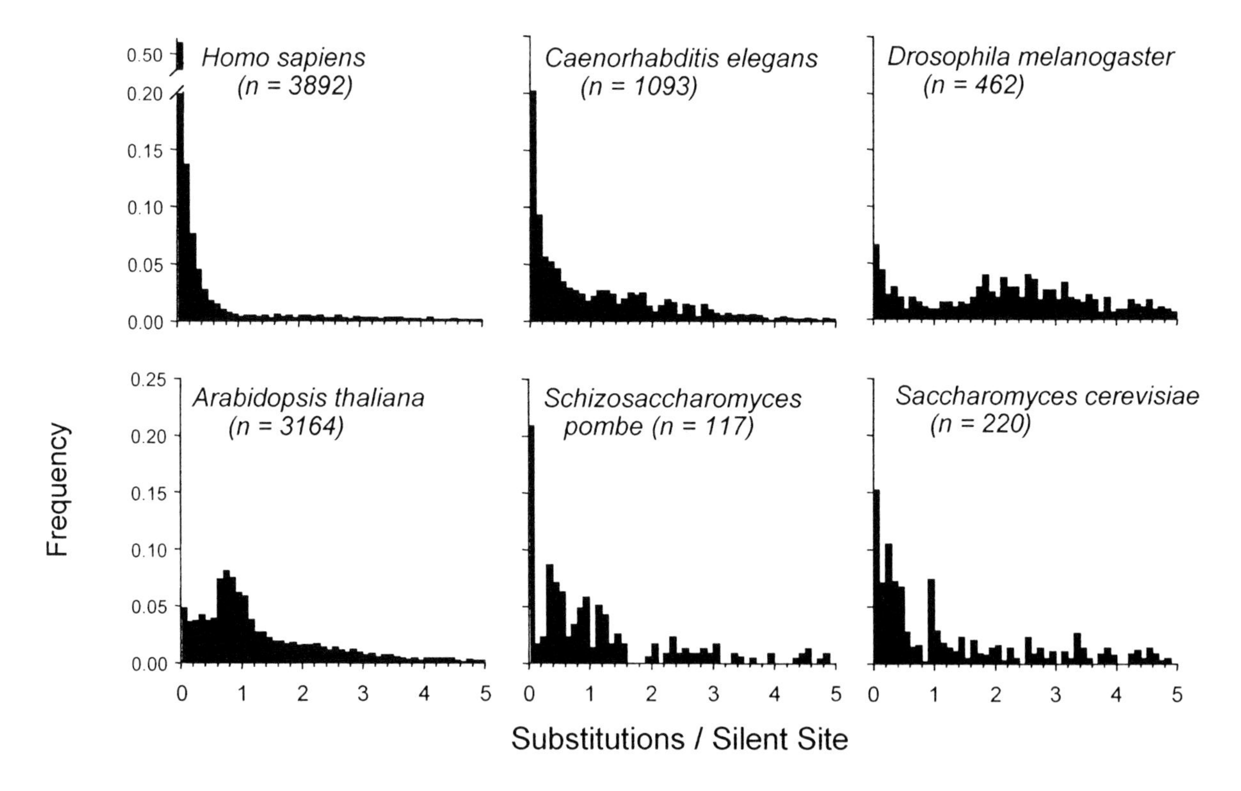

Figure 1. The age distribution of duplicate pairs of genes in six completely sequenced eukaryotic genomes. The total number of detected duplication events for each species is given in parentheses.

apparently reflects a 'baby boom' of genes that resulted from an ancient polyploidization event in the ancestor of *Arabidopsis* (Grant *et al.* 2000, Bevan *et al.* 2001). Using an estimated rate of silent-site substitution of 6.1/silent site/BY (an average of two independent estimates for vascular plants reported in Lynch (1997) and Li (1999), $S = 0.75$ is equivalent to approximately 60 million years of divergence.

Estimation of rates of birth and death

One interpretation of the taxonomically general patterns illustrated in Figure 1 is that of a birth-death process, with the very youngest age category representing newly arisen duplicates and the subsequent decline in frequency resulting from mutational processes that eliminate gene function (including large-scale deletions, frame shifts that introduce stop codons, etc.) Provided adequate numbers of young gene duplicates are available, estimates of the rates of birth and loss of such genes can be obtained directly from the observed age distribution, under the assumption that these rates have been essentially constant within the age classes employed in the analysis.

Letting n_t be the number of copies of a gene present at time t (in excess of the baseline number of one), then the dynamics of temporal change in the incidence of gene duplicates can be expressed as

$$n_t = n_{t-1} + B(1 + n_{t-1}) - Dn_{t-1}, \tag{1}$$

where we assume that only the excess copies of a gene are subject to loss. Setting $n_t = n_{t-1}$, the equilibrium number of additional copies per gene is found to be

$$n_{tot} = \frac{B}{D - B}. \tag{2}$$

Noting that the expected number of newborns per interval is $B(1 + n_{tot})$, the expected number of duplicates in the ith age category is

$$n_i = \frac{BD(1 - D)^i}{D - B}. \tag{3}$$

The slope of the least-squares regression of $\ln n_i$ on S provides an estimate of the instantaneous mortality rate d, defined such that the probability of duplicate-gene loss by the time divergence at silent sites has reached S is

$$\hat{D}=1-e^{-dS}. \quad (4)$$

The estimated half-life is then

$$\hat{S}_{0.5} = -\frac{\ln 0.5}{\hat{d}}. \quad (5)$$

Letting n_B be the number of duplicate pairs observed below some low level of S, then the birth rate over this span of divergence can be estimated as

$$\hat{B} = \frac{n_B \hat{d}S}{N(1-e^{-\hat{d}S})} \quad (6)$$

where N is the total number of genes in the analysis (not including the excess duplicates), and the term $dS/(1-e^{-dS})$ accounts for the loss of duplicates within the range of divergence up to S.

In the analyses reported here, we let $S=0.01$ in Equations (4) and (6), so $\hat{B}$ and $\hat{D}$ are estimated rates of birth and loss of duplicates over the time scale required for a pair of genes to diverge by 1% at silent sites. We also restricted our entire demographic analyses to duplication events with $S \leq 0.10$, so strictly speaking the assumptions regarding stationarity of rates are only relevant to this range of divergence. Consistent with this assumption, the log-arithmetic plots of n_i on S are approximately linear, although there is considerable scatter for the fungal genomes for which the numbers of duplication events are small (Figure 2).

The average half-lives of gene duplicates in the three fungal species are in the narrow range of $0.01 \leq \hat{S}_{0.05} \leq 0.02$, whereas those for the three metazoans are in the higher range of $0.04 \leq \hat{S}_{0.05} \leq 0.10$, and that for *A. thaliana* is still higher at $\hat{S}_{0.05} \simeq 0.21$ (Table 1). These half-lives in units of S can be crudely rescaled to absolute time by assuming an approximately constant rate of silent substitution. Using the rationale outlined in Lynch and Conery (2000), we assume a rate of 2.5/site/BY for human, 15.6/site/BY for invertebrates, 6.1/site/BY for *A. thaliana*, and 8.1/site/BY for fungi. Thus, the approximate half-lives for gene duplicates in humans, flies, and nematodes are 7.5, 3.2, and 1.7 MY, respectively, for an overall average of about 4 MY for animals. The average half-life for *A. thaliana* duplicates is much higher, on the order of 17.3 MY, whereas that for the three fungi is much lower, averaging to ~ 1.0 MY.

Table 1. Estimated mortality rates, half lives, and birth rates. $\hat{B}$ and $\hat{d}$ are estimated on a time scale for which silent-site divergence is 1%.

Species	$\hat{d}$ (SE)	$\hat{S}_{0.5}$	$\hat{B}$ (SE)
H. sapiens	18.4 (4.6)	0.038	0.0345 (0.0032)
C. elegans	13.0 (3.6)	0.053	0.0097 (0.0008)
D. melanogaster	6.9 (1.7)	0.100	0.0006 (0.0002)
A. thaliana	3.3 (2.0)	0.212	0.0032 (0.0005)
S. cerevisiae	30.5 (11.2)	0.023	0.0044 (0.0010)
S. pombe	42.6 (15.4)	0.016	0.0050 (0.0012)
E. cuniculi	62.3 (11.2)	0.011	0.0364 (0.0053)

Considerable interspecific variation also appears to exist for the rate of birth of duplicate genes, with the range being on the order of 0.001 to 0.04 over a time span equivalent to 1% divergence as silent sites (Table 1). At the high end, with essentially identical estimates of 0.035, are *H. sapiens* and *E. cuniculi*, whereas the remaining species fall in the range of 0.001 to 0.01. Using the molecular clocks noted above, these estimates translate into 0.009/gene/MY for humans, 0.016 for *C. elegans*, 0.001 for *D. melanogaster*, 0.002 for *A. thaliana*, 0.004 for both yeasts, and 0.030 for *E. cuniculi*. Thus, averaging over all taxa, the probability of duplication of a eukaryotic gene is at least 1% per million years. This rate is on the order of, if not greater than, most estimated rates of nucleotide substitution at silent sites (Li 1999). It is conceivable, however, that some of the duplicates that we have identified are 'dead-on-arrival.' For newborn duplicates in *C. elegans*, for example, we find that approximately one-third are complete over the entire coding region, with the remainder exhibiting one or more unique exons in one or both copies (Katju and Lynch, in prep.)

Given the potential errors in whole-genome sequences, all of the above estimates must be taken as provisional. However, there is no obvious reason to expect either the birth- or death-rate estimates to be upwardly biased, and in fact, the contrary may be true.

Whole-genome sequencing, particularly that using shot-gun approaches, is likely to lead to the exclusion of some of the youngest duplicates from the final genomic sequence, falsely interpreting them as simple redundant (or allelic) sequences. Such problems,

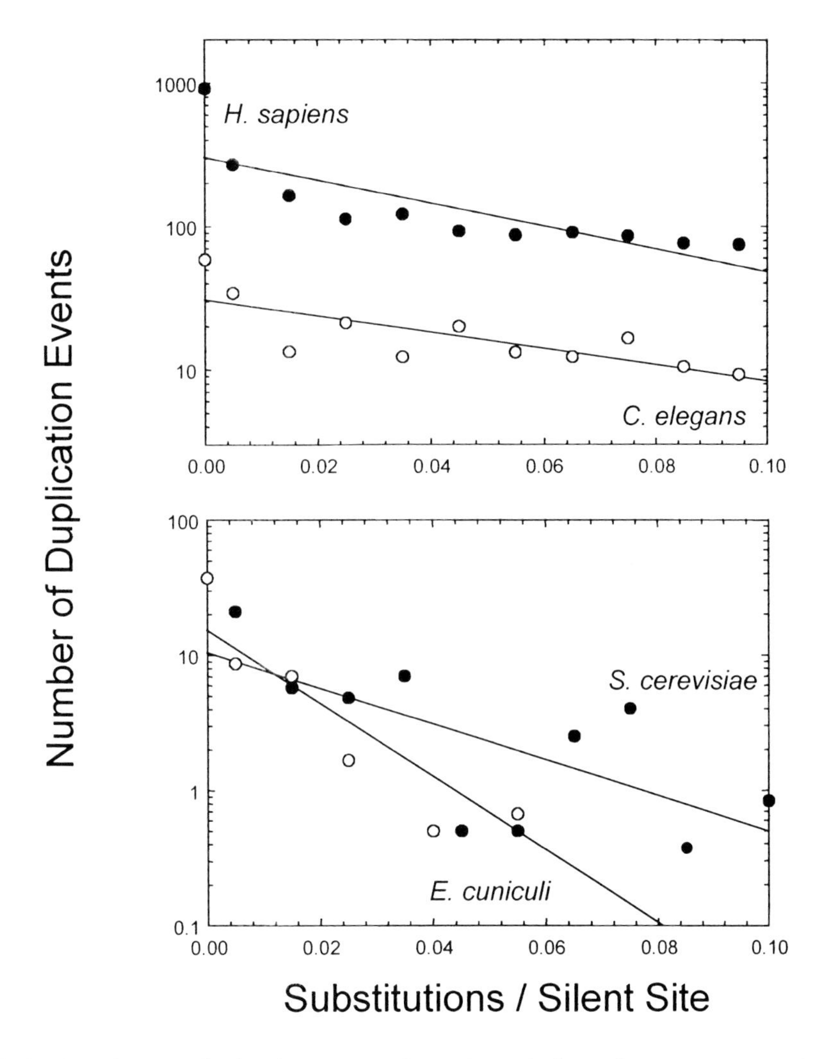

Figure 2. Survivorship curves fitted to the youngest cohorts of gene duplicates for two metazoans and two fungi.

which are most likely for the human and *Drosophila*genomes, would cause downward bias in the estimates of both the birth and death rates. In addition, our birth- and death-rate estimates may be somewhat downardly biased because we have ignored large multigene families.

Thus, the overall interpretation of these full-genome analyses, consistent with the earlier conclusions of Lynch and Conery (2000), is that the gene duplication is at least as significant as nucleotide substitution as an on-going contributor to genome evolution. Multiplying the species-specific duplication rates per gene by the genome size, the average number of newborn duplicates arising on a time scale of one million years is approximately 100 per genome. Roughly speaking, this implies that over a time span of 100 of 200 MY, nearly all of the genes within an average eukaryotic genome will have had an opportunity to duplicate. On the other hand, over 50% of such duplicates are likely to be silenced in only a few million years and most of the remainder shortly thereafter.

Evidence for genomic equilibrium

To test whether the incidences of gene duplicates in various genomes correspond to expectations under a long-term-state birth-death process, the predicted $\hat{n}_{tot}$, obtained by use of Equation (2), can be compared with the observed abundances of duplications. Such comparisons are somewhat subjective in that it is not

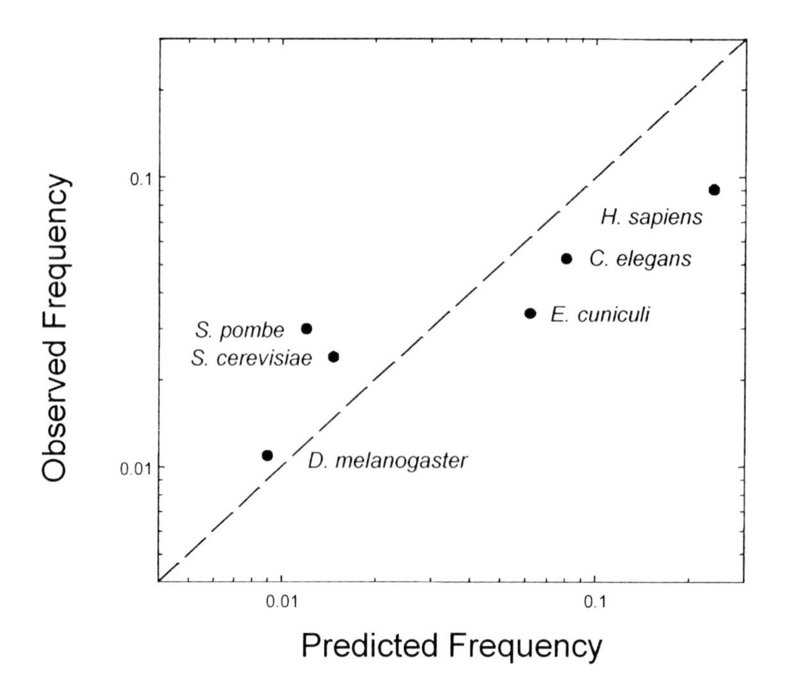

Figure 3. Comparison of the observed numbers of duplicates per gene with expectations based on a steady-state process, obtained by applying the demographic parameters in Table 1 to Equation (2). The diagonal dashed lines denotes the line of equivalence expected if the model perfectly fit the data.

entirely clear where to draw the upper boundary (with respect to S) for the subset of duplicates that are simply subject to the birth-death process (as opposed to the minority that are permanently preserved by positive selection). Nevertheless, there is a good correlation between observed and expected values for a broad range of cutoff values for S. We simply show the results for the pairs of duplicates with $S < 1.0$ in Figure 3, as estimates of S beyond this point are highly unreliable due to saturation effects. (We have excluded *A. thaliana* from this particular analysis because its ancestral polyploidization event clearly violates the assumption of equilibrium).

Note that because of its high rate of duplication, the microsporidian *E. cuniculi* actually has an equilibrium number of duplicates per gene approaching that of human and nematode, despite the fact that *E. cuniculi* has the smallest genome of any of the study species. On the other hand, because of its exceptionally low birth rate and moderately high loss rate, the *D. melanogaster* genome is rather depauperate with respect to gene duplicates.

Estimation of patterns of selective constraints

It is often assumed that redundancy of genes results in a relaxation of selection in one or both copies at least early in their history, and that this somehow enables one copy to take on a new function that would not otherwise be possible (e.g., Ohno 1970). From a population-genetics perspective, it is difficult to see how natural selection could isolate one gene for evolutionary exploration while keeping the other constant. This issue might be evaluated by considering the historical development of replacement- and silent-site substitutions in duplicates from the time of birth to the time of preservation or elimination, but the time scale of the mutational process necessitates an alternative approach. The problem that we are confronted with is that the observed estimates of R and S for any pair of extant duplicates are the cumulative outcomes of the joint evolutionary pressures operating on both loci since the initial duplication event. Such estimates potentially average over heterogeneous phases of molecular evolution, to an extent that increases with the age of the pair. However, some insight into the average temporal dynamics of selection can be acquired by examining the joint distribution of R and S for the

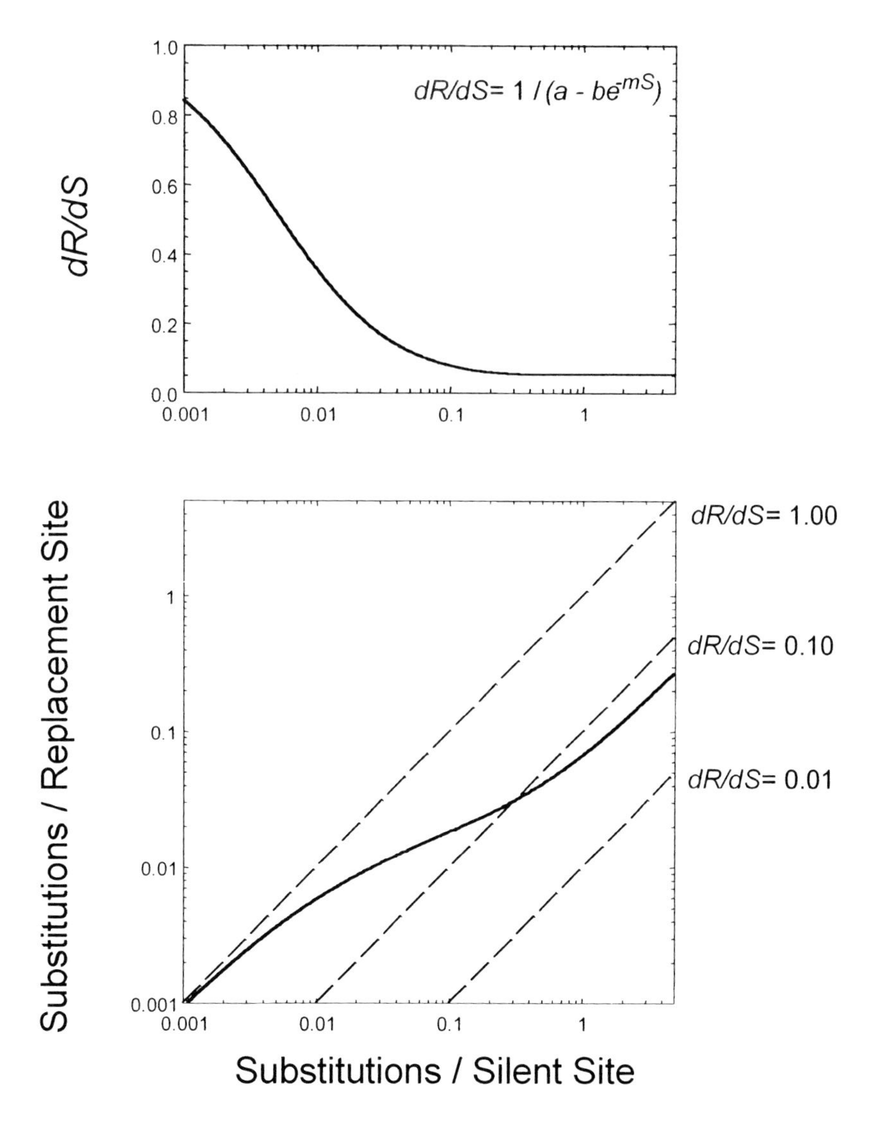

Figure 4. The change in the *R/S* ratio with increasing evolutionary time (measured in units of *S*). Upper panel: The instantaneous ratio of replacement to silent substitutions, defined by Equation (7) with $a = 20$, $b = 19$, and $m = 10$. Lower panel: The cumulative behavior of the *R* vs. *S* ratio, defined by Equation (8). Here, the dashed lines represent points of equal *R/S*. In this particular example, the ratio of replacement to silent substitutions initiates at 1.0 for newly arisen duplicates and gradually declines to a stable ratio of 0.05 as $S \to \infty$.

entire assemblage of gene duplicates within a species, under the assumption that approximately the same average temporal pattern of selection intensity operates on all cohorts of duplicates. If, for example, the intensity of purifying selection operating on duplicate genes typically increases with the age of a pair, this should be reflected in a reduction in the *R/S* ratio with increasing *S*, whereas the opposite is expected if selection is progressively relaxed.

To account for such behavior, we describe the ratio of the instantaneous rates of replacement and silent substitutions by the function.

$$\frac{dR}{dS} = \frac{1}{a - be^{-mS}}, \tag{7}$$

where a, b and m are constants (Lynch and Conery 2000). This function allows for two different phases of divergence, as well as a gradual transition between them. Assuming positive m, the ratio of rates of replacement to silent substitutions initiates with an expected value of $1/(a - b)$ at $S = 0$ (newly arisen duplicates) and declines to $1/a$ as $S \to \infty$ (ancient

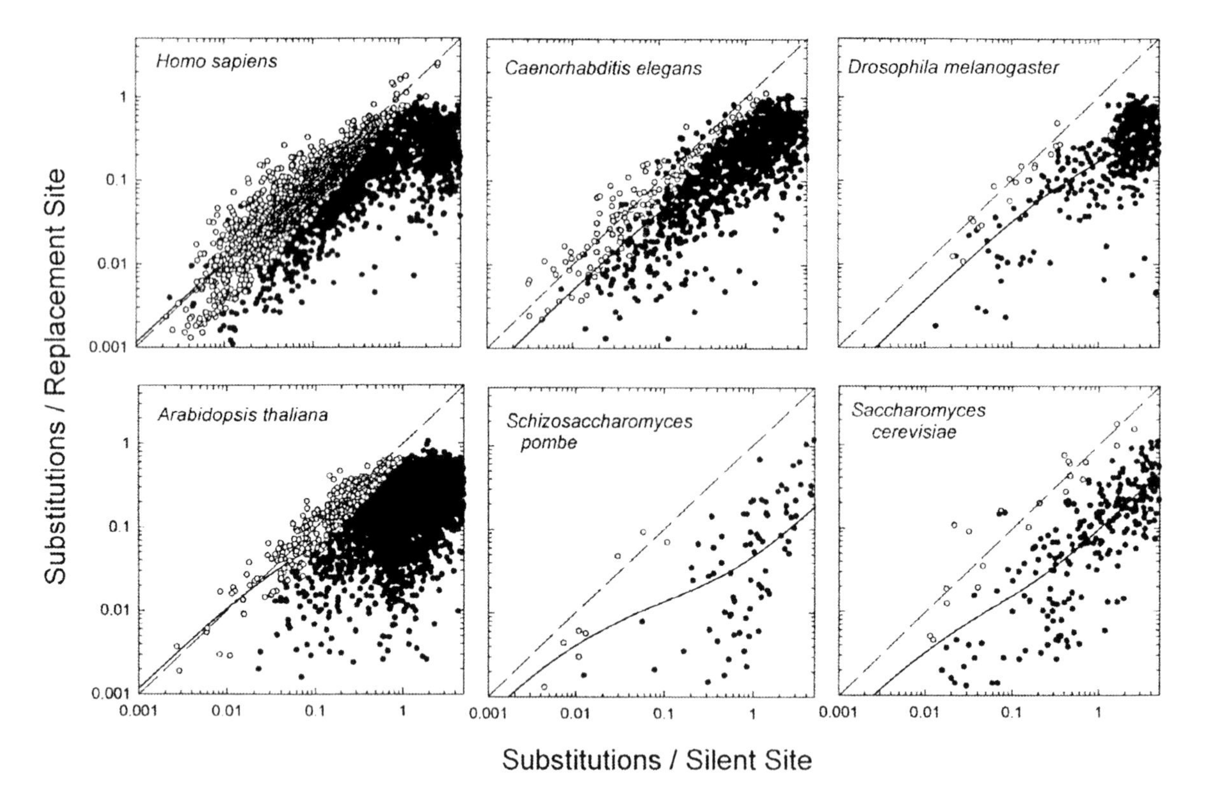

Figure 5. R vs. S plots for duplicate pairs of genes in six completely sequenced eukaryotic genomes. Open points denote pairs for which R is not significantly different from S.

duplicates). Other mathematical functions could be constructed to have qualitatively similar behavior, but Equation (7) is useful because it is readily integrated to yield a simple algebraic relationship between the cumulative number of substitutions per replacement site and the cumulative number of substitutions per silent site,

$$R = \frac{1}{am}\left[mS + \ln\left(\frac{a-b}{a-be^{-mS}}\right)\right]. \tag{8}$$

On a log-log plot, points with equal R/S ratios fall on a diagonal line, with the height of the line being defined by the magnitude of R/S (Figure 4). Thus, with the preceding model, a linear relationship appears between log S and log R when S is small enough, as $dR/dS \simeq 1/(a-b)$ (phase 1). The response of log R to log S then becomes shallower as a transition is made to lower dR/dS, until a slope of one is again arrived at as dR/dS approaches $1/a$ (phase 2). The coefficient determines the rate of the transition between these two extreme phases. During a period in which genes are evolving in a neutral fashion, the response will be coincident with the diagonal describing $R/S = 1.0$ (the main diagonal in Figure 4).

To obtain the parameter estimates for Equation (8), we performed least-squares analyses, using logarithms of observed and expected values, so as not to give undue weight to sequence pairs with large R. Pairs of sequences for which S or R are equal to zero cannot be included in such an analysis, and we excluded the few data in which either S or R were <0.001, as such estimates are highly unreliable. To test for the constancy of dR/dS, we also obtained fits for the reduced model with a single parameter (i.e., $R/S = 1/a$, independent of S). Letting n denote the number of gene pairs, and r_f and r_r denote the correlation coefficients for the full and reduced models, the relevant test statistic is

$$F = \frac{n(r_f^2 - r_r^2)}{1 - r_f^2} \tag{9}$$

with 1 and $n-2$ degrees of freedom (p. 633, Sokal and Rohlf 1995). (Although the models differ by two

Table 2. Estimated parameters relating replacement-site to silent-site divergence. Note that $(dR/dS)_{S \to 0} = 1/(a - b)$ and $(dR/dS)_{S \to \infty} = 1/a$. The one-parameter model was rejected for all species except *E. cunicluli*.

Species	m	$(dR/dS)_{S \to 0}$	$(dR/dS)_{S \to \infty}$	r^2	n
H. sapiens	0.31	0.898	0.029	0.671	3570
C. elegans	0.55	0.475	0.056	0.635	1253
D. melanogaster	0.54	0.360	0.050	0.524	445
A. thaliana	1.18	0.956	0.044	0.270	3790
S. cerevisiae	69.49	1.000	0.087	0.467	213
S. pombe	9.20	0.940	0.023	0.264	98
E. cuniculi	∞	0.296	0.296	0.919	27

parameters, m is effectively fixed at ∞ in the reduced model).

All of the eukaryotes examined except *E. cuniculi* exhibit a significant decline in dR/dS with increasing S (Table 2, Figure 5). The asymptotic values of dR/dS at low S are somewhat variable among species, with *D. melanogaster*, *C. elegans*, and *E. cuniculi* ranging between 0.3 and 0.5, the remaining species approaching 1.0, and an overall average of 0.70.

Excluding the estimate for *E. cuniculi*, the estimates of dR/dS at high S are much more homogeneous, ranging from 0.02 to 0.09, and averaging to 0.05. Thus, for most species, selection against amino-acid altering mutations is indeed relaxed early after duplication, in several cases approaching the neutral expectation. On average, the stringency of selection against replacement substitutions subsequently increases approximately 14-fold as a pair of duplicate genes ages, but the estimated values of m indicate that the transition between the two extreme phases occurs much more rapidly for the unicellular fungi than for metazoans and plants.

Discussion

These results largely corroborate the earlier conclusions in Lynch and Conery (2000). The estimated half-life of duplicate genes averaged over all species, approximately 4 MY, is identical in both studies, although it should be kept in mind that there is an order-of-magnitude range of variation among species. The average rate of origin of new duplicates in this study, 0.01/gene/MY, is about the same as that reported in our earlier study. Gu *et al.* (2002) also recently reported estimates of the rate of origin of duplicates genes in *D. melanogaster*, *C. elegans*, and *S. cerevisiae*. As in our study, these authors employed pre-screening devices to avoid pseudogenes, alternatively spliced variants, and annotation errors, but their additional methods of analysis deviated from those used herein in a number of ways. For example, Gu *et al.* (2002) did not make a distinction between the number of duplicate pairs and the actual (necessarily smaller) number of duplication events; they included very large multigene families, whereas we used an arbitrary cutoff for family size of five; they attempted to remove incomplete or chimeric duplicates; and they eliminated some duplicate pairs in which the nucleotide substitution levels in flanking regions exceeded those in silent sites. Despite these differences, it is comforting that estimates from both studies for B appear to deviate by no more than a factor of three.

Since decisions as to what constitutes a legitimate gene duplication will always involve a certain degree of subjectivity, and since the quality of current complete-genome databases is still in a state of flux, there is little question that the rates of birth and death of duplicate genes noted above will be subject to future modification. However, without a major upheaval, it appears difficult to avoid the conclusion that on-going gene turnover is a fundamental evolutionary feature of all eukaryotic genomes. Moreover, the smallest genomes may be kept small not by a low rate of origin of new duplicates but by a high rate of attrition. As a consequence of the duplication process, most species are expected to exhibit transient presence/absence polymorphisms at multiple loci (see Lynch 2003 for a review), and microchomosomal rearrangements among closely related species should commonly arise when the descendant members of duplicate pairs survive at the expense of their ancestral copy. Such reassignments of chromosomal locations provide a simple and powerful mechanism for the origin of post-zygotic isolating barriers that requires no intermediate stage of reduced fitness and no accumulation of negative epistatic interactions between heterospecific genes (Lynch and Force 2000). Thus, although gene duplication is often regarded primarily as a mechanisms for adaptive evolution, its role in the other major engine of evolution, speciation, may be even more significant.

Finally, we note that although the majority of duplicate genes appear to be transient, contributing little to long-term phenotypic evolution, a minority of such genes become preserved for very long periods, either by subfunctionalization of neofunctionalization, as illustrated by the long, shallow shoulders on the age distributions displayed in Figure 1 and by the strong purifying selection operating on such genes (Figure 5). Although such genes may be simply selected for purely on the basis of redundancy, this appears to be unlikely on theoretical grounds (Lynch *et al.* 2001), so they almost certainly contribute to adaptive evolution. Nevertheless, unless the eukaryotic genome is undergoing a gradual and prolonged expansion, even members of these pairs must ultimately be subject to loss.

Acknowledgments

We are extremely grateful to the large number of individuals who have contributed to the genome sequencing projects from which this study draws its analyses. Our work has been supported by NIH grant GM20887 and NSF grant DEB-0003920.

References

Altschul, S. F., T. L. Madden, A. A. Schaffer, J. Zhang, Z. Zhang, W. Miller, and D. J. Lipman. 1997. Gapped BLAST and PSI-BLAST: a new generation of protein database search programs. Nucleic Acids Res. 25: 3389–3402.

Bevan, M., K. Mayer, O. White, J. A. Eisen, D. Preuss, T. Bureau, S.L. Salzberg, H. W. Mewes. 2001. Sequence and analysis of the *Arabidopsis* genome. Curr. Opin. Plant Biol. 4: 105–110.

Conery, J. S., and M. Lynch. 2001. Nucleotide substitutions and the evolution of duplicate genes. Pacific Symp. Biocomput. 6: 167–178.

Cormen, T. H., C. E. Leiserson, and R. L. Rivest. 1990. Introduction to Algorithms. McGraw-Hill.

Grant, D., P. Cregan, and R. C. Shoemaker. 2000. Genome organization in dicots: genome duplication in *Arabidopsis* and synteny between soybean and *Arabidopsis*. Proc. Natl. Sci. USA 97: 4168–4173.

Gu, Z., A. Cavalcanti, F.-C. Chen, P. Bouman, and W.-H. Li. 2002. Extent of gene duplication in the genomes of *Drosophila*, nematode, and yeast. Mol. Biol. Evol. 19: 256–262.

Li, W.-H. 1999. Molecular Evolution. Sinauer Assocs., Sunderland, MA.

Lynch, M. 1997. Mutation accumulation in nuclear, organelle, and prokaryotic transfer RNA genes. Mol. Biol. Evol. 114: 914–925.

Lynch, M. 2003. Gene duplication and evolution. In A. Moya (ed.), Evolution: From Molecules to Ecosystems. Oxford University Press. (in press).

Lynch, M., and J. Conery. 2000. The evolutionary fate and consequences of duplicate genes. Science 290: 1151–1154.

Lynch, M., and A. Force. 2000. The origin of interspecific genomic incompatibility via gene duplication. Amer. Natur. 156: 590–605.

Lynch, M., M. O'Hely, B. Walsh, and A. Force. 2001. The probability of fixation of a newly arisen gene duplicate. Genetics 159: 1789–1804.

Ohno, S. 1970. Evolution by Gene Duplication. Springer-Verlag, Berlin.

Shapira, S. K., and V. G. Finnerty. 1986. The use of genetic complementation in the study of eukaryotic macromolecular evolution: rate of spontaneous gene duplication at two loci of *Drosophila melanogaster*. Mol. Biol. Evol. 23: 159–167.

Shioura, A., A. Tamura, and T. Uno. 1997. An optimal algorithm for scanning all spanning trees of undirected graphs. SIAM J. Comput. 26: 678–692.

Sokal, R. R., and F. J. Rohlf. 1995. Biometry. 3rd Ed. Freeman, Yew York.

Yang, Z. 1997. PAML: a program package for phylogenetic analysis by maximum likelihood. Comput. Appl. Biosci. 13: 555–556.

A. Meyer, Y. Van de Peer (eds.), Genome Evolution, 45-52.

Functional evolution in the ancestral lineage of vertebrates or when genomic complexity was wagging its morphological tail

Rami Aburomia[1], Oded Khaner[1,2] & Arend Sidow[1]*
[1]*Department of Pathology, Stanford University Medical Center, Room 248B, 300 Pasteur Drive, Stanford, CA 94305–5324, USA;* [2] *Present address: Department of Bio-Medical Sciences, Hadassah College, Jerusalem, Israel.*
* *Author for correspondence: E-mail: arend@stanford.edu*

Received 23.02.2002; accepted in final form 29.08.2002

Key words: cis-regulatory changes, early vertebrate evolution, Myb gene family, Pax 2/5/8

Abstract

Early vertebrate evolution is characterized by a significant increase of organismal complexity over a relatively short time span. We present quantitative evidence for a high rate of increase in morphological complexity during early vertebrate evolution. Possible molecular evolutionary mechanisms that underlie this increase in complexity fall into a small number of categories, one of which is gene duplication and subsequent structural or regulatory neofunctionalization. We discuss analyses of two gene families whose regulatory and structural evolution shed light on the connection between gene duplication and increases in organismal complexity.

Introduction

We are here concerned with the correlation of organismal complexity and new molecular functionality in the ancestral lineage of vertebrates. In principle, three kinds of molecular changes can give neomorphic phenotypes that could contribute to increased organismal complexity. They are, (1) cis-regulatory changes that affect timing and/or place of gene expression, (2) structural changes due to missense mutations or more severe lesions, and (3) entirely new proteins (or functional RNAs). Our analyses focus on the first two classes of changes, cis-regulatory and structural, for which duplication of an intact gene provides the facilitating raw material on which evolutionary pressures can act.

It has been clear for some time now that the ancestral lineage of vertebrates contains an excess of gene duplications in comparison to most other chordate lineages, excepting those that underwent recent genome duplications such as fish and amphibians. By contrast, the amount of morphological evolution that occurred in different lineages of chordates has not been quantified. It was therefore not known whether the increase in morphological complexity in the vertebrate ancestor was unusual or whether other vertebrate lineages underwent equally dramatic morphological changes. To add to the debate of gene duplication and its relationship with increases in organismal complexity at the origin of vertebrates, we present analyses that are intended to shed light on the interface of the two processes.

Results

Morphological complexity in early vertebrate evolution

We devised a method to estimate the amount of change in morphological complexity during all of vertebrate evolution (O. Khaner and A. Sidow, unpublished data). We first scored 21 extant higher-order chordate groups for the presence or absence of 479 morphological characters whose states for each group were obtained from the literature (Holland, 1996; Baker and Bronner-Fraser, 1997a, 1997b; Gilbert and Raunio, 1997; Kardong, 1997; Pough *et al.*, 1999;

Table 1. Number of characters scored in each subgroup of organismal traits

System	No. of Characters
Early embryonic development	47
Notochord, vertebrae, skull and jaws	12
Musculature	25
Cardiovascular and respiratory systems	45
Urogenital system	35
Integument	34
Nervous system	62
Neural crest	16
Sensory organs	70
Endocrine system	26
Digestive system	48
Appendages	29

Shimeld and Holland, 2000). Table 1 shows the breakdown of the morphological characters into different organ systems. State transitions were inferred from the resulting matrix and mapped onto the currently accepted phylogenetic tree with MacClade (Maddison and Maddison, 2001). We then defined the Morphological Complexity Index, $MCI_b = (G_b - L_b)/T$, where subscript b denotes the branch in question, G_b and L_b the number of gained and lost characters on that branch, and T the total number of characters (479 in this case). We calculated MCI_b for each branch of the tree and, upon connecting the nodes of the last common ancestors of the major chordate classes, found that there were two phases with dramatically different rates of increase of the MCI (Fig. 1). The first phase (I in Fig. 1), during which jawless and jawed vertebrates evolved, had an extremely high rate of increase. The second phase (II in Fig. 1), during which the major vertebrate classes evolved, had a ten-fold slower rate of increase. Thus, the comparatively short period of time — between 50 and 100 million years — in which many gene duplications occurred and many new genes arose, coincides with the greatest rate of increase in morphological complexity.

For phase I, we see no comparable increase in the average rate of point substitutional evolution as estimated from standard treeanalyses. In fact, some of the difficulties encountered in reconstruction of phylogenetic trees of gene families that duplicated during this critical period is due to a paucity of missense changes. We conclude that the molecular events that did happen had disproportionately large phenotypic and functional consequences. We believe that the challenge in this area lies in the identification of functionally important changes in gene families, and the mapping of those changes onto robust gene trees.

Gene duplications in early vertebrate evolution

Inferring functional evolution in a gene family requires robust gene trees; wrong trees cause misinterpretation of character evolution. Unfortunately for the inference of functional evolution of vertebrate gene families, the dissection of the exact branching patterns is fraught with difficulties that have been amply discussed in the '2R' debate (Sidow, 1996; Spring, 1997; Hughes, 1999; Meyer and Schartl, 1999; Martin, 2001). The large number of gene duplications during the short time of early vertebrate evolution suggest tetraploidization as a reasonable explanation for the majority of duplications observed. This led to the prediction that treeanalyses of families with four vertebrate paralogs should generate the branching pattern (A,B),(C,D), but many gene families do not follow this pattern (Gibson and Spring, 2000). Long branch attraction (LBA) has since been cited as one of the artefacts that may cause a failure to recover the symmetric pattern but it cannot explain all cases. Treeanalyses are therefore regarded as inconclusive at best with respect to mechanism underlying the gene duplications.

Ploidy increases are common in evolution, as evidenced by the recent tetra- or polyploidizations in the grasses (Ahn and Tanksley, 1993; Gale and Devos, 1997), the genus *Xenopus* (Kobel and Du Pasquier, 1986), and Salmonid fishes (Allendorf and Utter, 1976). Several ancient tetraploidizations are also known, including those of yeast (Wolfe and Shields, 1997), flowering plants (Doyle *et al.*, 1990), and teleost fish (Amores *et al.*, 1998). It is by far the most common mechanism for the duplication of many genes during a short amount of time.

Chromosomes of a genome undergo diploidization independently after an autotetraploidization (Fig. 2A). This process can take tens of millions of years, as evidenced by the Salmonid genome duplications. This is shown in Figure 2 for three hypothetical chromosomes which remained allelic for different amounts of time after the tetraploidization. As a consequence, the beginning of independent evolution of paralogs from alleles of a tetraploid will vary greatly across the genome. This is of importance for phylogenetic reconstruction because it is the beginning of

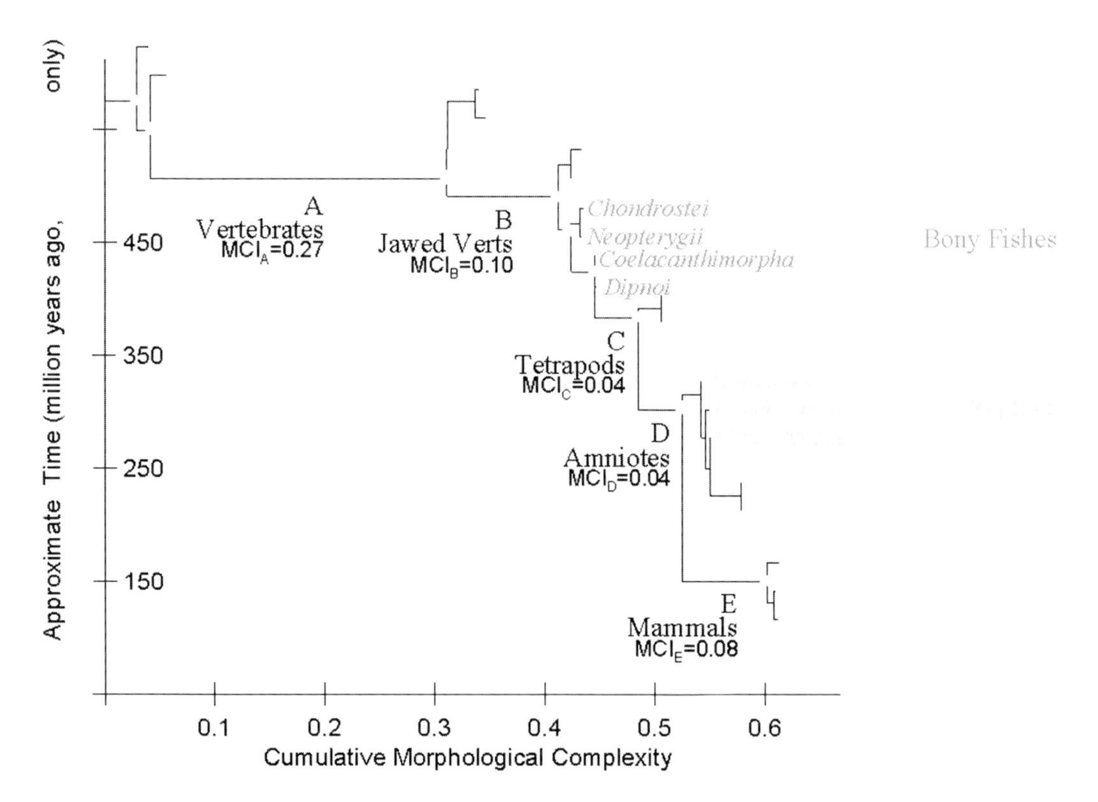

Figure 1. Phylogenetic tree of the groups analyzed in the study of morphological complexity. Branch lengths correspond to the values of the MCI for each branch and are drawn exactly to scale. The Y axis is approximate time with respect to the last common ancestors of the chordate subphyla and vertebrate classes, which are denoted by the red dots. Ancestral lineages with major transitions in morphology are labeled with their names and the values of the MCI. Red lines qualitatively illustrate the rate of increase of morphological complexity during two phases of vertebrate evolution. The rate of increase during Phase I was ten-fold higher than that of Phase II.

independent evolution of duplicates, not the autotetraploidization, that marks a gene duplication in a gene tree.

When one tetraploidization is rapidly followed by a second one (or by independent gene duplications), the time between diploidizations of different genes may vary considerably (for simplicity, Fig. 2 shows concurrent diploidizations after the second tetraploidization but this is unlikely to occur). In our example, chromosome I undergoes diploidization rapidly whereas chromosome III undergoes diploidization just before the next duplication; chromosome II is shown as intermediate. As a result, there is much phylogenetic signal that separates the duplications in the descendants of gene 1, but there is little or none for gene 3.

This situation is exacerbated when all four paralogs are retained (Fig. 2B). The times of diploidizations of the second round of paralogs, for example, 3a/3b and 3c/3d, are independent of each other. The phylogenetic signal for recovering the correct tree is small or even nonexistent if just one of the two diploidizations (for example, 3y to 3a and 3b) follows the first one (3 to 3y and 3z) rapidly. As a consequence, for many gene families in the genome, insufficient time may have elapsed between successive diploidizations to generate a phylogenetic signal that is detectable 500 million years later. A minimum of differences in evolutionary rates to avoid LBA, and sufficient time between successive diploidizations, seem to be necessary to recover the correct phylogenetic relationship. In the next section, we detail the evolution of two gene families and show how recovery of the correct tree is of paramount importance in the analysis of functional evolution of paralogs.

Functional evolution of the Pax2/5/8 gene family

Phylogenetic trees of the Pax2/5/8 gene family have been published in several reports (Heller and Brändli, 1999; Wada *et al.*, 1998; Kozmik *et al.*, 1999). All of them show a rooting of the vertebrate subtree on the

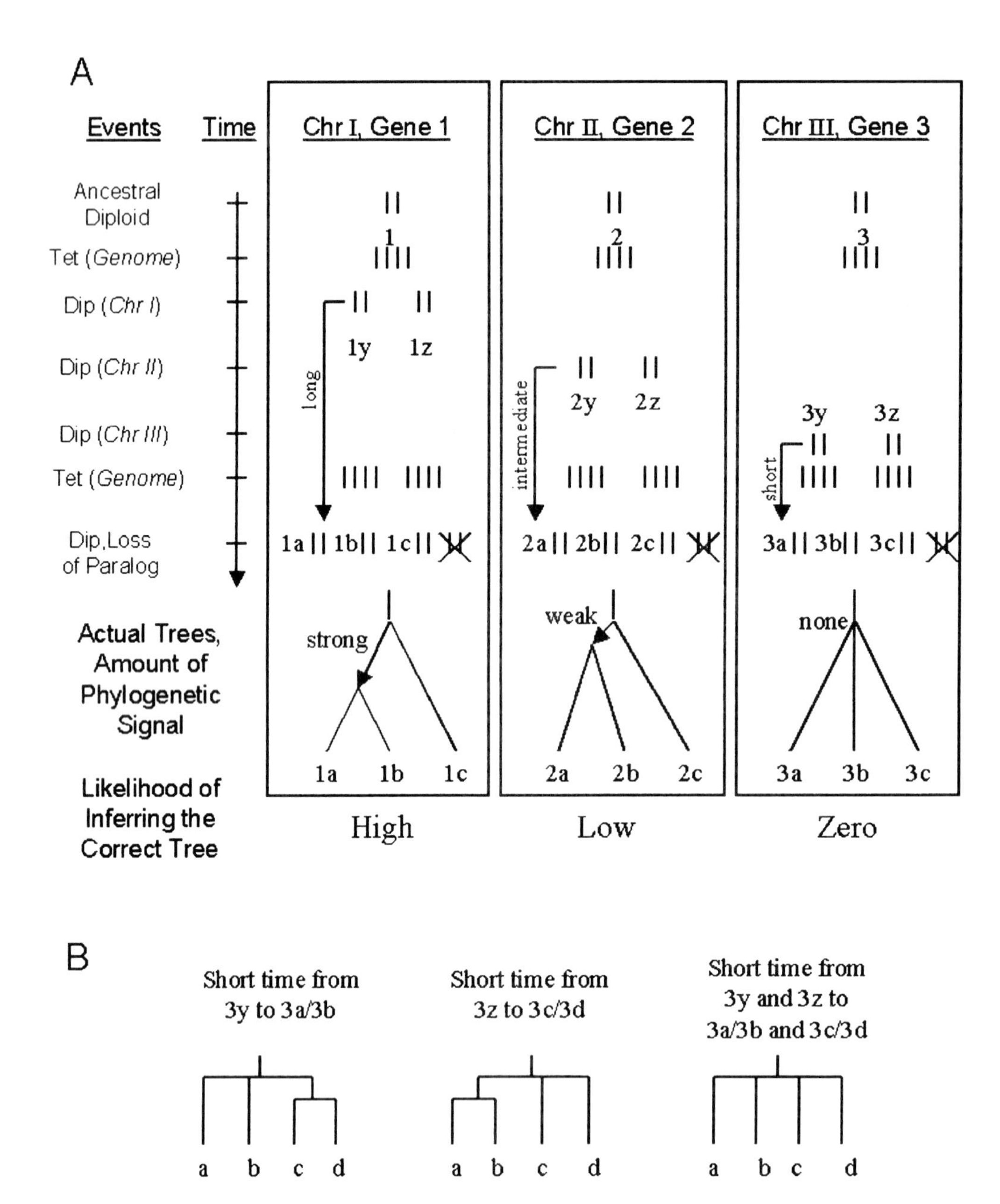

Figure 2. The relationship between timing of tetraploidization and chromosomal diploidization, and their effect on tree reconstruction. (A). Events are shown on the left and time progresses downward. For three chromosomes of a hypothetical genome, the tetraploidizations (Tet) happen at the same time, but each chromosome varies in the amount of time during which all four homologs remain allelic. Dip (diploidization) indicates the time at which the two pairs of homologs begin to evolve independently. Long, intermediate, and short is the amount of time between a chromosome's first and second diploidization. (B). Four-paralog cases analogous to Chr III, Gene 3. When all paralogs are retained, the phylogenetic signal may tend toward zero when the time between diploidizations is short for either pair of paralogs.

ancestral lineage of Pax8, with the Pax258 ancestor duplicating once to yield Pax8 and Pax25, and Pax25 duplicating again to give extant Pax2 and Pax5. We here suggest that this arrangement is erroneous, and that it is most likely due to LBA between invertebrate Pax258 and vertebrate Pax8: Pax8 evolves more than two-fold faster than Pax5 and nearly three-fold faster than Pax2 (Fig. 3A).

It is interesting to note that the rates of evolution of Pax2, Pax5, and Pax8 inversely correlate with the duration of the genes' expression and the importance of their functions as inferred from mouse knockout

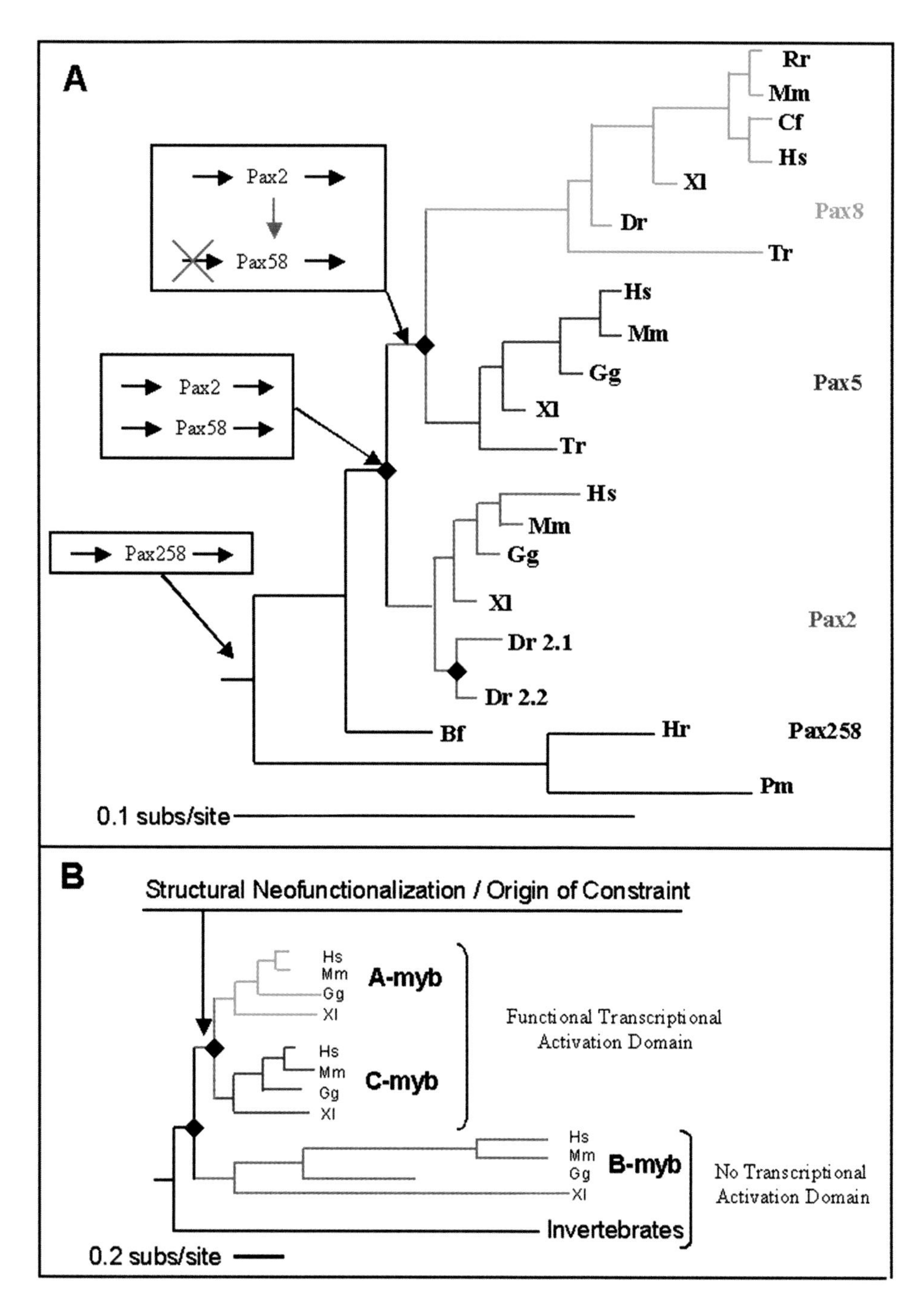

Figure 3. Regulatory and structural evolution in the Pax2/5/8 and Myb gene families. Diamonds are gene duplications. (A). Regulatory evolution of the Pax2/5/8 gene family. Arrows in boxes denote interactions in a simplified regulatory hierarchy. Arrows to the left of the Pax genes signify the initial signal that 'turns on' Pax expression. Each state is mapped onto the appropriate branch of the tree. (B). The Myb case study for illustration of protein neofunctionalization after gene duplication. Branch lengths of each Myb paralog are calculated from the positions corresponding to the transcriptional activation region only, and do not represent evolution of the entire protein. The transcriptional activation domain is constrained equally in A and C, but much less so in B. The arrow denotes the position of the origin of the constraint, and of the transcriptional activation function, which is shared between A and C, but not present in B and invertebrates. Maximum likelihood using PROTML was used to build the trees (Adachi and Hasegawa, 1992). Abbreviations: Bf, *Branchiostoma floridae*; Cf, *Canis familiaris*; Dr, *Danio rerio*; Gg, *Gallus gallus*; Hr, *Halocynthia roretzi*; Hs, *Homo sapiens*; Mm, *Mus musculus*; Rr, *Rattus rattus*; Pm, *Phallusia mammilata*; Tr, *Takifugu rubripes*; Xl, *Xenopus laevis*.

studies. Given that the physicochemical constraints on these closely related proteins are essentially equivalent (indeed, the protein coding regions can substitute for one another in knock-in experiments), the differences in their rates of evolution are a direct reflection of their differences in expression domains and timings.

We built the Pax2/5/8 tree in two steps in order to maximize the amount of usable sequence data. First, by excluding invertebrate chordates we were able to use 238 unambiguously homologous positions to build the best vertebrate-only tree. We then tested alternative rootings of the vertebrate tree, whose topology we fixed, with the invertebrate sequences. Only 136 positions, which are mostly in the highly conserved Paired domain, could be used for this analysis. The tree shown in Fig. 3A, in which the rooting is on the Pax2 lineage, has a log-likelihood value that is 3.8 points greater than the previously published arrangement in which the root falls on the Pax8 lineage.

In order to understand whether the duplications in this gene family facilitated the evolution of more complex regulatory hierarchies, we then asked whether any regulatory changes could be mapped to the periods after the gene duplications. We surveyed the extensive literature on the function and expression domains of each paralog and used the parsimony principle to map the regulatory changes onto the tree.

Pax258 of invertebrate chordates is expressed in a thin strip of cells in a region of the neural tube which may be homologous to the vertebrate mid-hindbrain (Wada *et al.*, 1998; Kozmik *et al.*, 1999). In vertebrates, the paralogs are expressed at the mid-hindbrain boundary (MHB) in spatial and temporal domains which are partially overlapping (Aasano and Gruss, 1992; Millet *et al.*, 1996; Murphy and Hill, 1991). In almost all vertebrate organs where these genes are expressed, Pax2 expression is initiated prior to that of Pax5 and Pax8 (Pfeffer *et al.*, 1998, 2002; Heller and Brändli, 1999; Bouchard *et al.*, 2000). It was also shown that Pax2 protein is required for maintaining Pax5 and Pax8 transcription, and that the mouse Pax2 knockout has a much more severe brain phenotype than those of Pax8 and Pax5.

Right after the duplication of Pax258 into Pax2 and Pax58, their regulation must have been equivalent. Then, on the ancestral lineage of Pax58, two events occurred that converted this equivalence into a hierarchical relationship: subfunctionalization, when Pax58 lost the ability to respond to the ancestral signal that initiated Pax258 gene expression; and neofunctionalization, when Pax58 gained the susceptibility to be controlled by Pax2. This combination of subfunctionalization and neofunctionalization is a likely hallmark of the increase in regulatory complexity that occurred in the vertebrate ancestor, and provides one molecular link to the increase in organismal complexity.

Functional evolution of the Myb gene family

Our analysis of the Pax2/5/8 gene family suggests how regulatory evolution may confer developmental, and therefore morphological, complexity. In this section, we use the Myb gene family to illustrate the evolution of a new biochemical function after an early vertebrate gene duplication.

Metazoan Myb genes have three highly conserved Myb repeats that function as DNA binding domains. Invertebrate Mybs and B-Myb function in the regulation of the cell cycle. A- and C-Myb are also known to have an additional independent transcriptional activation function that is missing in invertebrate Mybs and B-Myb. In A- and C-Myb, the second most-constrained region (after the Myb repeats) is the acidic domain, which has been shown to carry this transcriptional activation function.

We built the best tree of the Myb gene family, which agreed with previously published reports (Ganter and Lipsick, 1999). The vertebrate Myb tree is rooted on the B lineage, with A and C sharing a more recent common ancestor. LBA is not a problem because the average rate of evolution in those regions in which A-Myb, B-Myb, and C-Myb can be reliably aligned is very similar among the paralogs (Simon *et al.*, 2002). However, an analysis of the regions that correspond to the transcriptional activation domain revealed vastly different rates of evolution. B-myb evolves at a 6.6-fold faster rate than the average of the B-Myb protein (Fig. 3B; Simon *et al.*, 2002). In contrast, the corresponding region in both A-myb and C-myb evolves only 1.4 times faster than their average. Using the phylogeny of the Myb gene family, we can map the origin of this constraint onto the ancestral lineage of A- and C-myb, after the B-myb lineage diverged. This is also the most parsimonious placement of the origin of the transcriptional activation function.

Discussion

These case studies underline the potential pitfalls (wrong trees and their causes) in the interpretation of organismal and genomic evolution. Had we not taken special care to infer the best trees, our tree-based interpretation of the evolutionary events in the two gene families during this critical time period would have been misled. In our experience, building the best tree from the ingroup sequences first, and then rooting this tree with the outgroup sequences, consistently maximizes the sequence information used for building the tree.

The main purpose of this study was to explore whether connections could be made between the increase in genomic, developmental, and organismal complexity at the origin of vertebrates. Both regulatory evolution and protein structural evolution are likely to have contributed to an increase in complexity at the origin of vertebrates. We inferred potentially important changes in the regulation of the Pax gene family and in the function of the Myb proteins that occurred *between* consecutive duplication events.

The Pax gene family in particular illustrates the relationship between organismal and molecular complexity well, as it is involved in the patterning of the brain, one of the most important organs that underwent dramatic increases in morphological complexity. One of the characeristics that sets apart the brain of primitive chordates from that of vertebrates is the cerebellum, which forms at the MHB. Pax2 is necessary for its formation, whereas Pax5 and Pax8 have milder cerebellar phenotypes, as knockouts of the gene in mice and zebrafish show. Because the cerebellum is present in jawless vertebrates, but not in invertebrate chordates, the origin of the cerebellum appears to be broadly coincident with the molecular events that differentiated the function of these three Pax genes in brain patterning.

With the ever-increasing amount of genomic and developmental data from a broad range of chordate model organisms, the study of early vertebrate evolution is entering a new phase that uses robust gene trees as a basis for interpretation of functional evolution. Two such case studies are presented here, and we are looking forward to more.

References

Ahn, S. and Tanksley, S.D. (1993) Linkage maps of the rice and maize genomes. *Proc. Natl. Acad. Sci. USA*, **90,** 7980–7984.

Adachi, J. and Hasegawa, M. (1992) *MOLPHY, Programs for Molecular Phylogenetics, I. PROTML, Maximum Likelihood Inference of Protein Phylogeny* (Computer Science Monographs, Vol. 27). Institute of Statistical Mathematics, Tokyo, Japan

Allendorf, F.W. and Utter, F.M., (1976) Gene duplication in the family Salmonidae. III. Linkage between two duplicated loci coding for aspartate aminotransferase in the cutthroat trout (*Salmo clarki*). *Hereditas*, **82,** 19–24.

Amores, A., Force, A., Yan, Y.-L., Joly, L., Amemiya, C., Fritz, A., Ho, R.K., Langeland, J., Prince, V., Wang, Y.-L. *et al.* (1998) Zebrafish *hox* clusters and vertebrate genome evolution. *Science*, **282,** 1711–1714.

Asano, M. and Gruss, P. (1992) *Pax–5* is expressed at the midbrain-hindbrain boundary during mouse development. *Mech. Dev.*, **39,** 29–39.

Baker C.V.H. and Bronner-Fraser M. (1997a) The origins of the neural crest. Part I: Embryonic induction. *Mech. Dev.*, **69,** 13–29.

Baker, C.V.H. and Bronner-Fraser, M. (1997b) The origins of the neural crest. Part II: An evolutionary perspective. *Mech. Dev.*, **69,** 3–11.

Bouchard, M., Pfeffer, P. and Busslinger, M. (2000) Functional equivalence of the transcription factors Pax2 and Pax5 in mouse development. *Development*, **127,** 3703–3713.

Doyle, J.J., Doyle, J.L., Brown, A.H. and Grace, J.P. (1990) Multiple origins of polyploids in the *Glycine tabacina* complex inferred from chloroplast DNA polymorphism. *Proc. Natl. Acad. Sci. USA*, **87,** 714–717.

Gale, M.D. and Devos, K.M. (1997) Comparative genetics in the grasses *Proc. Natl. Acad. Sci. USA*, **95,** 1971–1974.

Ganter, B. and Lipsick, J.S. (1999) Myb and oncogenesis. *Adv. Cancer Res.*, **76,** 21–60.

Gibson, T.J. and Spring, J. (2000) Evidence in favour of ancient octaploidy in the vertebrate genome. *Biochem. Soc. Trans.*, **28,** 259–264.

Gilbert, S.F. and Raunio, A.M. (1997) *Embryology*, Sinauer, Sunderland, MA.

Heller, N. and Brändli, A.W. (1999) *Xenopus Pax–2/5/8* orthologues: novel insights into *Pax* gene evolution and identification of Pax–8 as the earliest marker for otic and pronephric cell lineages. *Dev. Genet.*, **24,** 208–219.

Holland P.W.H. (1996) Molecular biology of lancelets: insights into development and evolution. *Israel J. Zool.*, **42,** 247–272.

Hughes, A.L. (1999) Phylogenies of developmentally important proteins do not support the hypothesis of two rounds of genome duplication early in vertebrate history. *J. Mol. Evol.*, **48,** 565–576.

Kardong, K.V. (1997) *Vertebrates*, 2nd ed., McGraw-Hill, New York, NY.

Kobel, H.R. and Du Pasquier, L. (1986) Genetics of polyploid *Xenopus*. *Trends Genet.*, **2,** 310–315.

Kozmik, Z., Holland, N.D., Kalousova, A., Paces, J., Schubert, M. and Holland, L.Z. (1999) Characterization of an amphioxus paired box gene, *AmphiPax2/5/8*: developmental expression patterns in optic support cells, nephridium, thyroid-like structures and pharyngeal gill slits, but not in the midbrain-hindbrain boundary region. *Development*, **126,** 1295–1304.

Maddison, D.R. and Maddison, W.P. (2001) *MacClade Version 4.0*, Sinauer, Sunderland, MA.

Martin, A. (2001) Is tetralogy true? Lack of support for the 'one-to-four rule'. *Mol. Biol. Evol.*, **18,** 89–93.

Meyer, A. and Schartl, M. (1999) Gene and genome duplications in vertebrates: the one-to-four (-to-eight in fish) rule and the evolution of novel gene functions. *Curr. Opin. Cell Biol.*, **11,** 699–704.

Millet, S., Bloch-Gallego, E., Simeone, A. and Alvarado-Mallart, R.M. (1996) The caudal limit of *Otx2* gene expression as a marker of the midbrain/hindbrain boundary: a study using in situ hybridisation and chick/quail homotopic grafts. *Development*, **122,** 3785–3797.

Murphy, P. and Hill, R.E. (1991) Expression of the mouse labial-like homeobox-containing genes, *Hox 2.9* and *Hox 1.6*, during segmentation of the hindbrain. *Development*, **111,** 61–74.

Pfeffer, P.L., Gerster, T., Lun, K., Brand, M. and Busslinger, M. (1998) Characterization of three novel members of the zebrafish *Pax2/5/8* family: dependency of *Pax5* and *Pax8* expression on the *Pax2.1* (*noi*) function. *Development*, **125,** 3063–3074.

Pfeffer, P.L., Payer, B., Reim, G., di Magliano, M.P. and Busslinger, M. (2002) The activation and maintenance of *Pax2* expression at the mid-hindbrain boundary is controlled by separate enhancers. *Development*, **129,** 307–318.

Pough, F.H., Janis, C.M. and Heiser J.B. (1999) *Vertebrate Life*, 5th ed., Prentice Hall, Upper Saddle River, NJ.

Shimeld, S.M. and Holland, P.W.H. (2000) Vertebrate innovations. *Proc. Natl. Acad. Sci. USA*, **97,** 4449–4452.

Sidow, A. (1996) Gen(om)e duplications in the evolution of early vertebrates. *Curr. Opin. Genet. Dev.*, **6,** 715–22.

Simon, A.L., Stone, E.A. and Sidow, A. (2002) Inference of functional regions in proteins by quantification of evolutionary constraints. *Proc. Natl. Acad. Sci. USA*, **99,** 2912–2917.

Spring, J. (1997) Vertebrate evolution by interspecific hybridisation--are we polyploid? *FEBS Lett.*, **400,** 2–8.

Wada, H., Saiga, H., Satoh, N. and Holland, P.W. (1998) Tripartite organization of the ancestral chordate brain and the antiquity of placodes: insights from ascidian *Pax–2/5/8*, *Hox* and *Otx* genes. *Development*, **125,** 1113–1122.

Wolfe, K.H. and Shields, D.C. (1997) Molecular evidence for an ancient duplication of the entire yeast genome. *Nature*, **387,** 708–713.

A. Meyer, Y. Van de Peer (eds.), Genome Evolution, 53-63.

Numerous groups of chromosomal regional paralogies strongly indicate two genome doublings at the root of the vertebrates

Lars-Gustav Lundin*, Dan Larhammar & Finn Hallböök
Department of Neuroscience, BMC, Uppsala University, S–751 24 Uppsala, Sweden.
Author for correspondence: E-mail: lglundin@neuro.uu.se

Received 24.01.2002; accepted in final form 29.08.2002

Key words: conservation of chromosomal regions, genome doubling, paralogous chromosomal regions, vertebrate evolution

Abstract

The appearance of the vertebrates demarcates some of the most far-reaching changes of structure and function seen during the evolution of the metazoans. These drastic changes of body plan and expansion of the central nervous system among other organs coincide with increased gene numbers. The presence of several groups of paralogous chromosomal regions in the human genome is a reflection of this increase. The simplest explanation for the existence of these paralogies would be two genome doublings with subsequent silencing of many genes. It is argued that gene localization data and the delineation of paralogous chromosomal regions give more reliable information about these types of events than dendrograms of gene families as gene relationships are often obscured by uneven replacement rates as well as other factors. Furthermore, the topographical relations of some paralogy groups are discussed.

Introduction

For a long time a central and enigmatic question in evolutionary reasoning has been the possible differences in character between micro- and macroevolution. According to the Neo-Darwinian view, macroevolution is just the adding up of many small microevolutionary changes. Today it seems fairly obvious that the Neo-Darwinian explanation can be only part of the answer for drastic transitions such as the ones from unicellular protozoans to multicellular organisms, diploblasts to Bilateria or from simple chordates to vertebrates (Lundin, 1999). This is not to say that microevolutionary changes are not a necessary part of the macroevolutionary process but may not in many cases be the whole explanation. A priori, it seems reasonable to assume that a certain input of 'new' genes might be needed for the construction of more complicated organisms.

A decisive turning point, in our conception of the mechanisms underlying macroevolutionary events was the publication of articles by Susumu Ohno and coworkers (Ohno, 1969; Ohno *et al.*, 1968) followed by Ohno's book *Evolution by Gene Duplication* (Ohno, 1970). These changed the scene rather drastically in suggesting gene duplications as the necessary prerequisite for the evolution of new structures and functions. Ohno took a firm stand against the prevailing Neo-Darwinian view in his statement that 'allelic mutations of already existing gene loci cannot account for major changes in evolution'.

Two prominent processes leading to the presence of paralogous genes are regional duplications and total genome doubling (tetraploidization). The duplicated genes will slowly diverge in structure under the influence of mutation, selection and drift. Eventually one of them might have changed into a gene coding for a protein with a related but distinctly different function whereas the other gene in the pair will have kept the original function. In fact, numerous examples of this type of phenomenon have been described and gene duplications early in animal evolution often gave rise to separate gene families within superfamilies (Ono *et al.*, 1999; Suga *et al.*, 1999b, 1999c). Later duplications such as those found among vertebrate species have been described as isoforms of paralogous genes and their corresponding proteins (Suga *et al.*, 1999a). These isoform duplications most likely took place before the fish-tetrapod split and the corresponding proteins have similar activities but are very often tissue-specific (Iwabe *et al.*, 1996).

Basically the view on microevolution versus macroevolution boils down to a question of time perspective on the processes involved. Drastic changes such as major gene duplication events will not immediately lead to visible macroevolutionary changes in the species concerned but are in some cases necessary prerequisites for later dramatic changes in form and function. Within the metazoans the evolution of vertebrates is very well suited for further analyses of major macroevolutionary events. In molecular terms, with the likely genome doublings at the root of the vertebrates, this suphylum is probably the most rewarding to tackle, to a large extent because we have access to more genetic and biochemical data than for other major animal groups. Among the invertebrates we only have abundant molecular data from two protostomian species, *Caenorhabditis elegans* and *Drosophila melanogaster.*

In this article we review three groups of paralogous chromosomal regions in the human genome and discuss the indicative values of different parameters and approaches used to solve questions associated with gene and putative genome duplications in vertebrate evolution.

Paralogous chromosomal regions in vertebrate genomes

A parallel to the dramatic evolutionary events that took place in Cambrium, although possibly unfolding within a somewhat longer period of time, is the spectacular appearance of the different vertebrate groups. In this latter case a scenery is beginning to reveal its features. An hypothesis for the basis of these dramatic changes postulates that there were two major waves of gene duplications in an early chordate, most likely two rounds of genome doubling, leading to the evolution of jawed vertebrates (Ohno, 1970, 1998; Lundin, 1989, 1993; Holland *et al.*, 1994; Katsanis *et al.*, 1996; Sidow, 1996; Nadeau and Sankoff, 1997; Spring, 1997; Pebusque *et al.*, 1998; Meyer and Schartl, 1999; Popovici *et al.*, 2001). However, there seems to still be some hesitation to connect major evolutionary events, such as the occurrence of the neural crest or the central nervous system of the vertebrates, with possible or likely genome duplications (Skrabanek and Wolfe, 1998; Hughes, 1999; Martin, 2001).

In the present article different parameters relevant to the study of gene and genome duplications will be discussed and it is obvious that chromosomal locations of paralogous genes is a paramount variable to be considered in attempts to find paralogous chromosomal regions including groups of genes that have been transposed by inversions, fissions and translocations. For the locations of genes we have in most cases used data from OMIM (www3.ncbi.nlm.nih.gov/Omim/). In this process two central factors are number of paralogues in different gene families or subfamilies within a species and gene silencing. After two total genome doublings from an ancestral species, with a basic setup of single genes within families, four paralogues are to be expected in the ideal case. This is far from the real present situation due to gene silencing and often to the fact that all close paralogues in a gene family have not yet been found and mapped. Very old duplications will also allow for considerable changes in structure and function of genes involved.

Other important components in the analyses are identification of orthologous genes in other vertebrate species including cyclostomes and in different invertebrates as well as phylogenetic relationships between paralogous genes. It is also important to look out for large differences in the rate of evolutionary change in paralogues to be able to make an approximate estimate of the time of gene duplications.

One factor that has sometimes been overlooked is the time span between the putative genome doublings in the emerging vertebrates. The length of this span is likely to be of great importance for the information gained by dendrograms of gene families among other things.

Genes in different gene families or subfamilies with two, three or four members which are located on different chromosomal arms are useful to define conserved paralogous regions. These chromosomal regions are the ones of particular interest here and carry what Ono *et al.* (1999) and Suga *et al.* (1999b) have named isoform duplications. Gene families with five or more unlinked members can sometimes be split up in subfamilies where the gene duplications in a particular subfamily can be shown to have taken part within the vertebrate line. However, large gene families often contain members which are the results of very early duplications.

The paralogy groups shown here have been published earlier (Lundin, 1989, 1993; Spring *et al.*, 1994; Kasahara *et al.*, 1996; Katsanis *et al.*, 1996; Pebusque *et al.*, 1998) but are updated, expanded and graphically displayed (Figures 1–3). These figures show paralogous genes from different gene families primarily in four chromosomal regions but other regions are also included to demonstrate the likely outcome of chromosomal rearrangements. It should be noted that we have mainly chosen gene families with representatives in three or four chromosomal regions. However, a large number of gene families with genes mapped in the human gene map have only two known members. The majority of these genes, relevant for the paralogous regions, have been left out for technical reasons, not to overload the figures.

One of the paralogous groups, that has aroused much interest as well as criticism, consists of regions from chromosomes 1, 6p, 9 and 19p (Katsanis *et al.*, 1996) or part of this group (Kasahara *et al.*, 1996) (Fig. 1). Originally some paralogues, that stem from old branches before the separation of deuterostomes and protostomes, were included in this group but as discussed below this is nevertheless likely to reflect a conserved constellation of genes (Kasahara *et al.*, 1996; Smith *et al.*, 1999). Some of the gene families with early or very early branches on chromosomes 6 and 9 are ABC (ABC transporter), PSMB (Proteasome subunit, beta type) and HSPA (Heat-shock 70-kD protein). Many of the paralogous genes in this group are located around the MHC region.

The prime example of a group of paralogous regions in the human gene map contains genes from many families located in 2q, 7, 12q13 and 17q and is found in Figure 2 (Lundin, 1989, 1993). These regions contain representatives of some gene families that are found only in animals such as four clusters of genes for Hox proteins (HOX) and genes for voltage gated sodium channels (SCN) and for fibrillar collagens (COL) of a different subfamily than those found on chromosomes 1, 6 and 9 (Fig. 1).

In figure 3 are found regions from human chromosomes 4, 5q, 8, and 10q (Pebusque *et al.*, 1998). Some of the prominent gene families in these code for fibroblast growth factor receptors (FGFR), annexins (ANXA), glycine and GABA receptors (GLR and GABR), NPY receptors (NPY*R) and secreted frizzled-related proteins (SFRP). Paralogues missing in chromosomes 8 and 10 are frequently found in 2p and 13q.

Chromosomal rearrangements and gene silencing obscure original paralogous regions

Extensive comparisons of gene maps from many mammals demonstrate that chromosomal changes such as translocations and inversions have been frequent during evolution (Chowdhary *et al.*, 1998). Despite these chromosomal rearrangements it is possible to find a great number of orthologous chromosomal regions in comparisons between different mammalian gene maps (O'Brien *et al.*, 1997; Searle and Selley, 1997; Nadeau and Sankoff, 1998; Burt *et al.*, 1999) as well as comparisons based on ZOO-FISH staining (Scherthan *et al.*, 1994; Chaudhary *et al.*, 1996; Chowdhary *et al.*, 1996, 1998; Frönicke *et al.*, 1996; Raudsepp *et al.*, 1996). It is clear that rearrangements during mammalian evolution have been far more numerous in small mammals with short generation times such as the house mouse whereas in comparisons between larger mammals such as horses, cattle and man most often very large chromosomal regions have been conserved. Recently, an estimate was made of the relative roles of translocations and inversions in vertebrate evolution by comparing the zebrafish and human gene maps (Postlethwait *et al.*, 2000). The conclusion was that intrachromosomal rearrangements have been fixed more frequently than translocations.

Obviously, translocations and inversions make finding paralogous regions more difficult but it is often possible to trace such changes in the chromosomal regions included in the diagrams presented here.

Inversions

Chromosome 1 carries no less than 18 pairs or groups of paralogues located either on the two different chromosomal arms or in two cases far apart on one arm (Lundin, 1993; Lundin *et al.*, submitted). A reasonable explanation is that at least 11 of these groups result from one ancient, large regional duplication and that later on there was a pericentric inversion that split this duplicated region.This may be the reason why genes in the chromosome 1, 6p, 9 and 19p paralogy group (Fig. 1) are found both in 1p3, 1p2–1 and 1q2.

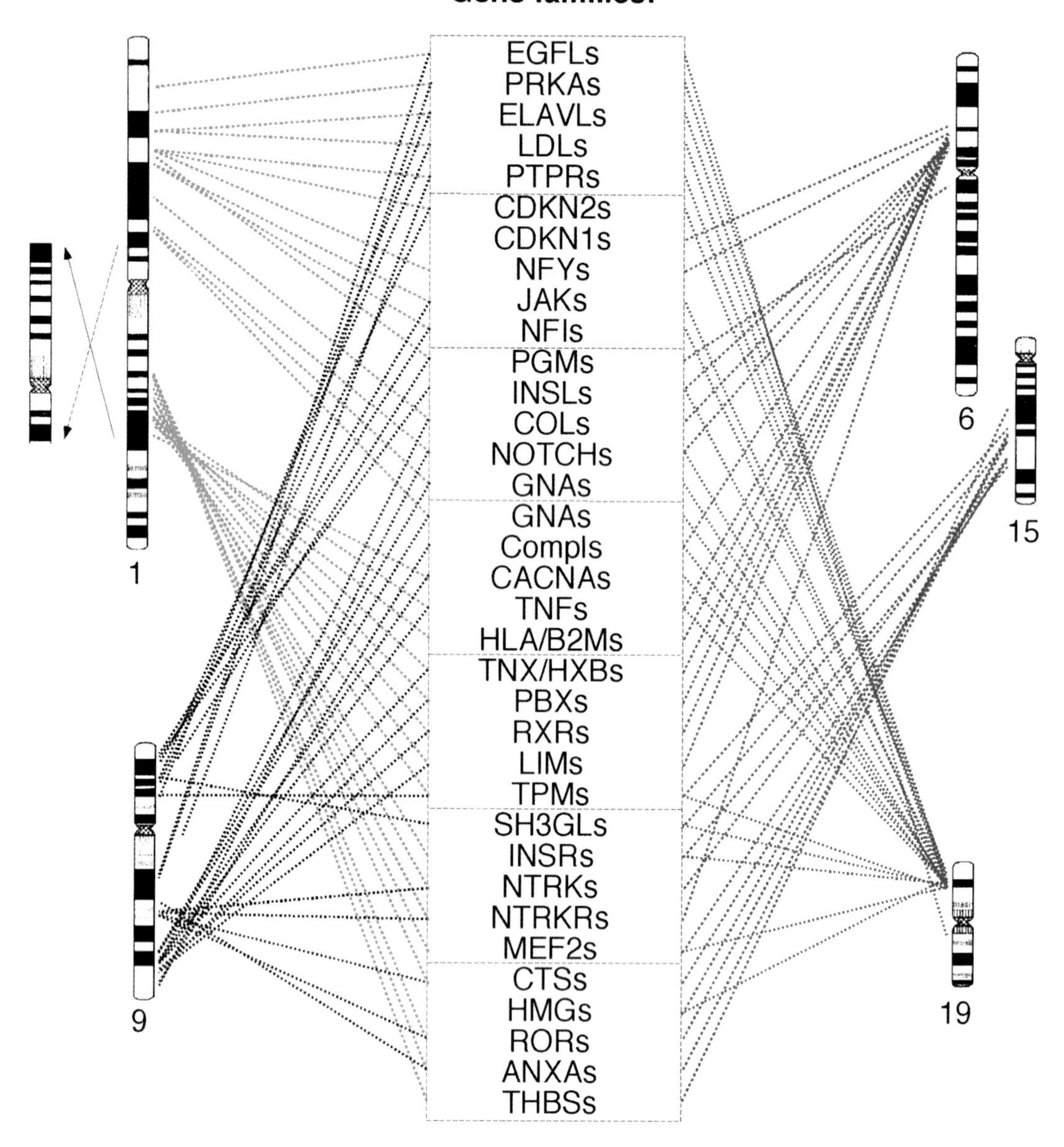

Figure 1. Gene family symbols and names.

EGFL, Epidermal growth factor-like; PRKA, cAMP-dependent protein kinase (catalytic); ELAVL, Embryonic lethal, abnormal vision (*Drosophila*, elav); LDL, LDL-receptor family, subfamily of short receptors; PTPR, Protein-tyrosine phosphatase, receptor-type, R2A subfamily; CDKN2, Cyclin-dependent kinase inhibitor 2; CDKN1, Cyclin-dependent kinase inhibitor 1; NFY, Transcription factor, NF-Y family; JAK, Non-receptor tyrosine kinase of the Janus/TYK2 family; NFI, Nuclear factor I; PGM, Phosphoglucomutase; INSL, Insulin-related family; COL, Fibrillar collagen; NOTCH, Homologue of *Drosophila* Notch; GNA, G-protein alpha family; Compl, Complement component genes C3, C4 and C5; CACNA, Calcium channel, voltage-dependent, alpha 1, non-L-type; TNF, Tumor necrosis factor ligand superfamily; HLA/B2M, MHC class I-like sequence/Beta–2-microglobulin; TNX/HXB, Tenascin/Hexabrachion; PBX, Pbx homeobox; RXR, Retinoid X receptor (Nuclear receptor, family II); LIM, LIM/homeodomain protein; TPM, Tropomyosin; SH3GL, SH3 domain/GRB2-like protein; INSR, Insulin/IGF1 receptor; NTRK, Neurotrophic receptor tyrosine kinase; NTRKR, Neurotrophic receptor tyrosine kinase-related; MEF2, MADS box transcription enhancer factor 2; CTS, Cathepsin, cysteine protease; HMG, High mobility group protein 20; ROR, RAR-related orphan receptor (Nuclear receptor, family I); ANXA, Annexin, first branch; THBS, Thrombospondin

Paralogous region 2 (3), 7, 12, 17

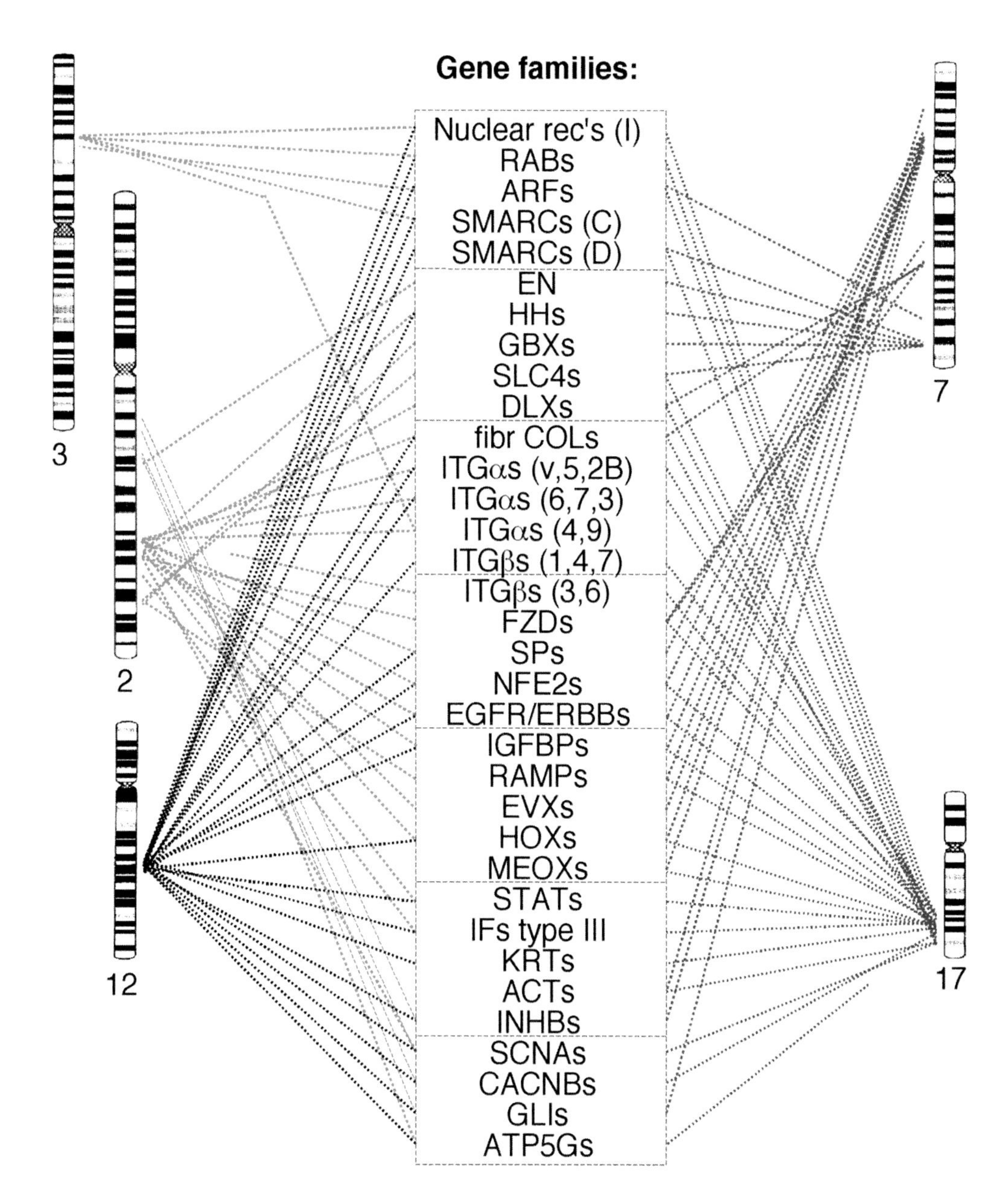

Figure 2. Gene family symbols and names.

Nuclear rec's (I), Nuclear receptor, family 1; RAB, Ras-related family Rab; ARF, ADP-ribosylation factor; SMARCC, SWI/SNF related, matrix-associated, actin-dependent regulator of chromatin, subfamily C; SMARCD, subfamily D; EN, Engrailed homeobox; *HH, Hedgehog family; GBX, Gastrulation brain homeobox; SLC4, Solute carrier family 4, anion exchanger subfamily; DLX, Distal-less homeo box; fibr COL, Fibrillar collagen; ITGα, Integrin alpha family; ITGβ, Integrin beta family; FZD, Frizzled (*Drosophila*); SP, Sp transcription factor (zinc-finger); NFE2, Erythroid-derived nuclear factor, family 2; EGFR/ERBB, Receptor tyrosine kinase, EGFR family; IGFBP, Insulin-like growth factor-binding protein; RAMP, Receptor activity-modifying protein; EVX, Even-skipped homeo box; HOX, HOX gene cluster; MEOX, Mesenchyme homeo box; STAT, Signal transducer and activator of transcription; IFs type III, Intermediate filament, type III; KRT, Keratin subfamily; ACT, Actin cytoplasmic subfamily; INHB, Activin/Inhibin subfamily; SCNA, Sodium channel, voltage-gated, alpha subunit; CACNB, Voltage-dependent calcium channel, beta subunit; GLI, GLI-Kruppel family; ATP5G, ATP synthase, H + transporting, mitochondrial F0 complex, subunit C

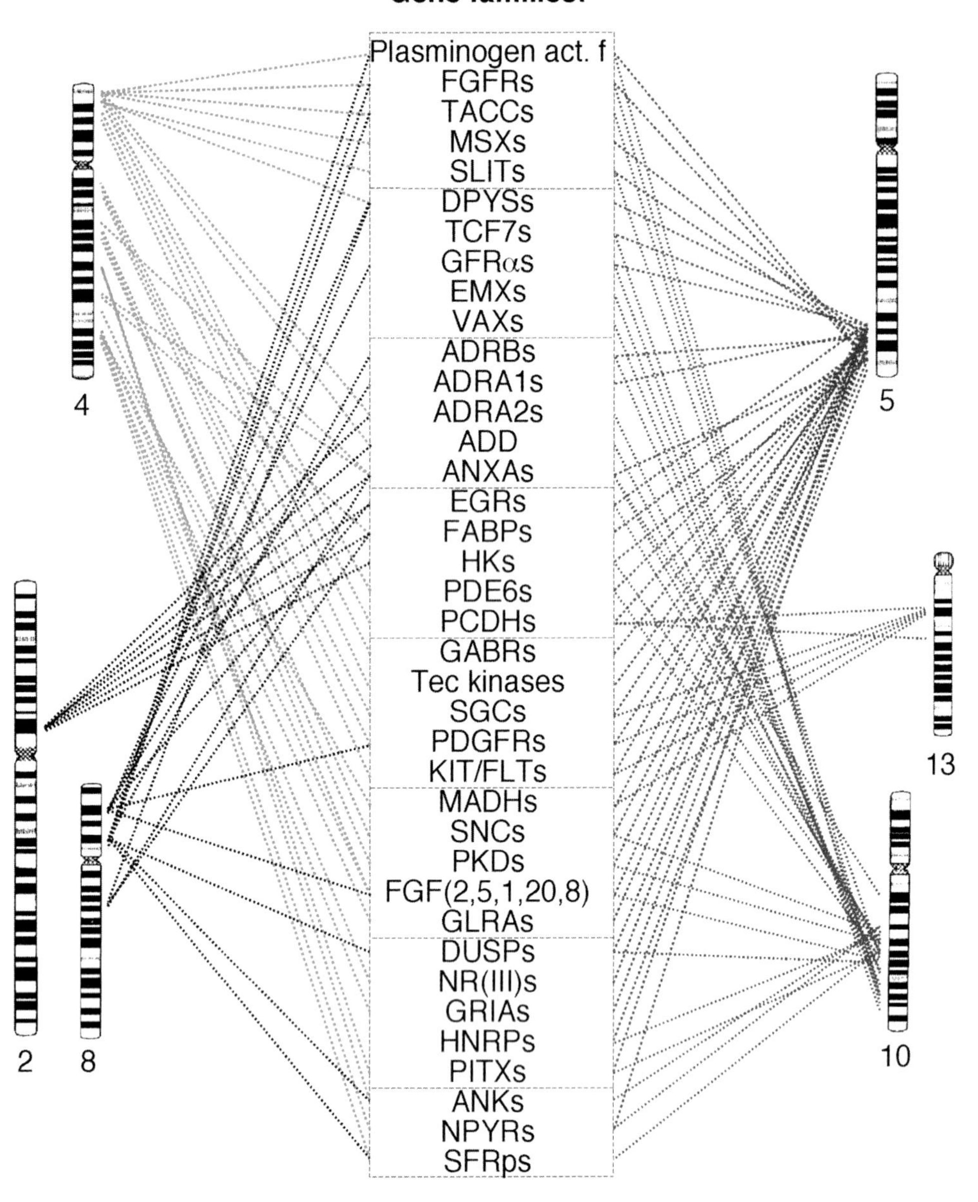

Figure 3. Gene family symbols and names.

Plasminogen act., Plasminogen activator serine protease family; FGFR, Fibroblast growth factor receptor (Tyrosine kinase); TACC, Transforming, acidic coiled-coil containing protein; MSX, Homeo box gene (*Drosophila* msh); SLIT, Homologue of Drosophila Slit; DPYS, Dihydropyrimidinase(-like); TCF7, Transcription factor–7(-like); GFRα, GDNF family receptor alpha; EMX, Empty spiracles (Drosophila); VAX, Ventral anterior homeo box; ADRB, Adrenergic receptor beta; ADRA1, Adrenergic receptor alpha–1; ADRA2, Adrenergic receptor alpha–2; ADD, Adducin; ANXA, Annexin A; EGR, Early growth response; FABP, Fatty acid-binding protein; HK, Hexokinase; PDE6, Phosphodiesterase 6, cGMP-specific; PCDH, Protocadherin; GABR, GABA A receptor cluster; Tec kinases, Non-receptor tyrosine kinase, Tec family; SGC, Sarcoglycan; PDGFR, KIT/FLT, Receptor tyrosine kinase, PDGFR family; MADH, Mothers against decapentaplegic (Drosophila), Subfamily C; SNC, Synuclein; PKD, Polycystin; FGF, Fibroblast growth factor; GLRA, Glycine receptor alpha; DUSP, Dual-specificity phosphatase, first subfamily; NR(III), Nuclear receptor family 3; GRIA, Glutamate receptor, ionotropic, AMPA; HNRP, Heterogeneous nuclear ribonucleoprotein; PITX, Paired-like homeodomain transcription factor; ANK, Ankyrin; NPY*R, Neuropeptide Y receptor; SFRP, Secreted frizzled-related protein

The effects of the putative inversion in chromosome 1 can also be seen in several regions of chromosome 1 in another paralogy group involving chromosomes 1, 11, 12 and 19q (Lundin *et al.*, submitted). The same reasoning can also be applied to chromosomes 11 and 12 in which an equal number of paralogues, belonging to this paralogy group, are located on both arms of these two chromosomes.

In the 2q, 7, 12q and 17q group, containing the four HOX clusters, chromosome 7 has most likely undergone a pericentric inversion as genes from relevant gene families are found in both chromosomal arms whereas the paralogous regions in the other three chromosomes are well kept together (Fig. 2). Likewise in the 4, 5q, 8 and 10q group the paralogous genes are spread out in chromosome 4 compared to the more compact groupings in chromosomes 5, 8 and 10 (Fig. 3). This fits well with a comparison between human chromosome 4 and the correspoding orthologous regions in the pig that led to the conclusion that two inversions can explain the different locations of the orthologous genes in the two species (Wraith *et al.*, 2000).

Translocations

In the diagrams presented here we find numerous examples of pieces of chromosomes that once must have belonged to some of the paralogous regions but which were translocated to another chromosomal region. Thus, in figure 1 the genes on 15q21–26 are likely to once have been syntenic to 6p21 in an ancestral mammal and those on 3p24–21 (Fig. 2) could have been translocated from either chromosome 2 or 7. Three genes in the 10p13–11 region probably also came from chromosome 7 (data not shown). In the 4, 5q, 8, 10q group (Fig. 3) genes originally on chromosomes 8 and 10 may have been translocated to 2p14–11 and 13q, respectively.

Gene silencing

Gene silencing is a phenomenon that is known to be frequent after gene and genome duplications (Li, 1997; Patton *et al.*, 1998; Wagner, 1998). The extent of silencing will no doubt vary for different gene families and in different lineages of various organismal groups but is estimated to be around 50% based on the hypothesis of genome duplications in vertebrate evolution (Nadeau and Sankoff, 1997).

A genome duplication was found to have taken place in an ancestor of *Saccharomyces cerevisiae* some 150 MYR ago (Wolfe and Shields, 1997). Around 15 % of the extra genome still seems to exist in this species which seems surprisingly little considering the fairly short time that has past since the putative duplication. As a comparison, an estimation of remaining gene pairs after the genome duplication that is likely to have taken place early in the evolution of the ray-finned fishes gave the result that at least 20% of the duplicated genes are still present in the zebrafish (Postlethwait *et al.*, 2000). It is to be noted that this tetraploidisation is likely to be considerably older than the one in yeast and that it took place in an organism that was already essentially octaploid compared to simple chordates, such as Amphioxus or *Ciona intestinalis*.

If we consider the Hox gene clusters it is evident that out of the 52 or 56 original genes (Ferrier *et al.*, 2000), that were likely to have resulted from the two genome doublings, there are now 39 still active in extant mammals (Holland, 1997). This means that 70–75% of the original Hox genes still play a role in mammals. What then about other gene families? With the two likely genome duplications in the vertebrate line in mind and the basic gene number among invertebrates being somewhere around 15,000–17,000, mammals and other vertebrates, except lineages known to have gone through another tetraploidisation, should have some 60,000 to 70.000 structural genes, unless gene silencing has been extensive. However, the real number may be considerably lower, maybe between 30,000 and 40,000 (Bork and Copley, 2001) meaning that perhaps up to around 50% of the genes were silenced in the line leading to mammals. This is reflected in the fact that paralogues in different gene families in the paralogy groups discussed here often do not occur as quadruplets but as triplets or doublets. No doubt there are gaps that will be filled when more genes are put on the map but it should also be remembered that many subfamilies that branched off before the separation of proto- and deuterostomes contain just one extant gene.

Why do phylogenetic trees give equivocal answers?

There are reasons to believe that the time span between the two putative genome doublings discussed here took place within a short time span (Gibson and Spring, 2000). It was further argued, and in our opinion rightly so, that 'whenever sequence trees for gene quartets lack informative data for resolving internal branches, the tree topology will be essentially random'. Gibson and Spring argue that the period between the genome duplications could have been as short as 10 million years or maybe even shorter. This contrasts to another attempt to estimate the time span between the putative genome doublings in the vertebrate line which gave figures between 90–106 million years based on 26 gene families (Wang and Gu, 2000). However, it should be noted that the span between the figures from the gene families is very large, from 9–295 million years and in one or two gene families the authors used regionally duplicated genes which will give misleading estimates. Furthermore, the FGF8 gene, used in one of the families, branched off before the separation of nematodes and vertebrates.

Another factor that is not always considered fully is the apparent large differences found in the rate of change for different genes within a gene family which may cause uncertainty about the reliability of trees. A rather drastic example is found among neurotrophins where the NTF5 gene has changed about four times faster than the BDNF gene and the Ntf6/7 from teleosts has had an even higher exchange rate (Hallböök *et al.*, 2001). This shows that the molecular clock is very variable and as pointed out there were often rapid changes in vertebrate genes after the extensive subtype and isoform duplications (Suga *et al.*, 1999a).

When phylogenetic trees based on gene sequences are used to test the validity of hypotheses about tetraploidisations it is obviously also necessary to avoid including paralogues that are the results of regional duplications. This rule has been violated at some occasions (Hughes, 1999; Wang and Gu, 2000; Martin, 2001) and it is certainly warranted to watch out for obvious anomalies, e.g. when there are large discepancies between the positions of genes from related groups within a tree (Hughes, 1999) or when the opposite effect is seen namely that genes from distantly related animal groups end up close to each other (Martin, 2001).

As discussed above gene silencing is likely to have been much more extensive than previously thought. This has to a large part been neglected in discussions and criticisms of the vertebrate genome duplication hypothesis. Taking gene silencing into account as a very real and frequent phenomenon (Page and Charleston, 1997) and to combine this with the rather frequent conservation of synteny between ancient paralogues as discussed below will add to our abilities in the analysis of genomic evolution. It is evident that in many situations the demonstration of conserved paralogous regions is often much more informative than phylogenetic trees based on gene sequences.

Presence of genes from ancient gene duplications in conserved regions

A considerable number of cases can be found where paralogous genes are still located in the same chromosomal region, often very close to each other, although they resulted from ancient duplications before the split between proto- and deuterostomes. Some examples are genes for gap junction alpha and beta polypeptides (GJA and GJB, one each in 1p35.1 and 13q11), G-protein alpha proteins (GNAI and GNAT, one each in 1p13 and 3p21) and mothers against decapentaplegic (MADH in 15q21–22 (2) and 18q21 (3)). Sometimes this is also the case for paralogues that stem from duplications before the separation of animals and fungi, e.g. proteasome subunit, beta (PSMB8 and 9 in 6p21.3) or animals and plants, e.g. voltage-gated chloride channel genes (CLCN*, three in 1p36). Among the first category of duplications are all those that gave rise to separate gene families within superfamilies early in animal evolution (Ono *et al.*, 1999; Suga *et al.*, 1999b, 1999c).

Not only is it possible to find close syntenies between pairs or groups of distant paralogues but this will also have an effect on the way we look upon the role of such genes in conserved paralogous regions. An interesting case was given by Patton *et al.* (1998) involving aromatic amino acid hydroxylase and insulin-related genes. In view of the frequent silencing of genes discussed above it is reasonable to assume that also paralogous genes resulting from ancient duplications may often give a good indication of very early syntenic regions. This has been elegantly demonstrated by Pollard and Holland (2000) for a large superfamily of homeobox genes. Very recently a similar way of reasoning was applied in order to tie together many paralogous chromosomal groups in a nested series of regions covering the entire human genome (Popovici *et al.*, 2001). This shows that we should not immediately dismiss the relevance of ancient paralogues in paralogy groups but to carefully consider them in their context.

Relationships between different groups of paralogous regions

In line with the observations on the distribution of homeobox families (Pollard and Holland, 2000; Popovici *et al.*, 2001) it is interesting to see that several gene families or superfamilies contain paralogues located in more than one of the paralogy groups discussed in this article and among these some of the most obvious cases are: receptor tyrosine kinases (Fig. 1, 2 and 3), family I of the nuclear receptors and fibrillar collagens (Fig. 1 and 2), annexins (Fig. 1 and 3) and frizzled and secreted frizzled and early growth response/SP family genes (Fig. 2 and 3). No doubt, this type of data will give us indications about the gene contents of prevertebrate chromosomal regions and hopefully also why so many of the syntenies were broken by chromosomal fission.

What independent support might there be for earlier synteny of chromosomal regions that are now found in different paralogy groups? One important indication would be if it is likely that an ancestral vertebrate genome was carried on a small number of chromosomes. In an analysis of the origins of vertebrate chromosomes, using the zebrafish genetic map as a comparative tool, it was suggested that the last common ancestor of rayfinned fishes and mammals had around 12 or 13 chromosomes in the haploid genome (Postlethwait *et al.*, 2000). This in effect means that the number of chromosomes in the chordate ancestor of the vertebrates could have been very small, indeed, in the order of 3–4 for the haploid set. An interesting conclusion was that there has been an extensive number of chomosome fissions in the tetrapod lineage. This may explain some intriguing features of the zebrafish and human gene maps.

In the human gene map three paralogy groups (Lundin *et al.*, submitted) have two regions in common, 1p3 and 6p2, which may give an indication as to how the ancestral regions of these three paralogy groups might have been syntenic or nested into each other at an earlier stage of vertebrate evolution (Lundin *et al.*, submitted). We suggest that parental genes, of the ones found in the 10 main regions included in these groups plus the likely translocated regions, once could have been syntenic in an early prevertebrate. According to this scenario this ancestral region duplicated twice as a result of the postulated genome doublings and these four newly formed regions must then successively have split up into a larger number of regions excepting the one in chromosome 1p3.

Furthermore, Hox clusters and other genes from gene families, which in the human map are found in the two paralogy groups 2, 7, 12, 17 and 1, 6, 9, 19, are found to be syntenic in linkage groups 3 and 19 in the zebrafish. The fact that this involves two of the zebrafish linkage groups is a strong indication that this is not a chance effect and that ancestral genes of those found in these two linkage groups may once have been syntenic.

Conclusion

As argued above, maybe the most decisive and powerful tool to find out if there has been one or more genome doublings is to identify conserved paralogous chromosomal regions within one or more species. Thus, due to the results presented in this article and earlier articles referred to above it is very likely that at the root of vertebrate evolution were two rounds of total genome doubling after an 'Amphioxus stage' but before the fish stage (Holland *et al.*, 1994). Numbers of active genes in invertebrate species support this hypothesis, e.g. *Drosophila*, an ecdysosoan animal, is estimated to have around 13,600 genes (Adams and al., 2000) and a closer relative of the vertebrates, the invertebrate chordate *Ciona intestinalis,* has around 15,500 protein-coding genes (Simmen *et al.*, 1998) whereas the human genome contains some 30,000 to 40,000 active genes (Bork and Copley, 2001). Taking the process of gene silencing into consideration we conclude that there were two genome doublings at the root of the vertebrates.

It has been suggested that the first of these genome doublings eventually was leading to the formation of neural crest structures, something that laid the ground for further evolution of the head region with a larger brain and paired sense organs. The second genome doubling might have given the potential for forming jaws and paired fins (Holland *et al.*, 1994; Sharman and Holland, 1996). However, it has proven difficult to find definite proof for candidate molecular changes in relation to the evolutionary appearance of the neural crest (Shimeld and Holland, 2000).

References

Adams, M.D., *et al.* (2000) The genome sequence of *Drosophila melanogaster*. *Science*, **287,** 2185–2195.

Bork, P. and Copley, R. (2001) Filling in the gaps. *Nature*, **409,** 818–820.

Burt, D.W., Bruley, C., Dunn, I.C., Jones, C.T., Ramage, A., Law, A.S., Morrice, D.R., Paton, I.R., Smith, J., Windsor, D., Sazanov, A., Fries, R. and Waddington, D. (1999) The dynamics of chromosome evolution in birds and mammals. *Nature*, **402,** 411–413.

Chaudhary, R., Raudsepp, T., Guan, X.-Y., Zhang, H. and Chowdhary, B.P. (1998) Zoo-FISH with microdissected arm specific paints for HSA2, 5, 6, 16, and 19 refines known homology with pig and horse chromosomes. *Mamm. Genome*, **9,** 44–49.

Chowdhary, B.P., Frönicke, L., Gustavsson, I. and Scherthan, H. (1996) Comparative analysis of the cattle and human genomes: Detection of ZOO-FISH and gene mapping-based chromosomal homologies. *Mamm. Genome*, **7,** 297–302.

Chowdhary, B.P., Raudsepp, T., Frönicke, L. and Scherthan, H. (1998) Emerging patterns of comparative genome organization in some mammalian species as revealed by zoo-FISH. *Genome Res.*, **8,** 577–589.

Ferrier, D.E.K., Minguillón, C., Holland, P.W.H. and Garcia-Fernández, J. (2000) The amphioxus Hox cluster: deuterostome posterior flexibility and *Hox14*. *Evol. Dev.*, **2,** 284–293.

Frönicke, L., Chowdhary, B.P., Scherthan, H. and Gustavsson, I. (1996) A comparative map of the porcine and human genomes demonstrates ZOO-FISH and gene mapping-based chromosomal homologies. *Mamm. Genome*, **7,** 285–290.

Gibson, T.J. and Spring, J. (2000) Evidence in favour of ancient octaploidy in the vertebrate genome. *Biochem. Soc. Trans.*, **28,** 259–264.

Hallböök, F., Kullander, K. and Lundin, L.-G. (2001) Evolution of the neurotrophins and their receptors. In *Neurobiology of the Neurotrophins* (Ed. Mocchetti, I.), FP Graham Publishing Co, Johnson City, TN, pp. 65–100.

Holland, P.W.H. (1997) Vertebrate evolution: Something fishy about *Hox* genes. *Curr. Biol.*, **7,** R570-R572.

Holland, P.W.H., Garcia-Fernández, J., Williams, N.A. and Sidow, A. (1994) Gene duplications and the origins of vertebrate development. *Development* Suppl., 125–133.

Hughes, A.L. (1999) Phylogenies of developmentally important proteins do not support the hypothesis of two rounds of genome duplication early in vertebrate history. *J. Mol. Evol.*, **48,** 565–576.

Iwabe, N., Kuma, K. and Miyata, T. (1996) Evolution of gene families and relationship with organismal evolution: Rapid divergence of tissue-specific genes in the early evolution of chordates. *Mol. Biol. Evol.*, **13,** 483–493.

Kasahara, M., Hayashi, M., Tanaka, K., *et al.* (1996) Chromosomal localization of the proteasome Z subunit gene reveals an ancient chromosomal duplication involving the major histocompatibility complex. *Proc. Natl. Acad. Sci. USA*, **93,** 9096–9101.

Katsanis, N., Fitzgibbon, J. and Fisher, E.M.C. (1996) Paralogy mapping: Identification of a region in the human MHC triplicated onto human chromosomes 1 and 9 allows the prediction and isolation of novel *PBX* and *NOTCH* loci. *Genomics*, **35,** 101–108.

Li, W.-H. (1997) *Molecular Evolution*. Sinauer Associates, Sunderland, MA.

Lundin, L.-G. (1989) Gene homologies and the nerve system. In *Genetics of Neuropsychiatric Diseases* (Ed. Wetterberg, L.), Vol. 51, MacMillan Press, London, pp. 43–58.

Lundin, L.-G. (1993) Evolution of the vertebrate genome as reflected in paralogous chromosomal regions in man and the house mouse. *Genomics*, **16,** 1–19.

Lundin, L.-G. (1999) Gene duplications in early metazoan evolution. *Sem. Cell Dev. Biol.*, **10,** 523–530.

Martin, A. (2001) Is tetralogy true? Lack of support for the 'one-to-four rule'. *Mol. Biol. Evol.*, **18,** 89–93.

Meyer, A. and Schartl, M. (1999) Gene and genome duplications in vertebrates: the one-to-four (-to-eight in fish) rule and the evolution of novel gene functions. *Curr. Opin. Cell Biol.*, **11,** 699–704.

Nadeau, J.H. and Sankoff, D. (1997) Comparable rates of gene loss and functional divergence after genome duplications early in vertebrate evolution. *Genetics*, **147,** 1259–1266.

Nadeau, J.H. and Sankoff, D. (1998) Counting on comparative maps. *Trends Genet.*, **14,** 495–501.

O'Brien, S. J., Wienberg, J. and Lyons, L.A. (1997) Comparative genomics: lessons from cats. *Trends Genet.*, **13,** 393–99.

Ohno, S. (1969) The role of gene duplication in vertebrate evolution. In *The Biological Basis of Medicine* (Eds., Bittar, E.D. and Bittar, N.), Vol. 4, Academic Press, London, pp. 109–132.

Ohno, S. (1970) *Evolution by Gene Duplication*, Springer-Verlag, Berlin.

Ohno, S. (1998) The notion of the Cambrian pananimalia genome and a genomic difference that separated vertebrates from invertebrates. In *Progress in Molecular and Subcellular Biology* (Ed., Müller, W.E.G.), Vol. 21, Springer-Verlag, Berlin, pp. 97–117.

Ohno, S., Wolf, U. and Atkin, N.B. (1968) Evolution from fish to mammals by gene duplication. *Hereditas*, **59,** 169–187.

Ono, K., Suga, H., Iwabe, N., Kuma, K. and Miyata, T. (1999) Multiple protein tyrosine phosphatases in sponges and explosive gene duplication in the early evolution of animals before the parazoan-eumetazoan split. *J. Mol. Evol.*, **48,** 654–662.

Page, R.D.M. and Charleston, M.A. (1997) From gene to organismal phylogeny: Reconciled trees and the gene tree/species tree problem. *Mol. Phylogenet. Evol.*, **7,** 231–240.

Patton, S.J., Luke, G.N. and Holland, P.W.H. (1998) Complex history of a chromosomal paralogy region: insights from Amphioxus aromatic amino acid hydroxylase genes and insulin-related genes. *Mol. Biol. Evol.*, **15,** 1373–1380.

Pebusque, M.J., Coulier, F., Birnbaum, D. and Pontarotti, P. (1998) Ancient large-scale genome duplications: phylogenetic and linkage analyses shed light on chordate genome evolution. *Mol. Biol. Evol.*, **15,** 1145–1159.

Pollard, S.L. and Holland, P.W.H. (2000) Evidence for 14 homeobox gene clusters in human genome ancestry. *Curr. Biol.*, **10,** 1059–1062.

Popovici, C., Leveugle, M., Birnbaum, D. and Coulier, F. (2001) Homeobox gene clusters and the human paralogy map. *FEBS Lett.*, **491,** 237–242.

Postlethwait, J.H., Woods, I.G., Ngo-Hazelett, P., Yan, Y.-L., Kelly, P.D., Chu, F., Huang, H., Hill-Force, A. and Talbot, W.S. (2000) Zebrafish comparative genomics and the origins of vertebrate chromosomes. *Genome Res.*, **10,** 1890–1902.

Raudsepp, T., Frönicke, L., Scherthan, H., Gustavsson, I. and Chowdhary, B.P. (1996) ZOO-FISH delineates conserved chromosomal segments in horse and man. *Chromosome Res.*, **4,** 218–225.

Scherthan, H., Cremer, T., Arnason, U., Weier, H.-U., Lima-de-Faria, A. and Frönicke, L. (1994) Comparative chromosome painting discloses homologous segments in distantly related mammals. *Nature Genet.*, **6,** 342–347.

Searle, A.G. and Selley, R.L. (1997) Tables of genetic homology. *Mouse Genome*, **95,** 106–160.

Sharman, A.C. and Holland, P.W.H. (1996) Conservation, duplication, and divergence of developmental genes during chordate evolution. *Netherlands J. Zool.*, **46,** 47–67.

Shimeld, S.M. and Holland, P.W.H. (2000) Vertebrate innovations. *Proc. Natl. Acad. Sci. USA*, **97,** 4449–4452.

Sidow, A. (1996) Gen(om)e duplications in the evolution of early vertebrates. *Curr. Opin. Genet. Dev.*, **6,** 715–22.

Simmen, M.W., Leitgeb, S., Clark, V.H., Jones, S. J. M. and Bird, A. (1998) Gene number in an invertebrate chordate, *Ciona intestinalis*. *Proc. Natl. Acad. Sci. USA*, **95,** 4437–4440.

Skrabanek, L. and Wolfe, K.H. (1998) Eukaryote genome duplication - where's the evidence? *Curr. Opin. Genet. Dev.*, **8,** 694–700.

Smith, N.G.C., Knight, R. and Hurst, L.D. (1999) Vertebrate genome evolution: a slow shuffle or a big bang? *BioEssays*, **21,** 697–703.

Spring, J. (1997) Vertebrate evolution by interspecific hybridisation - are we polyploid? *FEBS Lett.*, **400,** 2–8.

Spring, J., Goldberger, O.A., Jenkins, N.A., Gilbert, D.J., Copeland, N.G. and Bernfield, M. (1994) Mapping of the syndecan genes in the mouse: Linkage with members of the Myc gene family. *Genomics*, **21,** 597–601.

Suga, H., Hoshiyama, D., Kuraku, S., Katoh, K., Kubokawa, K. and Miyata, T. (1999a) Protein tyrosine kinase cDNAs from Amphioxus, hagfish, and lamprey: Isoform duplications around the divergence of cyclostomes and gnathostomes. *J. Mol. Evol.*, **49,** 601–608.

Suga, H., Koyanagi, M., Hoshiyama, D., Ono, K., Iwabe, N., Kuma, K. and Miyata, T. (1999b) Extensive gene duplication in the early evolution of animals before the parazoan-eumetazoan split demonstrated by G proteins and protein tyrosine kinases from sponge and Hydra. *J. Mol. Evol.*, **48,** 646–653.

Suga, H., Ono, K. and Miyata, T. (1999c) Multiple TGF-beta receptor related genes in sponge and ancient gene duplications before the parazoan-eumetazoan split. *FEBS Lett.*, **453,** 346–350.

Wagner, A. (1998) The fate of duplicated genes: loss or new function? *BioEssays*, **20,** 785–788.

Wang, Y. and Gu, X. (2000) Evolutionary patterns of gene families generated in the early stage of vertebrates. *J. Mol. Evol.*, **51,** 88–96.

Wolfe, K.H. and Shields, D.C. (1997) Molecular evidence for an ancient duplication of the entire yeast genome. *Nature*, **387,** 708–713.

Wraith, A., Törnsten, A., Chardon, P., Harbitz, I., Chowdhary, B.P., Andersson, L., Lundin, L.-G. and Larhammar, D. (2000) Evolution of the neuropeptide Y receptor family: Gene and chromosome duplications deduced from the cloning and mapping of the five receptor subtype genes in pig. *Genome Res.*, **10,** 302–310.

A. Meyer, Y. Van de Peer (eds.), Genome Evolution, 65-73.

Are all fishes ancient polyploids?

Yves Van de Peer[1]*, John S. Taylor[2†] & Axel Meyer[2]
[1] *Department of Plant Systems Biology, Vlaams Interuniversitair Instituut voor Biotechnologie (VIB), Ghent University, K.L. Ledeganckstraat 35, B-9000 Gent, Belgium;* [2] *Department of Biology, University of Konstanz, D-78457 Konstanz, Germany; † present address: Department of Biology, PO Box 3020, University of Victoria, Victoria, Canada, V8W 3N5*
* *Author for correspondence: E-mail: yves.vandepeer@gengenp.rug.ac.be*

Received 26.03.2002; accepted in final form 29.08.2002

Key words: genome duplication, gene evolution, subfunctionalization

Abstract

Euteleost fishes seem to have more copies of many genes than their tetrapod relatives. Three different mechanisms could explain the origin of these 'extra' fish genes. The duplicates may have been produced during a fish-specific genome duplication event. A second explanation is an increased rate of independent gene duplications in fish. A third possibility is that after gene or genome duplication events in the common ancestor of fish and tetrapods, the latter lost more genes. These three hypotheses have been tested by phylogenetic tree reconstruction. Phylogenetic analyses of sequences from human, mouse, chicken, frog (*Xenopus laevis*), zebrafish (*Danio rerio*) and pufferfish (*Takifugu rubripes*) suggest that ray-finned fishes are likely to have undergone a whole genome duplication event between 200 and 450 million years ago. We also comment here on the evolutionary consequences of this ancient genome duplication.

Introduction

Several authors have presupposed that major evolutionary transitions in biology have required the genetic raw material provided by gene, chromosome, and/or entire genome duplications (Ohno, 1970; Sidow, 1996; Spring, 1997; Holland, 1999; Lundin, 1999; Patel and Prince, 2000). Already about 30 years ago, Ohno (1970) presented comparative data on genome size and chromosome numbers to support his hypothesis that one or more genome duplications occurred during the evolution of vertebrates and made their diversification possible. Ohno hypothesized that big leaps in evolution – such as the transition from an invertebrate to a vertebrate – required the creation of new gene loci with previously non-existent functions and emphasized genome duplication via tetraploidy as the mechanism for the production of such new genes. Gene number comparisons do provide support for large-scale gene or genome duplication events in the vertebrate lineage. Spring (1997) uncovered an average of three homologous genes in humans for each of 52 genes of *Drosophila* and proposed that the additional human genes were produced during two rounds of entire genome duplications. However, Spring's hypothesis, later referred to as the 'one-to-four rule' (Ohno, 1999), or the '2R' hypothesis (Hughes, 1999) remains controversial (Hughes, 1999; Wang and Gu, 2000; Hughes *et al.*, 2001; other chapters in this issue).

Recently, an additional genome duplication event has been proposed in ray-finned fishes (Amores *et al.*, 1998; Wittbrodt *et al.*, 1998). The first indications for a fish-specific genome duplication came from studies based on *Hox* genes and *Hox* clusters. *Hox* genes encode DNA-binding proteins that specify cell fate along the anterior-posterior axis of bilaterian animal embryos and occur in one or more clusters of up to 13 genes per cluster (Gehring, 1998). It is thought that the ancestral *Hox* gene cluster arose from a single gene by a number of tandem duplications. Protostome invertebrates and the deuterostome cephalochordate *Amphioxus* possess a single *Hox* cluster, whereas Sarcopterygia, a monophyletic group including lobe-finned fish, such as the coelacanth and lungfishes, amphibians, reptiles, birds, and mammals, have four clusters (Holland and Garcia-Fernandez, 1996; Holland, 1997). This finding has been regarded as important support for the '2R' hypothesis of two rounds of entire genome duplications early in vertebrate evolution. Recently, extra *Hox* gene clusters discovered in the zebrafish (*Danio rerio*), medaka (*Oryzias latipes*), the African cichlid (*Oreochromis niloticus*) and the pufferfish (*Takifugu rubripes*) suggest an additional genome duplication in ray-finned fishes (Actinopterygii) before the divergence of most teleost species (Amores *et al.*, 1998; Wittbrodt *et al.*, 1998; Meyer and Schartl, 1999; Naruse *et al.*, 2000; Málaga-Trillo and Meyer, 2001; A. Amores, personal communication). In the meantime, comparative genomic studies have turned up many more genes and gene clusters for which two copies exist in fishes but only one copy in other vertebrates (e.g., Postlethwait *et al.*, 2000; Robinson-Rechavi *et al.*, 2001; Taylor *et al.*, 2001a; Van de Peer *et al.*, 2001; Woods *et al.*, 2001). The observations that different paralogous pairs originate at about the same time (Taylor *et al.*, 2001a), that they are found on different linkage groups, and that they show synteny with other duplicated genes (Gates *et al.*, 1999; Postlethwait *et al.*, 2000; Woods *et al.*, 2000) support the hypothesis that these genes arose through a complete genome duplication event (Fig. 1a). On the other hand, several well-supported trees show one of the fish genes as the sister sequence to a monophyletic clade that included the second fish gene and genes from frog, chicken, mouse, and human (Taylor *et al.*, 2001a; Robinson-Rechavi *et al.*, 2001a). These so-called 'outgroup' topologies (Fig. 1d) might suggest that the origin of many fish duplicates predates the divergence of the Sarcopterygii and Actinopterygii and that tetrapods lost duplicates retained in fish (Fig. 1c).

Robinson-Rechavi *et al.* (2001a, 2001b) argued that an ancestral whole-genome duplication event was not responsible for the abundance of duplicated fish genes. They counted orthologous genes in fish and mouse and, where extra genes were found in fish, compared the number of gene duplications occurring in a single fish lineage with that shared by more than one lineage. Most mouse genes surveyed were also found as single copies in fish. Duplicated fish genes were detected, but most were interpreted as the products of lineage-specific duplication events in fish and not as an ancient duplication event (Fig. 1b). Here, we provide further evidence for the ancient fish-specific genome duplication based on phylogenetic inference, including sequences from multiple fish lineages.

Material and methods

Sequence alignments

Homologous sequences were collected and aligned as described before (Taylor *et al.*, 2001a). In short, protein sequences were collected using BLASTp (Altschul *et al.*, 1997) and aligned with CLUSTALX (Thompson *et al.*, 1997). Sequence alignments were edited with BioEdit (Hall, 1999) and only unambiguously aligned regions were retained for further analysis. For this study, our aim was to collect homologous sets of genes that contained sequences from at least two different fish species. In most cases, genes from either zebrafish or pufferfish (*Takifugu rubripes*) were collected. Sequence alignments and additional data and information on sequence retrieval and analysis can be found in the Wanda database on duplicated fish genes (Van de Peer *et al.*, 2002a; http://www.evolutionsbiologie.uni-konstanz.de/Wanda/).

Phylogenetic tree construction

In general, phylogenetic trees were constructed by neighbor-joining (Saitou and Nei, 1987) based on Poisson-corrected distances, as implemented in TREECON (Van de Peer and De Wachter, 1994). Recently, we developed a software tool called ASaturA to detect and consider saturation in amino acid sequences (Van de Peer *et al.*, 2002b). When saturation is observed, evolutionary distances between sequences can be computed from the fraction of unsaturated sites only and evolutionary trees inferred by pairwise distance methods (for details, see Van de Peer *et al.*, 2002b).

Results and discussion

Tree topologies support an ancient fish-specific genome duplication

Previously, we have shown that third-codon positions were saturated for most zebrafish paralogs (Taylor *et al.*, 2001a; Van de Peer *et al.*, 2001). Together with the observation that duplicated genes were found on different linkage groups, the most parsimonious

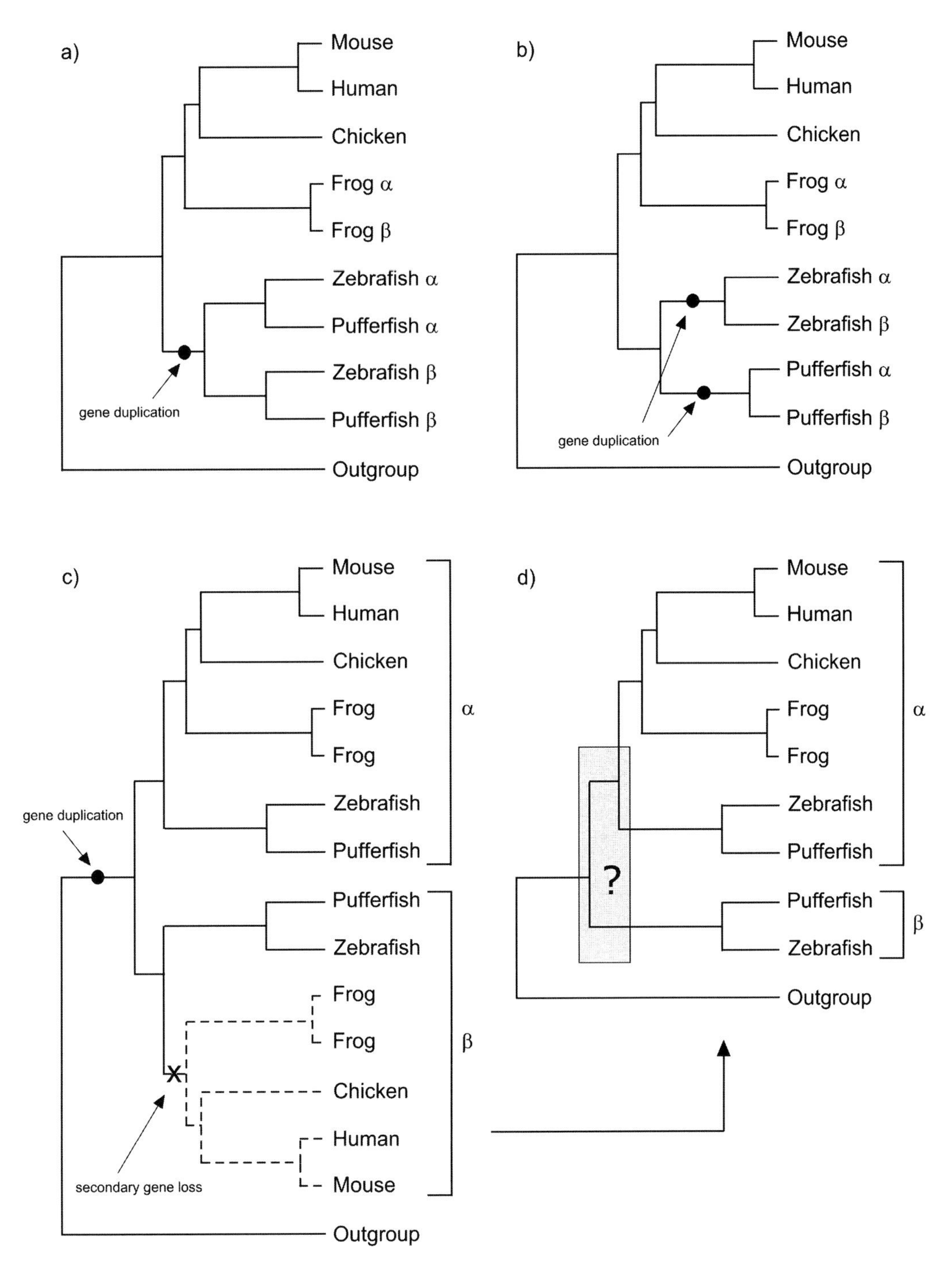

Figure 1. Different scenarios – and expected inferred tree topologies - to explain the presence of more genes in fish. (a) Duplicated fish genes are the result of a gene/genome duplication that preceded the divergence of zebrafish and pufferfish. (b) Duplicated genes are formed by independent gene duplications. The topology shown in d) is expected to be inferred when genes produced during a duplication event in the ancestor of Actinopterygii plus Sarcopterygii (shown in c) have been secondarily lost in the sarcopterygian lineage after the split of these two major lineages of jawed vertebrates. On the other hand, the tree topology shown in (d) might be an artifact in tree construction due to differences in evolutionary rates in the different duplicates (Taylor and Brinkmann, 2001) or due to saturation (Van de Peer *et al.*, 2002b), which often makes it difficult to infer the exact branching order for deeper regions in the tree as indicated by the gray border.

explanation is that all these 'old' paralogs originated by an ancient genome duplication, somewhere between 300 and 450 million years ago (Taylor *et al.*, 2001a). Because major teleost lineages are believed to have arisen between approximately 100 and 200 million years ago (Carroll, 1997; Lydeard and Roe, 1997), the working hypothesis was to assume that the genome duplication occurred in the ancestor of most (if not all) ray-finned fish. To find additional evidence, we compiled many vertebrate data sets, including the zebrafish genes described previously (Taylor *et al.*, 2001a) and, when available, their pufferfish homologs. The almost complete pufferfish genome sequence has been made available recently (http:// www.jgi.doe.gov/) and pufferfish orthologs could indeed be found for most of the zebrafish genes. Zebrafish and pufferfish both belong to the Euteleostei (a Subdivision of the Superorder Teleostei), together with at least 22,000 others species and are rather distantly related among Euteleost fish. Zebrafish and pufferfish are estimated to have diverged approximately 150 million years ago (Nelson, 1994).

Figure 2 shows some of the inferred tree topologies, including duplicated genes from both zebrafish and pufferfish. In general, the trees shown (and deposited in the Wanda database) are Poisson-corrected distance trees taking into account all sites of the alignment. However, in some cases, improved tree topologies with higher statistical support could be obtained by removing saturated sites from the sequence alignment. Saturation was detected with ASaturA, a software tool specifically developed for this purpose. AsaturA is a Java-based application that visualizes the amount of saturation in amino acid sequences by graphically displaying the number of observed frequent and rare amino acid replacements between pairs of sequences against their overall evolutionary distance. Discrimination between frequent and rare amino acid replacements is based on substitution probability matrices (e.g., PAM and BLOSUM). When amino acid sequences showed saturation for a fraction of the sites, evolutionary distances were computed from the fraction of unsaturated sites only (for details, see Van de Peer *et al.*, 2002b).

Figure 2a shows a tree for *Reggie*, a cell surface protein found in retinal ganglion cells during axon regeneration. For the *Reggie* gene, also two paralogs from the goldfish *Carassius auratus* (Schulte *et al.*, 1997) were included. In Figure 2b a tree topology is presented for *DLL1*, a homolog of the distal-less gene in *Drosophila*, which is the first genetic signal for limb formation to occur in a developing zygote. The tree topology for *FZD8*, a family of putative transmembrane receptors homologous to the product of the *Drosophila* tissue polarity gene *frizzled*, and that inferred from a sequence alignment of *RXRB*, the retinoid X receptor β gene, are seen in Figures 2c and 2d, respectively.

As can be observed, all the tree topologies shown, taking into account either all sites (*DLL1* and *RXRB*) or only unsaturated fractions of sites (*Reggie* and *FZD8*) are in perfect agreement with an ancient fish-specific genome duplication that occurred before the divergence of zebrafish and pufferfish (see also Fig. 1a). Many additional trees, including duplicated genes from zebrafish and pufferfish with similar topologies can be found in the Wanda database (Van de Peer *et al.*, 2002a).

It should also be noted that, in order to test whether two species experienced the same gene or genome duplication, it is not necessary to find two genes in both species. If a gene from one species clusters specifically with one of the two duplicates (i.e. paralogs) of a second species, then this can only be explained by a shared duplication event with a subsequent loss of one of the gene copies in one of the species. Examples are given for the *SHH* gene (Fig. 2e), which codes for a signal that is necessary in patterning the early embryo, and for the *BMP2* gene, a highly conserved member of the transforming growth factor β gene family (Fig. 2f). For instance, the *BMP2* gene of *Takifugu rubripes* clusters specifically with one of the *BMP2* paralogs of *Danio rerio*. The second *Takifugu BMP2* gene is probably waiting to be discovered or has been secondarily lost.

The evolutionary consequences of an ancient fish-specific genome duplication: Gene duplication, functional divergence of genes, and speciation

If a fish-specific genome duplication had occurred, fish genomes would be expected to contain more genes, at least initially, than the genomes of mammals. In our genome survey, we very often uncovered multiple gene copies in fish for single genes in other vertebrates, but almost never the opposite (see also Wittbrodt *et al.*, 1997: Robinson-Rechavi *et al.*, 2001a, 2001b). After at least 200 or more million years of evolution, these duplicated fish genes might be expected to have acquired quite different functions. Ohno's model, which Hughes (1994) first

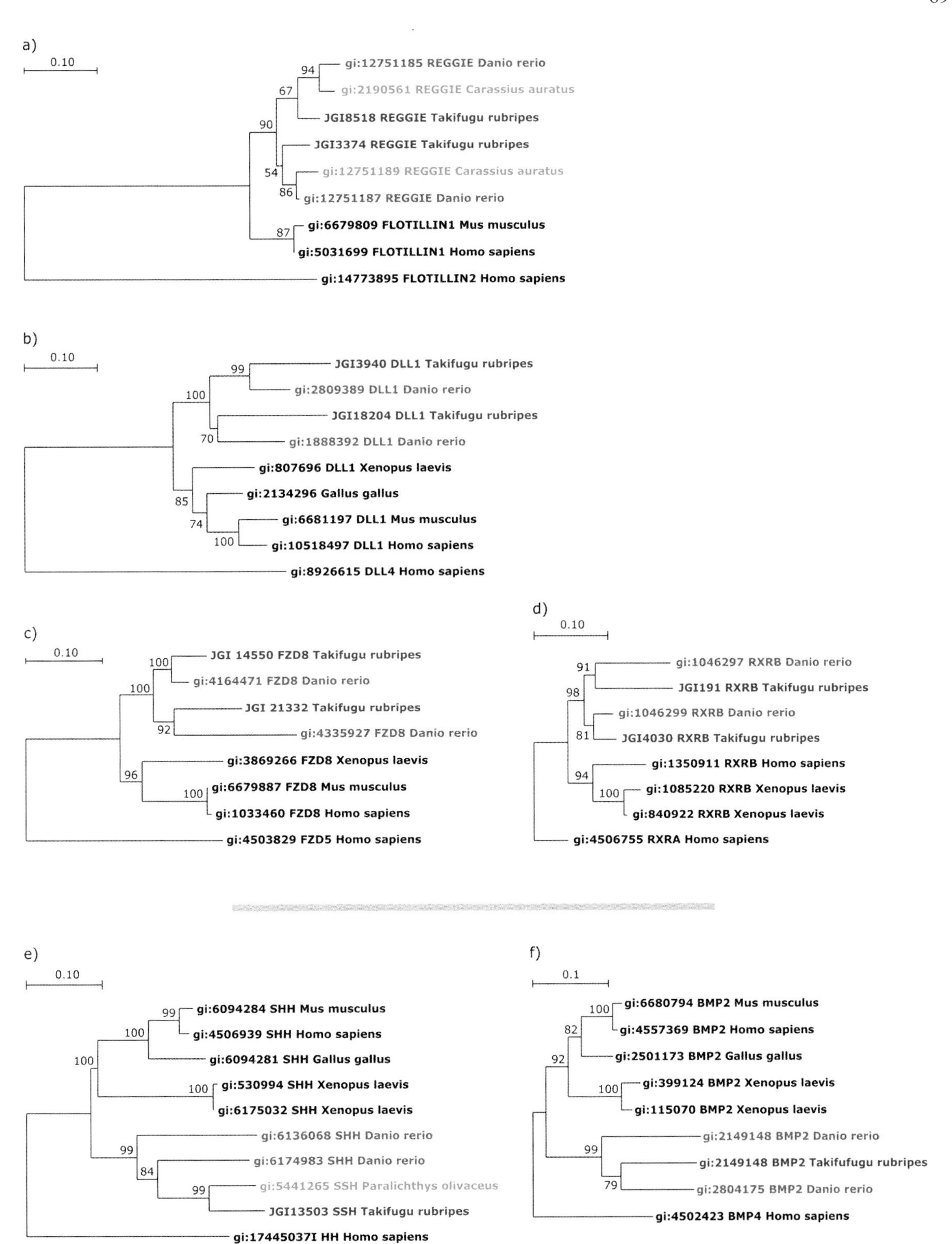

Figure 2. Selected phylogenies including genes of multiple fish lineages. All tree topologies, including either two or more pairs of duplicated fish genes (topologies a-d) or one pair of duplicated genes plus (a) single additional gene(s) from other fish species (topologies e-f) support an ancient fish-specific genome duplication. Additional trees supporting the fish-specific genome duplication can be found in the Wanda database (Van de Peer *et al.*, 2002a).

called the 'mutation during non-functionality' and later the 'mutation during redundancy' model (Hughes, 1999), predicts that, if genes do not get lost, by chance, a series of non-deleterious mutations might render the duplicate gene into a gene with a new function. This model has been widely adopted to explain the evolution of functionally novel genes, but little evidence can be found that new gene functions have evolved this way. Although it might seem unlikely that anciently duplicated genes perform completely redundant functions, redundancy has been shown to be widespread in genomes of higher organisms (Nowak *et al.*, 1997 and references therein; Gibson and Spring, 1998). Furthermore, many paralogous fish genes seem to have subdivided their functions rather than to have evolved novel functions. Recently, the 'duplication-degeneration-complementation' (DDC) model (Force *et al.*, 1999; Lynch and Force, 2000a) has been introduced to explain why duplicated genes might be retained. This model predicts that the likelihood of preservation is correlated with the number of 'subfunctions' that can be ascribed to a gene. The model starts from the assumption that a gene can perform several different functions, for instance, expression in different tissues and at different times during development, each of which may be controlled by different DNA regulatory elements. If duplicate genes lose different regulatory subfunctions, each affecting different spatial and/or temporal expression patterns, then they must complement each other by jointly retaining the full set of subfunctions that were present in the ancestral gene. Therefore, degenerative mutations facilitate the retention of duplicate functional genes, where both duplicates now perform different but necessary subfunctions. However, as predicted by the DDC model, the sum of the retained duplicates must be equal to the total number of subfunctions performed by the ancestral gene. Gene duplication then allows each daughter gene to specialize for one of the functions of the ancestral genes. Force *et al.* (1999) showed that this model might generally apply based on the *En1* genes in zebrafish. In mouse and chicken, *En1* is expressed in the developing pectoral appendage bud and in specific neurons of the hindbrain and spinal cord (Joyner and Martin, 1987; Davis *et al.*, 1991; Gardner and Barald, 1992). In zebrafish, however, one of the paralogs is expressed in the pectoral appendage bud, while the second paralog is expressed in the hindbrain/spinal cord neurons (Force *et al.*, 1999).

Possibly, retention of gene duplicates by subfunctionalization applies to many of the anciently duplicated fish genes. Besides *En1*, differences in the expression pattern of *Msx* zebrafish paralogs and homologous genes of other vertebrates also suggest subfunctionalization of the zebrafish genes after duplication (Ekker *et al.*, 1997). Similar conclusions can be drawn for *hedgehog* genes (Laforest *et al.*, 1998), *Bmp2* (Martinez-Barbera *et al.*, 1997), the transcription factors *mitfa* and *mitfb* (Mellgren and Johnson, 2002; Altschmied et al., 2002), *cyp19* (Chiang *et al.*, 2001), *GlyRalpha* genes (Imboden *et al.*, 2001), *Notch* and *Pax6* (Lynch and Force, 2000a). Models such as the DDC model may explain the retention and functional divergence of duplicated genes. However, when paralogs diverge in function mainly through subfunctionalization, functional divergence is probably limited to differences in timing and tissue specificity of expression. Until now, there is little evidence that the fish paralogs have changed functions completely in the course of evolution. Therefore, it is still an open question whether subfunctionalization of many duplicated genes resulting from the fish-specific genome duplication can be responsible for the large number of fish species and their tremendous morphological diversity, as suggested previously (Amores *et al.*, 1998; Wittbrodt *et al.*, 1998; Meyer and Schartl, 1999).

However, another phenomenon could explain the abundance of fish species we observe. Recently, a model called 'divergent resolution' has been proposed (Lynch and Conery, 2000; Lynch and Force, 2000a), in which the loss or silencing of duplicated genes might be more important to the evolution of species diversity than the evolution of new functions in duplicated genes. Divergent resolution occurs when different copies of a duplicated gene are lost in geographically separated populations and could genetically isolate these populations, should they become reunited (reviewed in Taylor *et al.*, 2001b; Fig. 3). Therefore, large-scale gene duplications and rapid speciation of organisms might be correlated. In this respect, it is noteworthy that also in plant evolution there is a strong indication for a polyploidy event that seems to coincide with a massive diversification of novel plant families (Raes *et al.*, this issue; Y. Van de Peer, unpublished data).

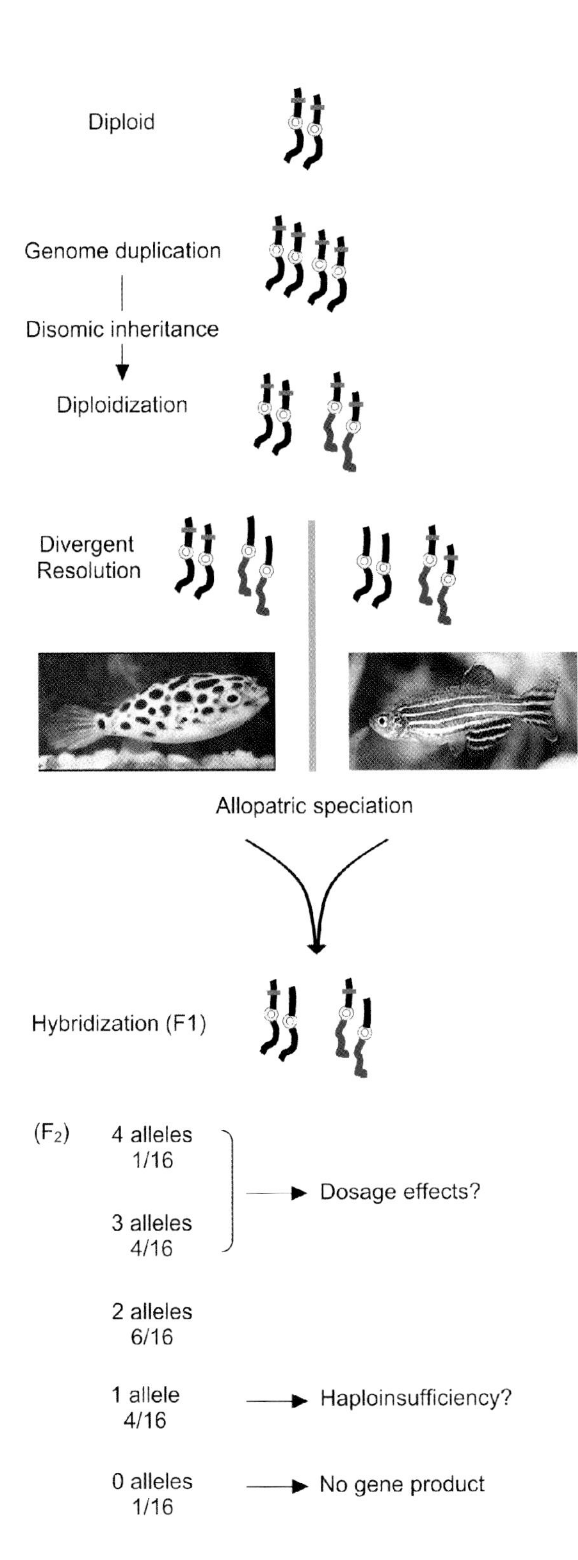

Figure 3. Speciation through genome duplication and divergent resolution. Natural selection will favour speciation over hybridization in populations fixed for different copies of a duplicated locus. Red bars represent a locus that is duplicated (along with all other loci) during a tetraploidy event. In this hypothetical example, diploidization is driven by a reciprocal translocation depicted by a change in chromatid colour. Pufferfish (left) and zebrafish (right) are shown as examples of the descendants of the two populations. If individuals from such populations mate, their 'hybrid' progeny would be heterozygous, possessing a functional allele and a pseudogene at each locus of the duplicated gene. Crosses between the F1 individuals produce some (about 6%) F2 individuals with only pseudogenes at both loci in question, and therefore lacking viability and/or fertility. Others would receive between one allele, which might lead to reduced function when the gene product from one functional allele is inadequate to support normal function (haploinsufficiency), to three or four functional alleles, which might have a negative dosage effect. All these might lead to postmating reproductive isolation (Lynch and Force, 2000b). Reprinted (and slightly adapted) from Trends in Genetics 17, pp. 299–301, © 2000, 'Genome duplication, divergent resolution, and speciation' by John S. Taylor, Yves Van de Peer, and Axel Meyer, with permission from Elsevier Science.

Acknowledgements

This work was supported by the German Science Foundation (DFG PE 842/2–1). J.S.T. is indebted to the National Sciences and Engineering Research Council of Canada for a Postdoctoral Fellowship. A.M. received support from the Deutsche Forschungsgemeinschaft (DFG), the University of Konstanz and the Fonds der Chemischen Industrie.

References

Altschmied, J., Delfgaauw, J., Wilde, B., Duschl, J., Bouneau, L., Volff, J.-N. and Schartl, M. (2002) Subfunctionalization of duplicate *mitf* genes associated with differential degeneration of alternative exons in fish. *Genetics* **161**, 259–267.

Altschul, S.F., Madden, T.L., Schaffer, A.A., Zhang, J., Zhang, Z., Miller, W. and Lipman, D.J. (1997) Gapped BLAST and PSI-BLAST: a new generation of protein database search programs. *Nucleic Acids Res.*, **25**, 3389–3402.

Amores, A., Force, A., Yan, Y.-L., Joly, L., Amemiya, C., Fritz, A., Ho, R.K., Langeland, J., Prince, V., Wang, Y.-L., Westerfield, M., Ekker, M. and Postlethwait, J.H. (1998) Zebrafish *hox* clusters and vertebrate genome evolution. *Science*, **282**, 1711–1714.

Aparicio, S., Hawker, K., Cottage, A., Mikawa, Y., Zuo, L., Venkatesh, B., Chen, E., Krumlauf, R. and Brenner, S. (1997) Organization of the *Fugu rubripes Hox* clusters: evidence for continuing evolution of vertebrate *Hox* complexes. *Nature Genet.*, **16**, 79–83.

Carroll, R.L. (1997) *Patterns and Processes of Vertebrate Evolution*, Cambridge University Press, Cambridge, UK.

Chiang, E.F., Yan, Y.L., Tong, S.K., Hsiao, P.H., Guiguen, Y., Postlethwaith, J. and Chung, B.C. (2001) Characterization of duplicated zebrafish *cyp19* genes. *J. Exp. Zool.*, **290**, 709–714.

Davis, C.A., Homyard, D.P., Millen, K.J. and Joyner, A.L. (1991) Examining pattern formation in mouse, chicken and frog embryos with an *En*-specific antiserum. *Development*, **2**, 287–298.

Ekker, M., Akimenko, M.A., Allende, M.L., Smith, R., Drouin, G., Langille, R.M., Weinberg, E.S. and Westerfield, M. (1997) Relationships among *msx* gene structure and function in zebrafish and other vertebrates. *Mol. Biol. Evol.*, **14,** 1008–1022.

Force, A., Lynch, M., Pickett, F.B., Amores, A., Yan, Y.-l. and Postlethwait, J. (1999) Preservation of duplicate genes by complementary, degenerative mutations. *Genetics*, **151,** 1531–1545.

Gardner, C.A. and Barald, K.F. (1992) Expression patterns of engrailed-like proteins in the chick embryo. *Dev. Dyn.*, **193,** 370–388.

Gates, M.A., Kim, L., Cardozo, T., Sirotkin, H.I., Dougan, S.T., Lashkari, D., Abagyan, R., Schier, A.F. and Talbot, W.S. (1999) A genetic linkage map for zebrafish: comparative analysis and localization of genes and expressed sequences. *Genome Res.*, **9,** 334–347.

Gehring, W.J. (1998). *Master Control Genes in Development and Evolution: the Homeobox Story*. Yale University Press, New Haven

Gibson, T.J. and Spring, J. (1998) Genetic redundancy in vertebrates: polyploidy and persistence of genes encoding multidomain proteins. *Trends Genet.*, **14,** 46–49.

Hall, T.A. (1999) BioEdit: a user-friendly biological sequence alignment editor and analysis program for Windows 95/98/NT. *Nucleic Acids Symp. Ser.*, **41,** 95–98.

Holland, P.W. (1997) Vertebrate evolution: something fishy about *Hox* genes. *Curr. Biol.*, **7,** R570-R572.

Holland, P.W.H. (1999) The effect of gene duplication on homology. In *Homology* (Eds., Bock, G.R. and Cardew, G.), Wiley, Chichester, UK, pp. 226–242.

Holland, P.W. and Garcia-Fernandez, J. (1996) *Hox* genes and chordate evolution. *Dev. Biol.*, **173,** 382–395.

Hughes, A.L. (1994) The evolution of functionally novel proteins after gene duplication. *Proc. R. Soc. Lond. B*, **256,** 119–124.

Hughes, A.L. (1999) Phylogenies of developmentally important proteins do not support the hypothesis of two rounds of genome duplication early in vertebrate history. *J. Mol. Evol.*, **48,** 565–576.

Hughes, A.L., da Silva, J. and Friedman, R. (2001) Ancient genome duplications did not structure the human Hox-bearing chromosomes. *Genome Res.*, **11,** 771–780.

Imboden, M., Devignot, V. and Goblet, C. (2001) Phylogenetic relationships and chromosomal location of five distinct glycine receptor subunit genes in the teleost *Danio rerio*. *Dev. Genes Evol.*, **211,** 415–422.

Joyner, A.L. and Martin, G.R. (1987) *En–1* and *En–2*, two mouse genes with sequence homolog to the *Drosophila engrailed* gene: expression during embryogenesis. *Genes Dev.*, **1,** 29–38.

Laforest, L., Brown, C.W., Poleo, G., Geraudie, J., Tada, M., Ekker, M. and Akimenko, M.-A. (1998) Involvement of the *Sonic Hedgehog*, *patched* 1 and *bmp2* genes in patterning of the zebrafish dermal fin rays. *Development*, **125**, 4175–4184.

Lundin, L.-G. (1999) Gene duplications in early metazoan evolution. *Cell Dev. Biol.*, **10,** 523–530.

Lydeard, C. and Roe, K.J. (1997) The phylogenetic utility of the mitochondrial cytochrome b gene for inferring relationships among actinopterygian fishes. In *Molecular Systematics of Fishes* (Eds., Kocher, T.C. and Stepien, C.A.), Academic Press, San Diego, CA, pp. 285–303.

Lynch, M. and Conery, J.S. (2000) The evolutionary fate and consequences of duplicate genes. *Science*, **290,** 1151–1155.

Lynch, M. and Force, A. (2000a) The probability of duplicate gene preservation by subfunctionalization. *Genetics*, **154,** 459–473.

Lynch, M. and Force, A. (2000b) The origin of interspecific genomic incompatibility via gene duplication. *Am. Nat.* **156**, 590–605.

Málaga-Trillo, E. and Meyer, A. (2001) Genome duplications and accelerated evolution of *Hox* genes and cluster architecture in teleost fishes. *Amer. Zool.*, **41**: 676–686.

Martinez-Barbera, J.P., Toresson, H., Da Rocha, S. and Krauss, S. (1997) Cloning and expression of three members of the zebrafish Bmp family: *Bmp2a*, *Bmp2b* and *Bmp4*. *Gene,* **198**, 53–59.

Mellgren E.M. and Johnson, S.L. (2002) The evolution of morphological complexity in zebrafish stripes. *Trends Genet.*, **18,** 128–134.

Meyer, A. and Schartl, M. (1999) Gene and genome duplications in vertebrates: the one-to-four (-to-eight in fish) rule and the evolution of novel gene functions. *Curr. Opin. Cell Biol.*, **11,** 699–704.

Naruse, K., Fukamachi, S., Mitani, H., Kondo, M., Matsuoka, T., Kondo, S., Hanamura, N., Morita, Y., Hasegawa, K., Nishigaki, R., Shimada, A., Wada, H., Kusakabe, T., Suzuki, N., Kinoshita, M., Kanamori, A., Terado, T., Kimura, H., Nonaka, M. and Shima, A. (2000) A detailed linkage map of medaka, *Oryzias latipes*: comparative genomics and genome evolution. *Genetics*, **154,** 1773–1784.

Nelson, J.S. (1994) *Fishes of the World*, 3rd ed., Wiley, New York, NY.

Nowak, M.A., Boerlijst, M.C., Cooke, J. and Maynard Smith, J. (1997) Evolution of genetic redundancy. *Nature*, **388,** 167–171.

Ohno, S. (1970) *Evolution by Gene Duplication*, Springer Verlag, New York, NY.

Ohno, S. (1999) The one-to-four rule and paralogues of sex-determining genes. *Cell. Mol. Life Sci.*, **55,** 824–830.

Patel, N.H. and Prince, V.E. (2000) Beyond the *Hox* complex. *Genome Biol.*, **1,** 1027.1–1027.4.

Postlethwait, J.H., Woods, I.G., Ngo-Hazelett, P., Yan, Y.-L., Kelly, P.D., Chu, F., Huang, H., Hill-Force, A. and Talbot, W.S. (2000) Zebrafish comparative genomics and the origins of vertebrate chromosomes. *Genome Res.*, **10,** 1890–1902.

Robinson-Rechavi, M., Marchand, O., Escriva, H., Bardet, P.-L., Zelus, D., Hughes, S. and Laudet, V. (2001a) Euteleost fish genomes are characterized by expansion of gene families. *Genome Res.*, **11,** 781–788.

Robinson-Rechavi, M., Marchand, O., Escriva, H. and Laudet, V. (2001b) An ancestral whole-genome duplication may not have been responsible for the abundance of duplicated fish genes. *Curr. Biol.*, **11,** R458-R459.

Saitou, N. and Nei, M. (1987) The neighbor-joining method: a new method for reconstructing phylogenetic trees. *Mol. Biol. Evol.*, **4,** 406–425.

Schulte, T., Paschke, K.A., Laessing, U., Lottspeich, F. and Stuermer, C.A. (1997) Reggie–1 and reggie–2, two cell surface proteins expressed by retinal ganglion cells during axon regeneration. *Development*, **124,** 577–587.

Sidow, A. (1996) Gen(om)e duplications in the evolution of early vertebrates. *Curr. Opin. Genet. Dev.*, **6,** 715–722

Spring, J. (1997) Vertebrate evolution by interspecific hybridisation — are we polyploid? *FEBS Lett.*, **400,** 2–8.

Taylor, J.S. and Brinkmann, H. (2001) 2R or not 2R. *Trends Genet.*, **17,** 488–489.

Taylor, J.S., Van de Peer, Y., Braasch, I. and Meyer, A. (2001a) Comparative genomics provides evidence for an ancient genome duplication in fish. *Phil. Trans. Roy. Soc. B*, **356,** 1661–1679.

Taylor, J.S., Van de Peer, Y. and Meyer, A. (2001b) Genome duplication, divergent resolution, and speciation. *Trends Genet.*, **17,** 299–301.

Thompson, J.D., Gibson, T.J., Plewniak, F., Jeanmougin, F. and Higgins, D.G. (1997) The CLUSTAL_X windows interface: flexible strategies for multiple sequence alignment aided by quality analysis tools. *Nucleic Acids Res.*, **25,** 4876–4882.

Van de Peer, Y., and De Wachter, R. (1994) TREECON for Windows: a software package for the construction and drawing of evolutionary trees for the Microsoft Windows environment. *Comput. Appl. Biosci.*, **10,** 569–570.

Van de Peer, Y., Taylor, J.S., Braasch, I. and Meyer, A. (2001). The ghost of selection past: rates of evolution and functional divergence in anciently duplicated genes. *J. Mol. Evol.*, **53,** 434–444.

Van de Peer, Y., Taylor, J.S., Joseph, J. and Meyer, A. (2002a) Wanda: A database of duplicated fish genes. *Nucleic Acids Res.*, **30,** 109–112.

Van de Peer, Y., Frickey, T., Taylor, J.S. and Meyer, A. (2002b) Dealing with saturation at the amino acid level: A case study based on anciently duplicated zebrafish genes. *Gene*, **295**, 205–211.

Wang, Y. and Gu, X. (2000) Evolution patterns of gene families generated in the early stage of vertebrates. *J. Mol. Evol.*, **51,** 88–96.

Wittbrodt, J., Meyer, A. and Schartl, M. (1998) More genes in fish? *BioEssays*, **20,** 511–512.

Woods, I.G., Kelly, P.D., Chu, F., Ngo-Hazelett, P., Yan, Y.-L., Huang, H., Postlethwait, J.H. and Talbot, W.S. (2000) A comparative map of the zebrafish genome. *Genome Res.*, **10,** 1903–1914.

A. Meyer, Y. Van de Peer (eds.), Genome Evolution, 75-84.

More genes in vertebrates?

Peter W.H. Holland
School of Animal & Microbial Sciences, The University of Reading, Whiteknights, PO Box 228, Reading RG6 6AJ, United Kingdom
E-mail: p.w.h.holland@reading.ac.uk
Present address: Department of Zoology, University of Oxford, South Parks Road, Oxford OX1 3PS, United Kingdom.
E-mail: peter.holland@zoology.oxford.ac.uk

Received 24.01.2002; Accepted for publication 29.08.2002

Key words: amphioxus, evolution, gene duplication, gene family, Hox, Parahox

Abstract

With the acquisition of complete genome sequences from several animals, there is renewed interest in the pattern of genome evolution on our own lineage. One key question is whether gene number increased during chordate or vertebrate evolution. It is argued here that comparing the total number of genes between a fly, a nematode and human is not appropriate to address this question. Extensive gene loss after duplication is one complication; another is the problem of comparing taxa that are phylogenetically very distant. Amphioxus and tunicates are more appropriate animals for comparison to vertebrates. Comparisons of clustered homeobox genes, where gene loss can be identified, reveals a one to four mode of evolution for Hox and ParaHox genes. Analyses of other gene families in amphioxus and vertebrates confirm that gene duplication was very widespread on the vertebrate lineage. These data confirm that vertebrates have more genes than their closest invertebrate relatives, acquired through gene duplication.

Abbreviations: IHGSC, International Human Genome Sequencing Consortium; TCESC, The C. elegans Sequencing Consortium.

Introduction

In several of his works, Aristotle divides the animals into sanguineous (blood–filled) and non-sanguineous (bloodless) forms. Presence of obvious blood is one basis for this distinction, but Aristotle draws attention to a second character: 'all sanguineous animals have a backbone of either one kind or other, that is composed either of bone or of spine' (Aristotle, 350 BC). It is clear, therefore, that his divisions equate rather neatly to our own concepts of vertebrate and invertebrate. Aristotle lists numerous other features that characterize either sanguineous or non-sanguineous animals; for the former he notes blood, lymph, fibre, flesh (?), bone (including fish-bone and gristle), skin, membrane, sinew, hair, nails (helpfully adding 'and whatever corresponds to these"), fat, suet, and some delightful excretions (dung, phlegm, yellow bile and black bile). This is not quite the same list that I would compile if drawing up differences between vertebrates and invertebrates; nonetheless, it does stress that there are numerous anatomical and physiological distinctions between vertebrates and invertebrates. Furthermore, this major dividing line in the animal kingdom has been noted for 2300 years.

Consideration of animal phylogeny indicates that vertebrates evolved from invertebrate ancestors (Figure 1); indeed, the phylum Chordata includes both invertebrate chordates (tunicates and cephalochordates) and vertebrates. It is interesting to ask, therefore, what was the basis for this major evolutionary transition from invertebrate to vertebrate?

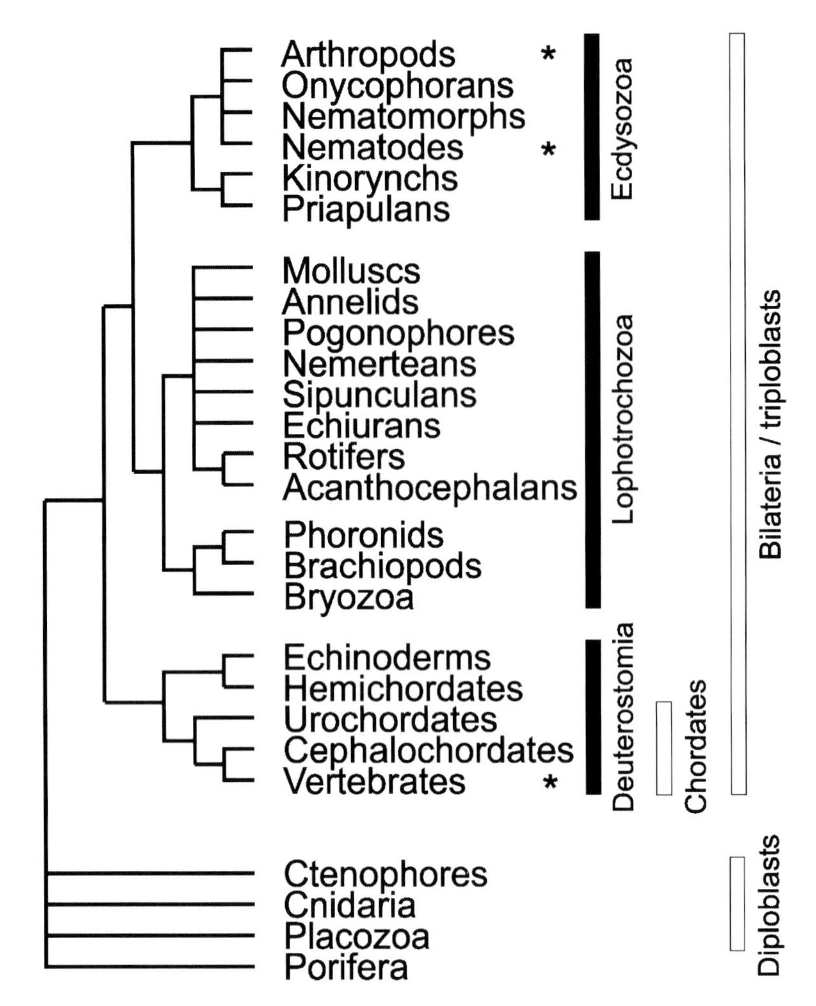

Figure 1. Proposed phylogeny of the animal kingdom based on molecular data. The primary subdivision into diploblasts and Bilateria is shown, as is subdivision of the latter into three great clades. Asterisks indicate the phylogenetic positions of the first three animal species to have their complete genome sequence determined (fly, nematode, human).

In his influential (and highly readable) 1970 book, Ohno made several arguments that are directly relevant to this question. First, he argued from a theoretical standpoint that gene duplication allows a 'new' gene to escape the purifying constraints of natural selection; new genes, therefore, have the potential to acquire new roles. Second, he suggested that whole genome doubling (by tetraploidy) would be a more effective route to achieve this than tandem gene duplication, partly because dosage problems would not be encountered. Third, he used empirical evidence to propose that vertebrates have more genes than some of their invertebrate relatives (Atkin and Ohno, 1967; Ohno 1970). The evidence included genome size (total DNA content), karyotype and allozyme complexity; with hindsight, we now realize that these lines of evidence are rather poor indicators of absolute gene number (Holland, 1999). In parallel with this proposal, Ohno noted that there was an increase in organismal complexity during vertebrate evolution, and speculated that an increase in gene number (probably effected by a combination of tandem gene duplication and genome doubling) may have facilitated this.

It is not the intention of this article to evaluate all these proposals. Instead, I will focus on the third of Ohno's points: was the origin of vertebrates accompanied by an increase in gene number?

Why not simply count genes?

How can we determine if gene number increased during vertebrate evolution? Without doubt, the ultimate answer to this question will come from comparison of complete genome sequences. Having said that, it is not a trivial task to extract the relevant information from genome sequences. At the time of writing, in the public domain are the complete genome sequences for just two invertebrates: a fly (*Drosophila melanogaster* and a nematode *Caenorhabitis elegans*) and, in draft form, the human genome (TCESC 1998; Adams et al., 2000; IHGSC, 2001; Venter *et al*., 2001). So, why not count up the total number of genes in each genome and simply compare invertebrate and vertebrate? This has been done by several authors; some concluding that genome duplication probably did not occur, and others giving a 'not proven' verdict. For example, Rubin (2001) states 'if such doubling did occur, the evidence for it has been obscured'. So have we no hope of answering the question? Or perhaps we can already say that Ohno was wrong? In fact, neither conclusion is sound, because the comparison undertaken is flawed. Before examining why this approach does not address the central question, consider the numbers.

Around 14,000 genes have been identified in the *Drosophila* genome sequence (Adams *et al*; 2000; Fly Base 1999) and about 19,000 in *C. elegans* (TCESC, 1998). These figures compare to 30,000 to 40,000 currently estimated in humans, although under 30,000 of these have actually been identified (Rubin, 2001; Baltimore, 2001). Many people have commented on the fact that the human figure is only a little larger than the invertebrate figures. Indeed, several people have suggested that 70,000 to 100,000 genes were 'expected' in the human genome, and that estimates of 30,000 to 40,000 are therefore surprisingly low (IHGSC, 2001; Venter *et al*., 2001; Claverie, 2001). It is not clear why the very high figures were ever expected, however. As commented by Claverie *et al.* (2001), the data from EST surveys and partial genome sequencing had been pointing towards a much lower figure for several years. So perhaps the 100,000 estimate came more from a deep-rooted desire to see humans possessing five or ten times more genes than a fly or a nematode, rather than accurate scientific reasoning? After all, isn't our own species immensely superior to all the others, in many possible measures of complexity? The problem with that line of argument is that a chimp, a squid, a parrot or a honeybee could probably make a similar claim. In passing, it is amusing to note that some commentators in the media have tried to suggest that the 'surprisingly low' number of genes implies that the nature versus nurture debate must now swing dramatically towards nurture, and that genes must play less of a role in our character. Such a conclusion is unjustified, of course, but it emphasizes just how confused the notions of genetic complexity and organismal complexity can become.

To return to evolutionary biology, the gene counting game suggests that humans have only 1.5 to 2 times as many genes as a nematode or a fly respectively. Despite claims to the contrary, this is not the death knell to Ohno's model of two rounds of genome doubling in vertebrate ancestry. In fact, I doubt that Ohno himself would have been at all surprised by these numbers. Towards the end of 1999, Susumu Ohno wrote an article for a journal issue I was editing; this was one of his last scientific articles before he died in January 2000. At the time, most estimates of human gene number ranged from 50,000 to 100,000. Ohno wrote 'I tend to favour an estimate towards the low end of the possible range, approximately 50,000' (Ohno, 1999).

Why is counting fly, nematode and human genes an inappropriate way to assess Ohno's hypothesis? The first problem is gene loss. After gene or genome duplication there is a finite probability that one of the daughter genes will be lost; the probability of loss may fluctuate widely depending on gene type or expression characteristics (Patton *et al*., 1998; Krakauer and Nowak, 1999). A clear example is given by the Hox gene clusters (Figure 2). There can be no doubt that the 39 human Hox genes, in four gene clusters, evolved by duplication and re-duplication of an ancestral cluster of (at least) 13 Hox genes. This implies a loss of 13 genes (13 x 4 = 52; 52 minus 39 = 13 genes lost). Other examples are more dramatic; for example, the Xlox gene duplicated from one to four in the vertebrate lineage, but gene loss has reduced this complement back to one again (Figure 3; Pollard and Holland, 2000). The implication is that a total genome doubling will not result in double the number of genes in a descendent lineage, particularly if tens of millions of years have elapsed since the genome doubling. If there was a 50% chance of retention (to pluck a figure from the air), then genome doubling would increase gene number by 1.5 times. Two rounds of genome doubling, with 50% loss after each, would cause gene number to increase by 2.25

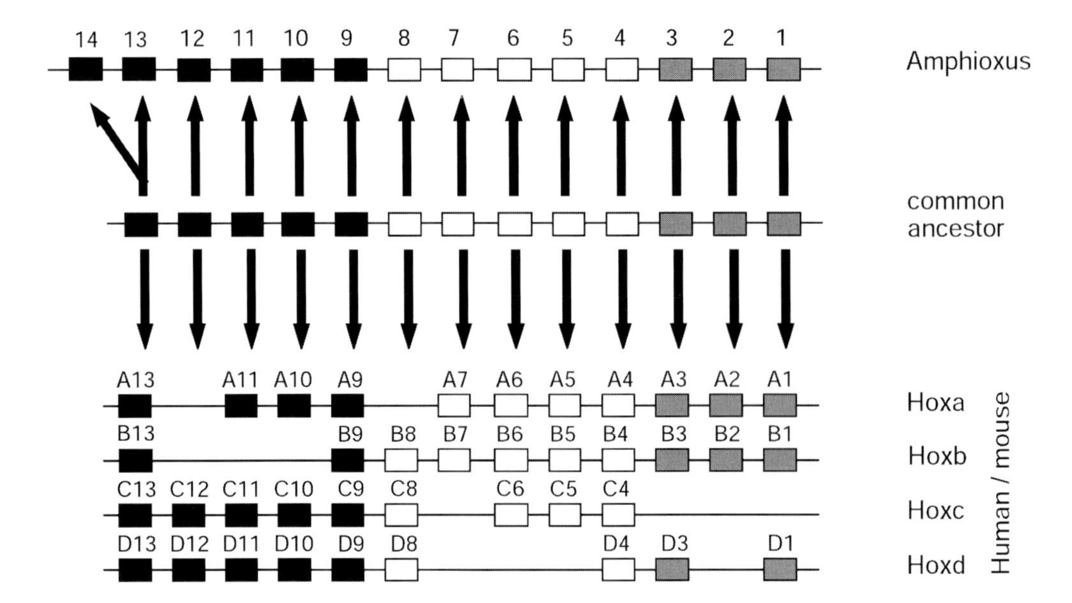

Figure 2. Evolutionary relationship between amphioxus and mammalian Hox gene clusters, showing deduced state for the last common ancestor of these taxa. Black boxes, posterior class genes; open boxes, middle class genes; grey boxes, PG3 genes; hashed boxes, anterior class genes.

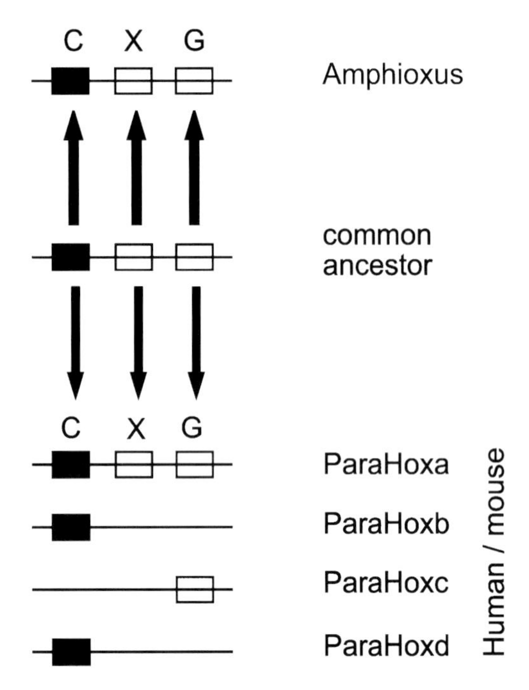

Figure 3. Evolutionary relationship between amphioxus and mammalian ParaHox gene clusters, showing deduced state for the last common ancestor of these taxa. Black boxes, posterior class genes; open boxes, middle class genes; grey boxes, PG3 or Xlox genes; hashed boxes, anterior class genes.

times. If, for the sake of argument, the invertebrate ancestor of the vertebrates possessed 16,500 genes (the arithmetic mean of the *Drosophila* and *C. elegans* gene numbers), then 16,500 times 2.25 would yield a mere 37,125 genes.

To some extent, the gene loss problem can be overcome by treating each gene family independently. If all gene families duplicated from one to four, then gene loss would reduce some families back to one gene, some to two genes and some to three, while a minority would remain as sets of four paralogues. Additional single gene duplication could complicate the picture, but if every gene family in the genome was examined, a pattern should be discernible. This type of analysis was attempted in both papers reporting the draft human genome sequence. Venter *et al.* (2001) used BLAST searches to group deduced proteins from the three species, and then plotted these according to the ratio of human to fly, or human to nematode, genes. The peak of this distribution lies over a 1:1 ratio, although the plot is markedly skewed in the direction of 'human predominance' (Figure 4; reproduced from Venter *et al.*, 2001). This indicates that 2:1, 3:1 and 4:1 ratios (human:fly or human:nematode) are relatively common. The International Human Genome Sequencing Consortium used a slightly different method (IHGSC, 2001). They reported finding 1,195 'orthologue groups' represented by single genes in all three animal taxa (that is, a 1:1 ratio); about the same number of orthologue

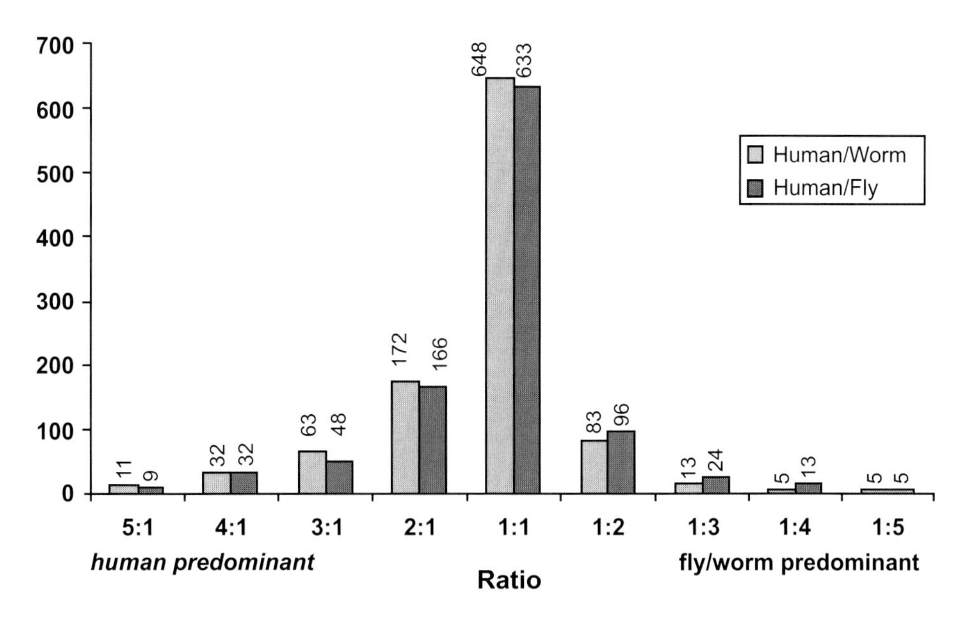

Figure 4. Gene number ratios for human to nematode (light bars), and human to fly (dark bars), within gene families as detected using a protein BLAST-based search strategy (Lek clustering) developed by Venter *et al.* (2001). Reproduced with permission from Venter *et al.* (2001). © 2001 American Association for the Advancement of Science.

groups had more vertebrate than invertebrate genes. The example shown in detail, nuclear hormone receptors, has six cases of 1:1, two of 2:1, four of 3:1 and one 4:1 in fourteen orthology groups detected. These patterns seem compatible with Ohno's model.

The second problem with comparing fly, nematode and human is grounded in phylogenetics. Quite simply, these are the wrong species to compare. It is becoming widely accepted that the 'higher' animals (Bilateria or triploblasts) diverged into three great lineages in evolution: the Ecdysozoa (moulting phyla), the Lophotrochozoa (spiral cleaving phyla plus lophophorates) and Deuterostomia, including vertebrates, other chordates, sea urchins and hemichordates (Aguinaldo *et al.*, 1997; de Rosa *et al.*, 1999; Figure 1). Flies and nematodes both fall into the first of these three great clades; vertebrates are in the third. To determine if extensive gene duplication accompanied the origin or evolution of vertebrates, we should focus on animals within the Deuterostomia. Genome evolution may have followed its own complex course in the Ecdysozoa, for example, which is not directly relevant to vertebrate origins. Ideally, we should compare vertebrate genomes with those of the non-vertebrate chordates (amphioxus and tunicates).

Ohno and amphioxus

There are three distinct taxonomic groups within the Phylum Chordata: the tunicates (also known as urochordates, including the ascidians), the cephalochordates (colloquially known as amphioxus or lancelets) and the vertebrates (used here as synonymous with craniates). It has long been thought that the amphioxus and vertebrate lineages are sister lineages, and that the tunicate linage is more distant. Indeed, this contention gains strong support from rDNA sequence analysis (Wada and Satoh, 1993). By comparing the genomes of these three taxa, therefore, it may be possible to get some insight into the intermediate states during chordate evolution. Atkin and Ohno (1967) determined the genome sizes of various chordates. They noted that the haploid genome size of a tunicate, *Ciona intestinalis*, is around 0.2 pg, whereas the genome size of amphioxus, *Branchiostoma lanceolatum,* is considerably larger, at 0.6 pg. Indeed, 0.6 pg is approximately the same size as the smallest vertebrate genomes; for example, several puffer fish and their relatives in the Tetraodontiformes have genome sizes in the order of 0.5 to 0.6 pg (Ohno, 1970; Hinegardner and Rosen, 1972). This was one line of reasoning that lead Ohno to suggest that gene number had increased on the stem lineage of amphioxus plus vertebrates, that is, before the divergence of amphioxus and vertebrates. He speculated that the ascidians retain the ancestral gene number for chordates, while the amphioxus and vertebrates share a derived condition with a greater total number of genes. Ohno (1970) notes 'Whether this increase was accomplished exclusively by tandem duplication or by a combination of tandem duplication and tetraploidy cannot be resolved at the moment'. Similar reasoning, incorporating other taxa, lead Ohno to suggest there was also a second round of genome expansion, within the vertebrate lineage.

Implicit in this line of argument is the assumption that the small genome sizes of tetraodontiform fish represent the primitive condition for vertebrates. This seems extremely unlikely, however, since it would demand independent increase in vertebrate genome size on multiple vertebrate lineages, leaving a few rather insignificant branches with the ancestral condition. These 'small genome' fish include puffer fish, sticklebacks, seahorses, flatfish, blue-striped grunt and Siamese algae eater (Ohno, 1970; Hinegardner and Rosen 1972; Arai *et al.*, 1988). Furthermore, it is noteworthy that representatives of the most basal vertebrate lineages, lampreys and hagfish, have genome sizes significantly larger than that of amphioxus. For example, the hagfish *Eptatretus* has a haploid genome size of 2.7 pg (Atkin and Ohno, 1967); lampreys are generally around 1.5 pg (Robinson *et al.*, 1975).

Overall, it is actually rather hard to discern any simple pattern in the genome sizes of chordates. If certain assumptions were made about the relative likelihood of increase or decrease in genome size (and whether small or large incremental changes were likely), it should be possible to model the pathways of genome size evolution in chordates, and deduce likely ancestral states. Until such an analysis is performed, it is difficult to argue strongly for one pattern of genome size evolution over another. Nonetheless, from cursory examination of genome sizes across the vertebrates, I will make the rash suggestion that the original vertebrate genome size was around 1.5 to 2 pg, and that this was independently increased in (for example) the hagfish, the mammals and the amphibians, but decreased in the Tetraodontiformes and some other fish lineages. Such a scheme differs from that deduced by Ohno (1970).

What do patterns of genome size evolution mean for total gene number? Possibly very little. This point was first demonstrated in relation to these taxa by Schmidtke *et al.* (1977). These authors used allozyme electrophoresis to estimate the 'complexity' of particular enzyme systems in a tunicate and in amphioxus. Although there are many caveats to this application of allozymes, if sufficient enzyme families are examined it should be possible to gain a low resolution look at gene number differences (in brief, if a gene family has multiple members, this should be reflected in a greater diversity of allozyme bands). Schmidtke and colleagues found little difference between the tunicate and amphioxus species, contrary to the prediction of Ohno (1970). Hence, even though amphioxus has a genome three times larger than *Ciona* (0.6 pg versus 0.2 pg), this was not accompanied by a three-fold increase in gene number (at least in the enzyme systems examined). Similar conclusions have been drawn from many other studies in many other taxa, and it is now accepted that genome size is not an accurate estimator of total gene number. This does not mean that there is no correlation at all, only that variation is extensive.

Schmidtke *et al.* (1977) concluded that their data do not support Ohno's 'polyploidation hypothesis'. They also state, however, that they 'do not contradict it' either! At best, the authors' data indicated that there was probably no whole genome doubling before the common ancestor of amphioxus and vertebrates; the data say absolutely nothing about duplication on the vertebrate-specific lineage (after divergence from the non-vertebrate chordates). In this regard, it is very unfortunate that Schmidtke and colleagues did not extend their analyses to lampreys, hagfish and other vertebrates. If they had, they may well have detected the true pattern of gene duplication in chordates: a pattern of gene duplication on the vertebrate lineage, after it had diverged from tunicates *and* amphioxus. This pattern was not clearly demonstrated until more than 15 years later.

Do Hox genes encapsulate Ohno's central tenet?

In 1992, my laboratory published the first Hox gene from a species of amphioxus, and two years later a physical map of the majority of the amphioxus Hox gene cluster (Holland *et al.*, 1992; Garcia-Fernandez and Holland, 1994). These data proved definitely that amphioxus has a single Hox gene cluster (correcting a previous suggestion of two clusters; Pendleton *et al.*, 1993), as compared to the four Hox gene clusters of mammals. Subsequent work added detail to this picture, confirming that all vertebrates have multiple

Hox gene clusters; all the other deuterostomes (including non-vertebrate chordates) have a single cluster (for review, see Holland and Garcia-Fernandez, 1996). It is clear that a single cluster is the ancestral condition, and that duplication (or multiple duplication) of this cluster has occurred on the vertebrate lineage. This pattern of evolution is compatible with that observed for enzyme systems by Schmidtke *et al.* (identity between tunicate and amphioxus), yet is also consistent with the central tenet of Ohno's hypothesis (more genes in vertebrates).

Careful comparison of the deduced protein coding sequences of the amphioxus and vertebrate Hox genes suggested a remarkably simple pathway of evolution. These analyses suggested the amphioxus Hox gene cluster retains the ancestral condition inferred for the last common ancestor of amphioxus as in vertebrates, whereas in the vertebrate lineage the cluster underwent cluster duplication and loss of numerous individual genes (Figure 2; Garcia-Fernandez and Holland, 1994). This pattern is particularly striking when one remembers that both lineages have been evolving for the same period of time since their last common ancestor (estimated at 550 million years). Furthermore, this highly asymmetrical pattern of molecular evolution has a morphological correlate. The anatomy of modern cephalochordates is somewhat similar to that reconstructed for some Cambrian chordates such as *Pikaia* and *Cathymyrus* (Shu *et al.*, 1996) and comparable in many respects to the hypothetical pre- or proto-vertebrate condition (e.g., Jefferies and Lewis, 1978). In contrast to relative stasis in the cephalochordate lineage, the vertebrate lineage has seen the acquisition of a large suite of lineage-specific characters (Shimeld and Holland, 2000). Is this a molecular incarnation of Ohno's ultimate vision, that vertebrate morphological evolution follows the constraints set by gene duplication?

The correlation between Hox genes and morphology is neat and captivating, but ultimately too simplistic. First, there are indeed anatomical specializations on the amphioxus lineage. Not as many as on the vertebrate lineage, for sure, but living members of the genera *Branchiostoma* and *Epigonichthys* certainly have derived characters. These include rostral extension of the notochord, unpaired nature of the frontal eye, and additional complexity of the branchial basket. Second, it has recently been found that the amphioxus Hox gene cluster is not completely prototypical in its arrangement. In our original physical map (Garcia-Fernandez and Holland, 1994), for technical reasons we did not characterize genes more 5' to paralogy group 10. Hox genes of paralogy group 9 (PG 9) and above are collectively termed posterior genes; there are five paralogy groups in this category in mammals and other vertebrates (PG9 to PG13) patterning the most posterior structures of the body axis. When we extended our characterization to this region of the amphioxus Hox gene cluster, we detected six posterior class genes: the genes already designated *AmphiHox–9* and *AmphiHox–10*, and four additional genes (Ferrier *et al.*, 2000). It has proved extremely difficult to deduce the relationship of *AmphiHox–9*, *–10*, *–11*, *–12*, *–13* and *–14* to the vertebrate PG9 to PG13 genes, possibly due to rapid sequence divergence in these genes. The most parsimonious model is that the common ancestor of vertebrates and amphioxus had a Hox gene cluster with 13 genes; on the amphioxus lineage the terminal gene (13) underwent an extra tandem gene duplication to yield *AmphiHox–13* and *AmphiHox–14* (Figure 2). Just to weaken the molecules-morphology correlation yet further, most of the derived morphological features of amphioxus are anterior, while the derived Hox gene arrangements affect the posterior genes. Despite these complications, however, there remains a broad correlation between complexity of the Hox gene system and complexity of the body plan, at least during the transition from invertebrates to vertebrates.

A paradigm paralleled by ParaHox

Brooke *et al.* (1998) reported that three homeobox genes outside the Hox gene cluster are arranged into a separate gene cluster, at least in the amphioxus genome. These three genes, *Cdx* (or caudal), *Xlox* (also known as *pdx* or *ipf*) and *Gsx* (or *Gsh*), form the ParaHox gene cluster. Furthermore, phylogenetic analysis suggests that the ParaHox gene cluster is as ancient as the Hox gene cluster, the two having originated early in animal evolution by duplication of an ancestral, and as yet hypothetical, ProtoHox gene cluster. Recent analyses date the origin of the Hox and ParaHox gene clusters to before the separation of the cnidarian (sea anemones, coral, hydroids, jellyfish) and bilaterian lineages (for review, see Ferrier and Holland, 2001). The relevance of these findings from the present perspective is that the ancestor of the vertebrates possessed a single ParaHox gene cluster in its genome. As with the Hox gene cluster, amphioxus retains a single ParaHox gene cluster inherited from the common ancestor of amphioxus and vertebrates.

What happened to the ParaHox gene cluster during vertebrate evolution? Pollard and Holland (2000) collated the chromosomal map positions of the human and mouse ParaHox genes. They report that the human homologues of *Cdx* (three genes), *Xlox* (one gene) and *Gsx* (two genes) map to four chromosomal locations; furthermore, analysis of neighbouring genes indicate that these four chromosomal regions are evolutionarily related. These data indicate that the single ParaHox gene cluster in the ancestor of the vertebrates duplicated to give four clusters; these were then subject to very extensive gene loss (Figure 3). This pattern of evolution is strikingly reminiscent to the paradigm established by Hox genes: the two gene clusters evolved in a parallel manner. It is interesting to note that gene loss was even more extensive than seen in the Hox clusters. In the human ParaHox clusters, 50% of genes have been lost after vertebrate-specific duplication (6/12 lost) compared to 25% loss in the Hox clusters (13/52 lost).

Massive gene duplication on the vertebrate lineage?

When compared between amphioxus and vertebrates, do other gene families show a similar pattern of expansion? This is actually rather hard to ascertain, because of the gene loss complication alluded to earlier. The advantage of the Hox gene clusters in this regard is that their stereotyped organization permits simple recognition of gaps produced by gene loss. The same advantage was afforded to the ParaHox analysis by use of genes linked to ParaHox genes. Without this sort of information, we would get a quite different insight into their mode of evolution. For example, if the Hox genes were dispersed rather than in clusters, we might conclude that only the PG4 and PG9 genes duplicated in a 1 to 4 pattern; in contrast, PG1, PG3, PG5, PG6, and PG8 duplicated from 1 to 3 genes, while PG2 and PG7 only went from one gene to two. In the case of ParaHox, we might conclude that none of the genes duplicated from 1 to 4, and that the *Xlox* gene did not duplicate at all! These conclusions would be erroneous, and again emphasize the need for care in interpreting gene numbers.

We can take two approaches to overcome this problem. First, we could restrict our consideration only to amphioxus genes that have vertebrate homologues mapping to well characterized paralogy regions in the human genome. In this way, we could detect the gaps left by missing genes as well as the extant genes. For example, Msx, NK1, NK3, NK4 and Emx homeoboxes, plus the FGR receptor gene, have all been cloned from amphioxus (Pollard and Holland, 2000; G.N. Luke and P.W.H. Holland, unpublished data). Their homologues map to a fourfold paralogy region in the human genome, comprising 4p16, 2p13–14/8p21, 10q24–26, 5q34 (Pollard and Holland, 2000). We can deduce that all these genes duplicated in the one to four pattern (just like Hox and ParaHox), but the final numbers in the human genome have been reduced by gene loss.

The second approach is to examine a diversity of amphioxus and human gene families, and ask whether they show a spectrum of duplication patterns ranging from 1:1 to 1:4. In other words, to perform an analysis similar to that performed by Venter *et al.* (2001) or the IHGSC (2001) when comparing fly, nematode and human gene families (Figure 4). If the genes were evolving in a similar way to Hox and ParaHox, few should show duplication patterns in excess of 1:4. The precision of the analysis can be improved markedly by drawing phylogenetic trees for each gene family. In this way, any independent gene duplication on the amphioxus lineage can be detected; such events would otherwise complicate the detection of gene duplication on the vertebrate lineage. (For example, two myogenic basic helix-loop-helix genes have been cloned from amphioxus, and four from mammals. A simple ratio would describe this as 1:2. In fact, phylogenetic analysis reveals that a single gene in the common ancestor of the two lineages duplicated independently to give two genes in amphioxus and four in human. The true pattern of gene duplication on the vertebrate lineage is therefore 1:4; Araki *et al.*, 1996).

This methodology was applied to 83 gene families with representatives cloned from amphioxus and mammals (Furlong and Holland, 2002). For this study, a gene family is defined as the inclusive set of genes that descend from a single ancestral gene in the common ancestor of amphioxus and vertebrates. Two exceptions to this definition are the Hox and ParaHox clusters, which are each treated here effectively as single genes. Approximately 14 gene families showed a 1 to 1 pattern, 34 families showed a 1 to 2 pattern, 20 showed a 1 to 3 pattern and 10 showed a 1 to 4 pattern of duplication on the vertebrate lineage. Just five families showed duplication patterns in excess of

1 to 4. (These figures are approximate because a small number of trees had ambiguous nodes, and some gene families have additional genes cloned from other vertebrates). Hence, 69 out of 83 gene families showed gene duplication on the vertebrate lineage. By comparison, just 13 of these genes showed duplication on the sister lineage, leading to amphioxus. Gene duplication is indeed widespread on the vertebrate lineage.

Conclusion

The central question addressed here is whether gene number increased during the transition from invertebrates to vertebrates. I have argued that comparing total gene complements between fly, nematode and human genome sequences is not appropriate for addressing this question. Even if comparisons are made at the level of gene families, there is still the complicating factor of gene loss to take into account. More crucially, these are simply the wrong species to compare. The ideal data to use for addressing this question would be the complete genome sequences for non-vertebrate chordates (tunicate and amphioxus) and for basal vertebrates (hagfish, lampreys, cartilaginous fish). At current estimates, it would cost around one billion dollars to acquire this information. Since this is rather excessive, we must find ways to extract maximal information from more limited data sets. A sampling of amphioxus genes has proved extremely informative for this purpose.

The best studied regions of the amphioxus genome are the Hox and ParaHox gene clusters. Both genomic regions duplicated from one copy to four during vertebrate evolution, probably before vertebrate radiation. Extending the comparative analyses to other gene families provides inconvertible evidence for extensive gene duplication on the vertebrate lineage. Most genes examined show duplication in a 1:2 or 1:3 pattern of duplication, with some cases of 1:4 (and very few cases in excess of this). These data strongly support Ohno's central tenet of extensive gene duplication close to vertebrate origins. Furthermore, if gene loss after duplication is frequent (as is suggested by the Hox and ParaHox gene studies), then these figures are also compatible with two whole genome doublings on the vertebrate lineage, followed by some gene loss.

Acknowledgements

I thank Rebecca Furlong, Graham Luke, Sophie Pollard and other members of my laboratory for helpful discussions. This paper is based on a presentation given at the Jacques Monod Conference on gene and genome duplication held at Aussois, France (2001), funded by C.N.R.S; I thank Axel Meyer, Hervé Philippe and Yves Van de Peer for the invitation to participate. The author's research on gene duplication is funded by the BBSRC.

References

Adams, M.D. *et al.* (2000) The genome sequence of *Drosophila melanogaster*. *Science*, **287,** 2185–2195.

Aguinaldo, A.M., Turbeville, J.M., Linford, L.S., Rivera, M.C., Garey, J.R., Raff, R. and Lake, J.A. (1997) Evidence for a clade of nematodes, arthropods and other moulting animals. *Nature*, **387,** 489–493.

Arai, R., Suzuki, A. and Akai Y. (1988) The karyotype and DNA value of a cypriniform algae eater, *Gyrinocheilus aymonieri*. *Jpn. J. Ichthyol.*, **34,** 515–517.

Araki, I., Terazawa, K. and Satoh, N. (1996) Duplication of an amphioxus myogenic bHLH gene is independent of vertebrate myogenic bHLH gene duplication. *Gene*, **171,** 231–236.

Aristotle (350 BC) *History of Animals* (Translation, D'Arcy Wentworth Thompson). Library of the Future, Second Series 1991, World Library, Inc., Garden Grove, CA.

Atkin, N.B. and Ohno, S. (1967) DNA values of four primitive chordates. *Chromosoma*, **23,** 10–13.

Baltimore, D. (2001) Our genome unveiled. *Nature*, **409,** 814–816.

Brooke, N.M., Garcia-Fernàndez, J. and Holland, P.W.H. (1998) The ParaHox gene cluster is an evolutionary sister of the Hox gene cluster. *Nature*, **392,** 920–922.

Claverie, J.-M. (2001) What if there are only 30,000 genes? *Science*, **291,** 1255–1257.

de Rosa, R., Grenier, J.K., Andreeva, T., Cook, C.E., Adoutte, A., Akam, M., Carroll, S.B. and Balavoine, G. (1999) Hox genes in brachiopods and priapulids and protostome evolution. *Nature*, **399,** 772–776.

Ferrier, D.E.K. and Holland, P.W.H. (2001) Ancient origins of the Hox gene cluster. *Nature Rev. Genet.* **2,** 33–38.

Ferrier, D.E.K., Minguillon, C., Holland, P.W.H. and Garcia-Fernandez, J. (2000) The amphioxus Hox cluster: deuterostome posterior flexibility and Hox14. *Evol. Dev.*, **2,** 284–293.

Flybase (1999) The Flybase database of the *Drosophila* genome projects and community literature. *Nucleic Acids Res.*, **27,** 85–88.

Furlong, R.F. and Holland, P.W.H. (2002) Were vertebrates octoploid? *Phil. Trans. R. Soc. Ser. B*, **357**, 531–544.

Garcia-Fernández, J. and Holland, P.W.H. (1994) Archetypal organization of the amphioxus *Hox* gene cluster. *Nature* **370,** 563–566 .

Hinegardner, R. and Rosen, D.E. (1972) Cellular DNA content and the evolution of teleostean fishes. *Amer. Naturalist*, **106,** 621–644.

Holland, P.W.H. (1999) Gene duplication: past, present and future. *Semin. Cell Dev. Biol.*, **10,** 541–547.

Holland, P.W.H. and Garcia-Fernàndez, J. (1996) Hox genes and chordate evolution. *Dev. Biol.*, **173,** 382–395.

Holland, P.W.H., Holland, L.Z., Williams, N.A. and Holland, N.D. (1992) An amphioxus homeobox gene: sequence conservation, spatial expression during development and insights into vertebrate evolution. *Development*, **116,** 653–661.

International Human Genome Sequencing Consortium (2001) Initial sequencing and analysis of the human genome. *Nature*, **409,** 860–921.

Jefferies, R.P.S. and Lewis, D.N. (1978) The English Silurian fossil *Placocystites forbesianus* and the ancestry of the vertebrates. *Phil. Trans. R. Soc. Ser. B*, **282,** 205–323.

Krakauer, D.C. and Nowak, M. (1999) Evolutionary preservation of redundant duplicated genes. *Semin. Cell Dev. Biol.*, **10,** 555–559.

Ohno, S. (1970) *Evolution by Gene Duplication*, Springer-Verlag, Heidelberg.

Ohno, S. (1999) Gene duplication and the uniqueness of vertebrate genomes circa 1970–1999. *Semin. Cell Dev. Biol.*, **10,** 517–522.

Patton, S.J., Luke, G.N. and Holland, P.W.H. (1998) Complex history of a chromosomal paralogy regions: insights from amphioxus aromatic amino acid hydroxylase genes and insulin-related genes. *Mol. Biol. Evol.*, **15,** 1373–1380.

Pendleton, J.W., Nagai, B.K., Murtha, M.T. and Ruddle, F.H. (1993) Expansion of the *Hox* gene family and the evolution of chordates. *Proc. Natl. Acad. Sci. USA*, **90,** 6300–6304.

Pollard, S.L. and Holland, P.W.H. (2000) Evidence for fourteen homeobox gene clusters in human genome ancestry, *Curr. Biol.*, **10,** 1059–1062.

Robinson, E.S., Potter, I.C. and Atkin, N.B. (1975) The nuclear DNA content of lampreys. *Experientia*, **31,** 912–913.

Rubin, G.M. (2001) Comparing species. *Nature*, **409,** 820–821.

Shimeld, S.M. and Holland, P.W.H. (2000) Vertebrate innovations. *Proc. Natl. Acad. Sci. USA* **97,** 4449–4452.

Schmidtke, J., Weiler, C., Kunz, B. and Engel, W. (1977) Isozymes of a tunicate and a cephalochordate as a test of polyploidisation in chordate evolution. *Nature*, **266,** 532–533.

Shu, D.-G., Conway Morris, S. and Zhang, X.-L. (1996) A *Pikaia*-like chordate from the Lower Cambrian of China. *Nature*, **384,** 157–158.

The *C. elegans* Sequencing Consortium (1998) Genome Sequence of the nematode *Caenorhabditis elegans*: a platform for investigating biology. *Science*, **282,** 2012–2018.

Venter, J.C., Adams, M.D., Myers, E.W. *et al.* (2001) The sequence of the human genome. *Science*, **291,** 1304–1351.

Wada, H. and Satoh, N. (1993) Details of the evolutionary history from invertebrates, as deduced from the sequences of 18S rDNA. *Proc. Natl. Acad. Sci. USA*, **91,** 1801–1804.

A. Meyer, Y. Van de Peer (eds.), Genome Evolution, 85-93.

2R or not 2R: Testing hypotheses of genome duplication in early vertebrates

Austin L. Hughes* & Robert Friedman
Department of Biological Sciences, University of South Carolina, Columbia, SC 29208, USA
**Author for correspondence: E-mail:austin@biol.sc.edu*

Received 6.04.2002; Accepted in final form 29.08.2002

Key words: gene duplication, gene number, genome size, polyploidy, vertebrate evolution

Abstract

The widely popular hypothesis that there were two rounds of genome duplication by polyploidization early in vertebrate history (the 2R hypothesis) has been difficult to test until recently. Among the lines of evidence adduced in support of this hypothesis are relative genome size, relative gene number, and the existence of genomic regions putatively duplicated during polyploidization. The availability of sequence for a substantial portion of the human genome makes possible the first rigorous tests of this hypothesis. Comparison of gene family size in the human genome and in invertebrate genomes shows no evidence of a 4:1 ratio between vertebrates and invertebrates. Furthermore, explicit phylogenetic tests for the topology expected from two rounds of polyploidization have revealed alternative topologies in a substantial majority of human gene families. Likewise, phylogenetic analyses have shown that putatively duplicated genomic regions often include genes duplicated at widely different times over the evolution of life. The 2R hypothesis thus can be decisively rejected. Rather, current evidence favors a model of genome evolution in which tandem duplication, whether of genomic segments or of individual genes, predominates.

Introduction

Both philosophers of science and practicing scientists agree that science progresses through the formulation and testing of hypotheses. Sometimes a hypothesis lives a fruitful life, in the sense that it provides a major stimulus to scientific research, before eventually being discarded. Though eventually shown to be implausible, such a hypothesis can play an important role in the scientific process, stimulating lines of research that lead to a more accurate understanding of nature.

The hypothesis that vertebrates underwent two rounds of genome duplication by polyploidization early in their history (the 2R hypothesis) will probably be seen by future historians of science as an example of this process. This hypothesis has stimulated numerous avenues of research in vertebrate evolutionary biology, particularly in the early days of genomic studies prior to the completion of a draft sequence of the human genome. Now, however, as we enter the genomic era, evidence has accumulated that casts serious doubt on the plausibility of the 2R scenario. Because the 2R hypothesis is a hypothesis regarding unique historical events within a single lineage of organisms, it is probably safe to say that it can never be falsified to everyone's satisfaction (Makałowski, 2001). However, we can now say for certain that, if these events did occur, they had minimal impact on the genomes of vertebrates alive today.

Much of the theoretical background for the 2R hypothesis derives from Ohno (1970). Ohno (1970) believed that tandem duplication of genes rarely if ever gave rise to new functional genes. The mechanism of gene expression in eukaryotes was unknown at that time, and Ohno apparently believed that each structural gene in a eukaryotic genome had a unique regulatory protein encoded by a gene which might not be linked to the structural gene (Hughes, 2000). Tandem duplication of the structural gene would thus lead to an imbalance between structural and regulatory genes, and thus could not lead to evolutionary advances (Ohno, 1970). We know now of course that Ohno's model of gene expression is incorrect, and we know of many cases where tandem duplication has lead to new genes with novel functions (Hughes, 1999a).

Even though the original rationale for Ohno's emphasis on polyploidization is no longer applicable, the hypothesis of ancient polyploidization has continued to be very popular in the genomic literature. In the case of the vertebrates, one polyploidization event is typically hypothesized to have occurred prior to the divergence of Agnatha (jawless vertebrates, represented by lamprey and hagfish) and Gnathostomata (jawed vertebrates), while another round is hypothesized to have occurred after the divergence of Agnatha but before the divergence of Chondryichthyes (cartilaginous fishes) (Sidow, 1996). This is the most popular form of the 2R hypothesis. Other authors have hypothesized a single round of genome duplication early in vertebrate history, a hypothesis that might be called the 1R hypothesis (Guigo *et al.*, 1996).

Our purpose here is, first, to review the lines of evidence that have been put forward in the past in favor of the 2R hypothesis and to review recent studies showing that these lines of evidence are no longer tenable, if indeed they ever were. The past few years have been particularly important in the life of this hypothesis, which was widely accepted for many years without being subjected to any critical testing. Only recently have the data been available to test the 2R hypothesis, and it has dramatically failed every test. Second, we summarize results from recent phylogenetic analyses regarding the timing of gene duplication in families of different sizes in the three complete (or nearly complete) animal genomes now available to us. While our primary focus will be on the 2R hypothesis, the results we present have implications for the 1R hypothesis as well.

Evidence from genome size

Ohno (1970) invoked genome size in support of the hypothesis of polyploidization. However, genome sizes are now known to vary greatly within different taxonomic groups and may be subject to selective constraints (Hughes, 1999a). Therefore, genome sizes are unlikely to reflect ancient polyploidization events. Furthermore, available data are not consistent with widely cited hypotheses such as the 2R hypothesis. For example, The lamprey has a genome size about 40% that of human, while the hagfish genome is about 80% that of human, while genome sizes of bony fishes vary from 11–4088% of human genome size (Hughes, 2000).

Evidence from relative gene number

Previously, it was conjectured that the number of protein-coding genes in vertebrates is about 4 times that in invertebrates such as *Drosophila*, consistent with the 2R hypothesis (Sidow, 1996). Similarly, the number of genes in the urochordate *Ciona intestinalis* was estimated at about 15,000, believed to be one quarter the number in a typical vertebrate, again as expected under the 2R hypothesis (Simmen *et al.*, 1998). With the completion or near-completion of genome sequencing projects, the picture appears less straightforward. The human genome is now predicted to have about 32,000–39,000 protein-codong genes, or 2.3–2.9 times as many genes as *Drosophila melanogaster* and only 1.7–2.0 times as many genes as another invertebrate animal, *C. elegans* (Bork and Copley, 2001).

It has been observed that certain gene families include four members in a typical vertebrate but only one in *Drosophila* or other invertebrates, a pattern that has been cited as evidence for the 2R hypothesis (Sidow, 1996). Despite talk of a 'four-to-one rule' (Meyer and Schartl, 1999) in comparison of human to *Drosophila* family sizes, the number of gene families for which a 4 : 1 ratio is observed is in fact quite low. In a recent survey of all gene families, we found that only 4.9% showed a ratio of 4 : 1 between the number of members in human and the number of members in *Drosophila* (Friedman and Hughes, 2001).

We computed the distribution of family size ratios (i.e., the ratio of gene number in homologous families) for human: *Drosophila* and C. elegans: *Drosophila*. Clearly, the human:*Drosophila* ratios differ from the *C. elegans*:*Drosophila* ratios in the occurrence of higher peaks in the human distribution around 2 : 1, 3 : 1, and 4 : 1 (Figure 1). But in the human: Drosophila comparison, the peaks around 2 : 1 and 3 : 1 are much more pronounced than that around 4 : 1, contrary to the prediction of the 2R hypothesis.

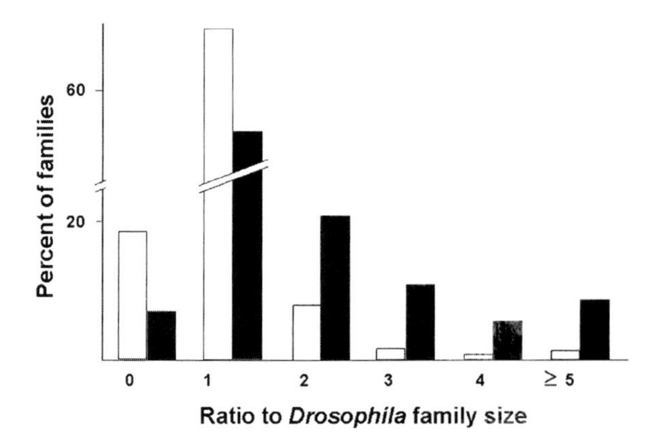

Figure 1. Frequency distributions of the ratios of gene numbers in shared gene families in human: Drosophila (solid bars) and *C. elegans*: *Drosophila* (open bars) comparisons. Each vertical bar shows the percentage of families having ratios greater than or equal to the value on the *X* axis underneath that bar and less than the value to the immediate right.

Furthermore, it is important to realize that a pattern of 4 genes in vertebrates and one in *Drosophila* does not support the 2R hypothesis unless certain conditions are met. Phylogenetic analysis is required to test whether these gene families are in fact consistent with the 2R hypothesis (Hughes, 1999b). If phylogenetic analysis shows that one or more gene duplications within a four-member vertebrate family has occurred prior to the origin of vertebrates, that phylogeny is clearly not consistent with the 2R hypothesis (Figure 2c). Furthermore, even if the vertebrate genes have been duplicated after the origin of vertebrates, there is only one topology of the phylogenetic tree that is consistent with the 2R hypothesis; that is, a topology showing two clusters of two genes, decribed as (AB) (CD) (Figure 2a). Although an ad hoc story can always be told to reconcile alternative topologies (Figure 2b) with the 2R hypothesis, on the face of it a topology other than (AB) (CD) does not support this hypothesis (Hughes, 1999b).

We searched the *Drosophila* genome and the available portion of the human genome for all four-member families, using a strict criterion of homology and examined the topologies of these families, as well as of four-member clusters in larger families (Friedman and Hughes, 2001). For 4-member families in human and *Drosophila*, topologies of trees were categorized as follows: (1) supporting duplication of at least one gene pair prior to the protostome-deuterostome divergence (Figure 2c); (2) supporting duplication after the deuterostome-protostome divergence and having a

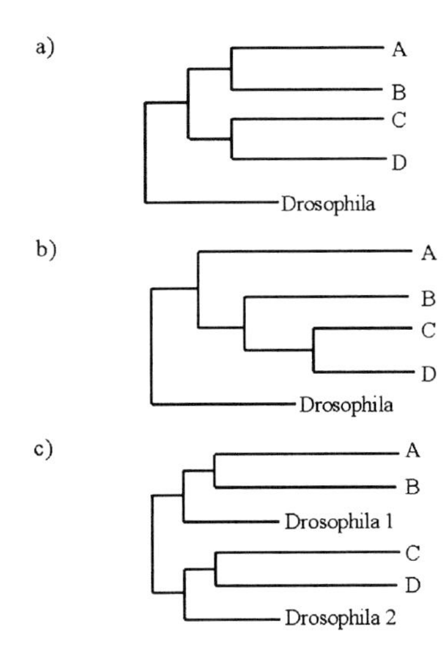

Figure 2. (a) Hypothetical 4-member vertebrate gene family having a topology of the form (AB) (CD) consistent with the hypothesis of two rounds of genome duplication (the 2R hypothesis). (b) Hypothetical 4-member vertebrate gene family having a topology of the form (A) (BCD) inconsistent with the 2R hypothesis. (c) Hypothetical 4-member vertebrate gene family in which one gene duplication predates the divergence of deuterostomes (including vertebrates) and protostomes (including *Drosophila*).

topology of the form (AB) (CD) (Figure 2a); and (3) supporting duplication after the deuterostome-protostome divergence and having a topology of the form (A) (BCD) (Figure 2b). In the case of the human genome, 32 of 92 4-member families for which the phylogeny resolved the topology showed a topology supporting duplication of one or more genes prior to the deuterostome-protostome divergence, and in 25 of these the relevant internal branch received significant support (Table 1). In 38 of the remaining families, the topology was of the form (A) (BCD), and in 17 of these the internal branch establishing this topology received significant support (Table 1). Thus, 70 of 92 human 4-member families (76.1%) showed topologies different from that predicted by the 2R hypothesis (Table 1).

Likewise, in 4-gene clusters within 5–8-member families, the (A) (BCD) topology occurred more frequently than (AB)(CD) (Table 1). Of 42 such clusters in which the topology was resolved, 25 (59.5%) showed topologies inconsistent with the 2R hypothesis (Table 1). Thus, of a total of 134 resolved 4-member phylogenies, 95 (70.9%) were not consistent with the 2R hypothesis. Similar results were reported for a smaller number of families by the International Human Genome Sequencing Consortium (2001).

Table 1. Topologies of 4-member families and clusters of 4 genes in 5–8-member families of human and *Drosophila*

	Human	*Drosophila*
4-member families		
• At least one duplication before deuterostome-protostome divergence	32 (25)[a]	16 (11)[a]
• All duplications afterdeuterostome-protostome divergence		
Topology (AB)(CD)	22 (11)[b]	4 (2)[b]
Topology (A) (BCD)	38 (17)[b]	2 (1)[b]
• Unresolved	14	2
• Total	106	24
4-member clusters in 5–8-member families		
• Topology (AB) (CD)	17 (5)[b]	5 (3)[b]
• Topology (A) (BCD)25 (10)[b]6 (2)[b]		
• Unresolved	8	0
• Total	50	11

Numbers are numbers of families showing a given topology. Phylogenetic trees were reconstructed by the maximum-likelihood quartet-puzzling method (Strimmer and von Haeseler, 1996). An internal branch was considered to be significantly supported if it was supported in 95% or better of 10,000 quartet-puzzling steps. [a] Numbers in parentheses are numbers of families in which the branch establishing duplication prior to the deuterostome-protostome divergence received significant support. [b] Numbers in parentheses are cases where the branches establishing the topology received significant support.

Interestingly, the patterns seen in Drosophila were quite similar to those seen in humans. In *Drosophila* 16 of 22 4-member families for which the topology was resolved (72.7%) showed topologies different from that predicted by the 2R hypothesis. These results thus suggest that the hypothesis of two rounds of genome duplication is no more likely to be true of vertebrates than of *Drosophila*. And we doubt that even the most ardent advocates of polyploidization would hold that *Drosophila* has undergone two rounds of polyploidization.

Putatively duplicated genomic regions

A number of studies have identified genomic regions in vertebrate genomes which share members of multiple gene families. Particularly if four such regions are found in the genome, they have been taken as providing support for the 2R hypothesis. Examples include homologous genes on human chromosomes 1, 6, 9, and, 19; and on human chromosomes 2, 7, 12, and 17 (Lundin, 1993; Kasahara *et al.*, 1997). However, in the case of such putatively duplicated gene clusters, it is important to use phylogenetic analysis to test the hypothesis that all the genes in a given cluster duplicated simultaneously, as is predicted if these regions are the remnants of ancient polyploidization events (Hughes, 1998).

When such analyses have been conducted, the results have shown that the gene families involved were duplicated at widely different times over the course of the evolution of life. For example, members of the heat shock protein 70 family on human chromosomes 6 and 9 arose by a duplication that occurred before the divergence of animals and fungi (Figure 3). This conclusion is independent of the rooting of the tree and does not depend on the assumption of a constant rate of molecular evolution ('molecular clock'). We can have a good deal of confidence in this conclusion because the internal branch of the tree on which it depends is strongly supported. Thus, the duplication of these genes could not have been part of a polyploidization event early in vertebrate history. The same is true of several other gene families on human chromosomes 1, 6, 9, and, 19 (Hughes, 1998; Yeager and Hughes, 1999). Thus, the history of the paralogous genes on these chromosomes is quite complex. The same gene families have evidently been assembled together independently in different chromosomal locations (Hughes, 1998).

Analysis of the gene families on human chromosomes 2, 7, 12, and 17 tells a similar story (Hughes *et al.*, 2001). These chromosomes are of particular interest because they bear the human Hox clusters. The existence of four *Hox* clusters in humans but only one in *Drosophila* has long been pointed out as evidence in support of the 2R hypothesis. Hughes *et al.*

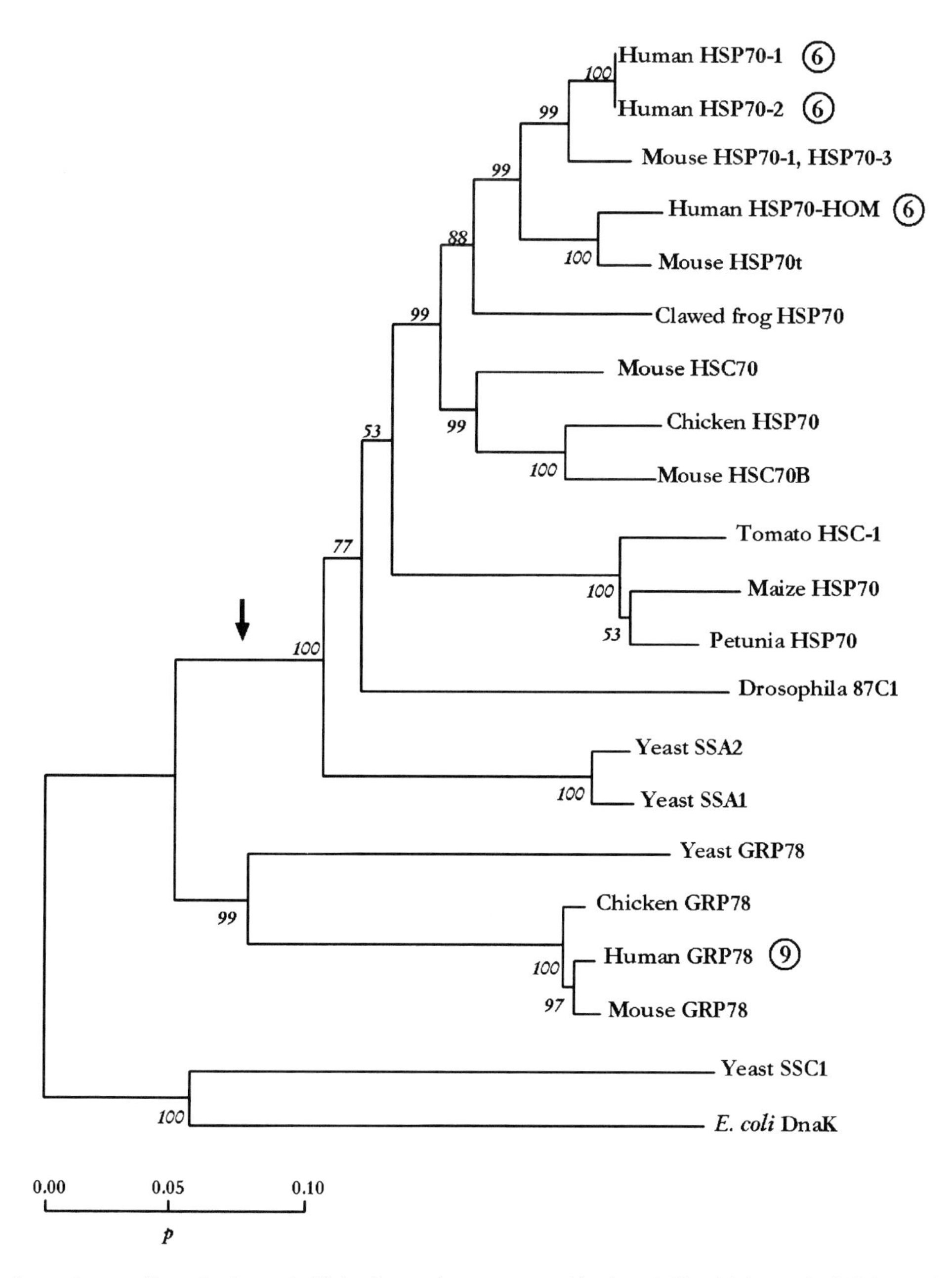

Figure 3. Phylogenetic tree of heat shock protein 70 family members constructed by the neighbor-joining method (Saitou and Nei, 1987) on the basis of the proportion of amino acid difference (p). Numbers on the branches are the percentage of 1000 bootstrap samples (Felsenstein, 1985) supporting each branch; only values > 50% are shown. Chromosomal locations of human genes are circled. The *arrow* marks the key branch for establishing that human genes on chromosomes 6 and 9 diverged before the animal-fungus divergence. For further details, see Hughes (1998).

(2001) conducted phylogenetic analyses of 42 gene families including members on at least two of the human Hox-bearing chromosomes (2, 7, 12, and 17). Phylogenetic trees showed that genes in these families were duplicated at very different times over the history of life (Hughes *et al.*, 2001).

The α and β integrins provide an interesting example of the complexity one encounters in analyzing families found on the human *Hox*-bearing chromosomes. Integrins are adhesion receptors involved in cell-cell and cell-matrix interactions in animals (Sonnenberg, 1993). The integrin receptor is a het-

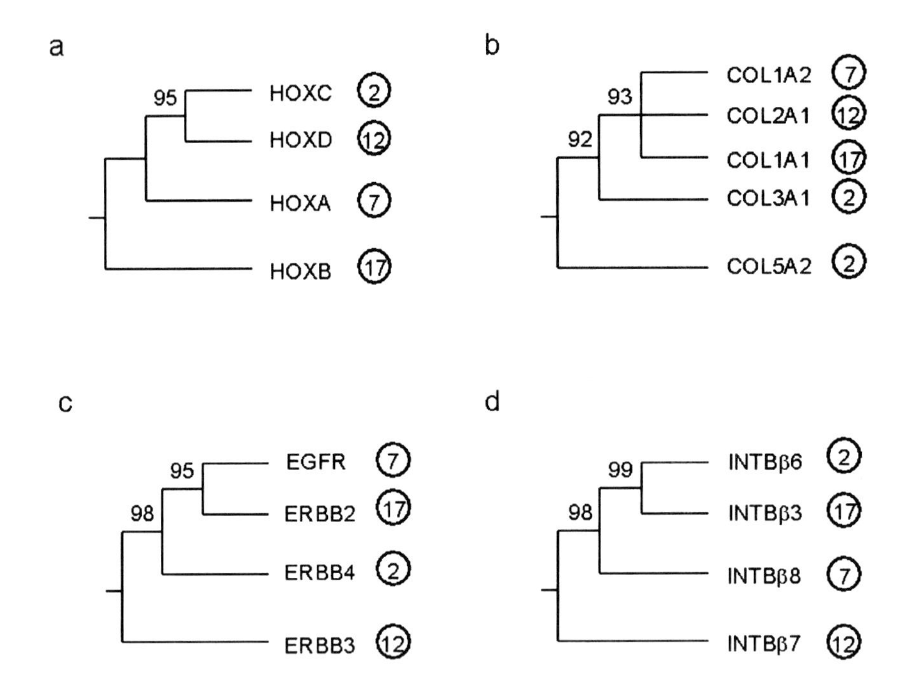

Figure 4. Schematic representation of phylogenetic trees (with bootstrap support for key internal branches) of gene families on human chromosomes 2, 7, 12, and 17: (a) Hox clusters; (b) collagen; (c) ERBB family; (d) α integrin. See Hughes *et al.* (2001) and Hughes (2001).

erodimer of noncovalently linked α and β chains encoded by two evolutionarily unrelated gene familes. Phylogenetic analysis suggests that the α integrin family includes three distinct subfamilies that were present in the common ancestors of deuterostomes and protostomes (Hughes, 2001). Two of these families, designated PS1 and PS2 after their *Drosophila* members (Hughes, 2001), include pairs of genes found on human chromosomes 2, 12, and 17, closely linked to each other and to the *Hox* clusters. It is plausible that these linked pairs of genes duplicated along with the *Hox* clusters, although the cluster linked to *HOXA* on chromosome 7 was evidently deleted. However, there are other integrin α genes on the human Hox-bearing chromosomes that duplicated at completely different times (Hughes, 2001). Likewise there are integrin β chain genes on these chromosomes that duplicated at still other times (Hughes, 2001).

Interestingly, even among genes on the human *Hox*-bearing chromosomes that appear to have duplicated around the same time as the *Hox* clusters, there is often evidence against the hypothesis that they duplicated along with the *Hox* clusters. Figure 4 shows topologies of four families on these chromosomes, including the Hox clusters themselves and integrin α chain genes. All four show different topologies of the chromosomes that are not easily reconciled. We are forced to conclude that duplications in these different gene families occurred by different processes even though they all occurred early in vertebrate evolution. These results are very strong evidence against any hypothesis of polyploidization early in vertebrate history as an explanation for linkage patterns in the chromosomes of living vertebrates.

Timing of gene duplications

In order to provide further evidence regarding the 2R hypothesis, Friedman and Hughes (2001) examined the timing of gene duplications in all 2–8-member families in human, *Drosophila*, and *C. elegans*, using phylogenetic analysis. The human genome differed from that of *Drosophila* in having significantly lower proportions of gene duplication events in 2–8-member families which could be dated by a significant internal branch prior to the animal fungus divergence, the coelomate-nematode divergence, or the deuterostome-protostome divergence (Figure 5). By contrast, the proportions of genes that could be dated prior to the animal-fungus divergence or the coelomate-nematode divergence did not differ significantly between *Drosophila* and *C. elegans* (Figure 5).

Before Animal-Fungus

C. elegans 5.0%
Drosophila 5.6%
Human 1.6% ***

Before Coelomate-Nematode

C. elegans 18.4% ***
Drosophila 26.3%
Human 13.9% ***

Before Deuterostome-Protostome

Drosophila 33.5%
Human 9.9% ***

Figure 5. Numbers of gene duplications in 2–8-member families. A duplication was dated prior to one of the three major cladogenetic events (the animal-fungus divergence, the coelomate-nematode divergence, and the deuterostome-protostome divergence) if its occurrence prior to the event was supported by a significant internal branch in phylogenetic trees constructed by maximum likelihood quartet puzzling (Strimmer and von Haeseler, 1996). A branch was considered significant if it was supported by 95% or better of 10000 quartet-puzzling steps. Chi-square tests of the hypothesis that the proportion of duplications prior to a cladogenetic event differed from that in Drosophila: $^{***}P < 0.001$. Numbers of duplication events were as follows: *C. elegans*, 463; *Drosophila* 567; human, 1760. For further detail, see Friedman and Hughes (2001).

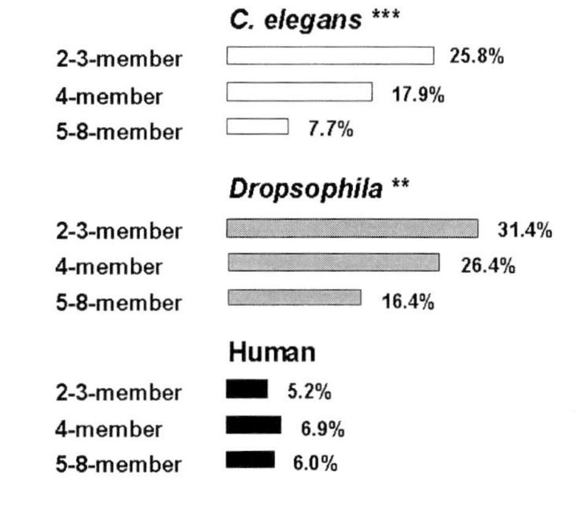

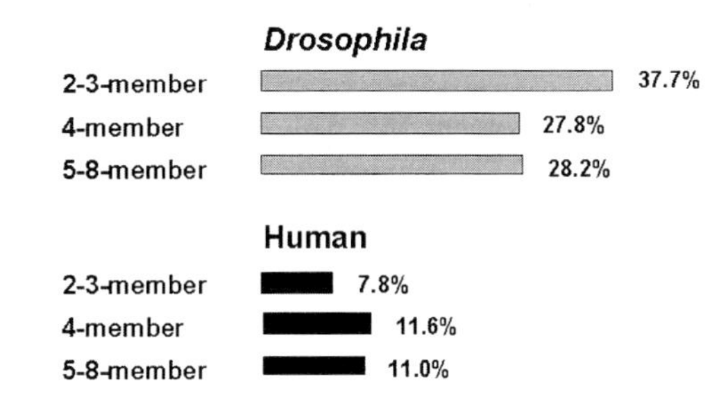

Figure 6. Numbers of gene duplications in three family size classes (2–3-member families, 4-member families, 5–8-member families). A duplication was dated prior to one of two major cladogenetic events, (a) the coelomate-nematode divergence, and (b) the deuterostome-protostome divergence, if its occurrence prior to the event was supported by a significant internal branch. Phylogenetic tree reconstruction and test of internal branches were as in Figure 5. Chi-square tests of the uniformity across family size classes of the proportion of duplications prior to the cladogenetic event : $^{**}P < 0.01$; $^{***}P < 0.001$.

In both *C. elegans* and *Drosophila* genomes, there was a significant non-uniformity among family size classes with respect to the proportion of duplications that could be dated prior to the coelomate-nematode divergence (Figure 6). In both of these species, 2–3-member families included the highest proportion of duplications that could be dated prior to the coelomate-nematode divergence, whereas the proportion was lower in 4-member families and lower still in 5–8-member families (Figure 6). By contrast, in the human genome, the proportions of duplications that could be dated prior to the coelomate-nematode divergence was remarkably constant across 2–3-member families, 4-member families, and 5–8-member families (Figure 6). In neither Drosophila nor human was there significant non-uniformity among family size classes with respect to the proportion of duplications that could be dated prior to the deuterostome-protostome divergence (Figure 6).

On the 2R hypothesis, we might expect to see evidence of a major burst of duplication in 4-member gene families of vertebrates after the deuterostome-protostome divergence but not in families with more or fewer members. Contrary to this expectation, the proportion of duplications in human gene families that could be dated prior to the deuterostome-protostome divergence was remarkably constant across family size categories (Figure 6).

Conclusions

A number of different analyses using a number of different methods have been used to test hypotheses of polyploidization early in vertebrate history. At the very least we can conclude that no strong signal consistent with the 2R hypothesis is present in the human genome. It might be argued that the data are consistent with the 1R hypothesis rather than the 2R hypothesis. However, the data are problematic for the 1R hypothesis as well. A high proportion (42.7%) of families shared between human and *Drosophila* were found to be represented by a single human gene (Friedman and Hughes, 2001). Furthermore, in 85.7% of families, the human:*Drosophila* ratio of gene number was less than 4:1 (Friedman and Hughes, 2001). Given these data, if early vertebrates underwent even a single polyploidization event, it must have been followed by deletion of the vast majority of duplicated genes.

Proponents of the 2R or 1R hypotheses might argue that the effects of polyploidization events are difficult to detect because they are obscured by hundreds of millions of years of subsequent evolution. This is certainly possible, but there are philosophical reasons for questioning the invocation of hypothetical past events whose traces are undetectable today. Consider the following two hypotheses: (1) 'Event *x* happened but it had no lasting detectable effects.' (2) 'Event *x* did not happen.' Clearly, from the point of view of empirical science, these two hypotheses are equivalent. There is no conceivable experiment or observation that can be used to decide between them. And in that case, the criterion of parsimony obliges us to favor hypothesis (2) because it is simpler.

Because the effects of a very ancient polyploidization event may be quite subtle, one might conclude that the 1R and 2R hypothesis are difficult (or even impossible) to falsify (Makałowski, 2001). On the other hand, emphasizing falsification of the polyploidization hypotheses misses a crucial point; namely, that in science it is always preferable to accept the null hypothesis (or the hypothesis of no effect) unless the observed data are extremely unlikely under the null hypothesis. In the case of polyploidization, the null hypothesis must be that polyploidization did not happen. We should only reject this null hypothesis if there is compelling evidence of polyploidization, which is certainly not the case with vertebrates.

It is often assumed that polyploidization, because it duplicates numerous genes simultaneously, is a more parsimonious explanation of an increase in gene number than multiple independent events of tandem duplication, but this is not necessarily the case (Hughes *et al.*, 2001). In the case of vertebrates, the numbers of events of gene deletion that must be assumed under either the 1R or the 2R hypothesis far exceeds the number of events of tandem duplication that must be assumed if polyploidization is not invoked. Hughes *et al.* (2001) showed this in the case of gene families where gene duplication occurred prior to the origin of vertebrates. In these familes, because duplicate genes were already present at the time of the alleged polyploidizations, a large number of deletions must be assumed in order to account for both polyploidization and current vertebrate gene numbers. Likewise, as mentioned above, there are a large number of single-copy genes that must have been present in the ancestor of vertebrates because they are shared with invertebrates such as *Drosophila*. Here again, the polyploidization hypotheses require extensive gene deletion. Therefore, the hypothesis that the increase in gene number in vertebrates occurred as a result of multiple independent gene duplications, as well as occasional duplication of chromosomal blocks, is far more parsimonious than any hypothesis invoking polyploidization.

It is worth emphasizing that the origin of new genes having new functions by tandem duplication is a well understood phenomenon. This process has occurred repeatedly in the history of the mammals, and these events are recent enough to be reconstructed without difficulty. For example, the type I interferon (IFN) genes constitute a small gene family encoding cytokines that regulate immune responses in vertebrates. In humans, there are 15 functional genes in this family, all located in a cluster on chromosome 12 (Diaz *et al.*, 1994). Mammalian type I IFNs can be divided into a number of subfamilies; three of these, IFN-α (13 genes), IFN-β (1 gene), and IFN-γ (1-gene) are present in the human type I IFN cluster. Phylogenetic analyses show that these subfamilies arose by gene duplication after mammals diverged from birds but before the radiation of the eutherian (placental) orders of mammals (Hughes, 1995; Hughes and Roberts, 2000). The 13 IFN-α genes arose by gene duplications that occurred in the primate lineage after the radiation of the eutherian orders (Hughes, 1995). Many similar examples could be cited. By contrast, in the case of vertebrate genes duplicated by polyploidization, evidence for evolution of new function has been much harder to find (Hughes and Hughes, 1993).

Reconstructing the genetic events that gave rise to the unique adaptations of vertebrates remains a formidable challenge. One of the ironic legacies of Ohno (1970) is that vertebrate genome biologists have struggled mightily and in vain to fit the data to a model (polyploidization) whose ability to account for evolutionary novelty remains uncertain at best, while discounting a mechanism (tandem duplication) for which we have abundant evidence. The result is an approach to genome evolution which might be called 'genomic catastrophism' by analogy with catastrophism in geology (Hughes, 1999c). We recommend that vertebrate biologists, rather than concocting scenarios involving low-probability events early in evolutionary history, at least initially adopt a position of 'genomic uniformitarianism.' In other words, wherever possible, we should attempt to explain ancient genomic events by the same mechanisms that are familiar to us from recent genome evolution (Hughes, 1999c).

References

Bork, P. and Copley, R. (2001) Filling in the gaps. *Nature*, **409,** 818–820.

Diaz, M.O., Pomykala, H.M., Bohlander, S.K., Maltepe, E., Malik, K., Brownsein, B. and Olapede, O.I. (1994) Structure of the human type-I interferon gene cluster determined from a YAC clone contig. *Genomics*, **22,** 540–552.

Felsenstein, J. (1985) Confidence limits on phylogenies: an approach using the bootstrap. *Evolution*, **39,** 783–791.

Friedman, R. and Hughes, A.L. (2001) Pattern and timing of gene duplication in animal genomes. *Genome Res.*, in press.

Guigo, R., Muchnik, I. and Smith, T.F. (1996) Reconstruction of ancient molecular phylogeny. *Mol. Phyl. Evol.*, **46,** 189–213.

Hughes, A.L. (1995) The evolution of the type I interferon gene family in mammals. *J. Mol. Evol.*, **41,** 539–548.

Hughes, A.L. (1998) Phylogenetic tests of the hypothesis of block duplication of homologous genes on human chromosomes 6, 9, and 1. *Mol. Biol. Evol.*, **15,** 854–870.

Hughes, A.L. (1999a) *Adaptive Evolution of Genes and Genomes*, Oxford University Press, New York, NY.

Hughes, A.L. (1999b) Phylogenies of developmentally important proteins do not support the hypothesis of two rounds of genome duplication early in vertebrate history. *J. Mol. Evol.*, **48,** 565–576.

Hughes, A.L. (1999c) Genomic catastrophism and the origin of vertebrate immunity. *Arch. Immunol. Ther. Exper.*, **47,** 347–353.

Hughes, A.L. (2000) Polyploidization and vertebrate origins: a review of the evidence. In *Comparative Genomics* (Eds., Sankoff, S. and Nadeau, J.H.). Kluwer, Dordrecht, pp. 493–502.

Hughes, A.L. (2001) Evolution of the integrin α and β protein families. *J. Mol. Evol.*, **52,** 63–72.

Hughes, M.K. and Hughes, A.L. (1993) Evolution of duplicate genes in a tetraploid animal, *Xenopus laevis*. *Mol. Biol. Evol.*, **10,** 1360–1369.

Hughes, A.L. and Roberts, R.H. (2000) Independent origin of IFN-α and IFN-β in birds and mammals. *J. Interferon Cytokine Res.*, **20,** 737–739.

Hughes, A.L., da Silva, J. and Friedman, R. (2001) Ancient genome duplications did not structure the human *Hox*-bearing chromosomes. *Genome Res.*, **11,** 771–780.

International Human Genome Sequencing Consortium. (2001) Initial sequencing and analysis of the human genome. *Nature*, **409,** 860–891.

Kasahara, M., Nayaka, Y., Satta, Y. and Takahata, N. (1997) Chromosomal duplication and the emergence of the adaptive immune system. *Trends Genet.*, **13,** 90–92.

Lundin, L.G. (1993) Evolution of the vertebrate genome as reflected in paralogous chromosome regions in man and the house mouse. *Genomics*, **16,** 1–19.

Makałowski, W. (2001) Are we polyploids? A brief history of one hypothesis. *Genome Res.*, **11,** 667–670.

Meyer, A. and Schartl, M. (1999) Gene and genome duplication in vertebrates: the one-to-four (-to-eight in fish) rule and the evolution of novel gene functions. *Curr. Opin. Cell Biol.*, **11,** 699–704.

Ohno, S. (1970) *Evolution by Gene Duplication*, Springer, New York, NY.

Saitou, N. and Nei, M. (1987) The neighbor-joining method: a new method for reconstructing phylogenetic trees. *Mol. Biol. Evol.*, **4,** 406–425.

Sidow, A. (1996) Gen(om)e duplications in the evolution of early vertebrates. *Curr. Opin. Genet. Dev.*, **6,** 715–722.

Simmen, M.W., Leitger, S., Clark, V.H., Jones, S.J.M. and Bird, A. (1998) Gene number in an invertebrate chordate, *Ciona intestinalis*. *Proc. Natl. Acad. Sci. USA*, **95,** 4437–4440.

Sonnenberg, A. (1993) Integrins and their ligands. *Curr. Top. Microbiol. Immunol.*, **184,** 7–35.

Strimmer, K. and von Haeseler, A. (1996) Quartet puzzling: a quartet maximum-likelihood method for reconstructing tree topologies. *Mol. Biol. Evol.*, **13,** 964–969.

Yeager, M. and Hughes, A.L. (1999) Evolution of the mammalian MHC: natural selkection, recombination, and convergent evolution. *Immunol. Rev.*, **167,** 45–58.

A. Meyer, Y. Van de Peer (eds.), Genome Evolution, 95-110.

The 2R hypothesis and the human genome sequence

Karsten Hokamp[†], Aoife McLysaght[†] & Kenneth H. Wolfe[*]
Department of Genetics, Smurfit Institute, University of Dublin, Trinity College, Dublin 2, Ireland;
[*]*Author for correspondence: E-mail: khwolfe@tcd.ie*
[†]*These authors contributed equally to this work.*

Received 02.04.2002; accepted in final form 29.08.2002

Key words: human genome, paralogon, polyploidy, 2R hypothesis

Abstract

One theory formalised in 1970 proposes that the complexity of vertebrate genomes originated by means of genome duplication at the base of the vertebrate lineage. Since then, the theory has remained both popular and controversial. Here we review the theory, and present preliminary results from our analysis of duplications in the draft human genome sequence. We find evidence for extensive duplication of parts of the genome. We also question the validity of the 'parsimony test' that has been used in other analyses.

The 2R hypothesis

In his 1970 book Susumu Ohno proposed that there may have been one or more whole genome duplications in the lineage leading to vertebrates. He postulated that genome duplication in the vertebrate lineage provided a platform for increasing the sophistication of the vertebrate genome and thus increasing morphological complexity. Genome duplication may be particularly powerful because all genes in a biochemical pathway will be duplicated simultaneously. Ohno was not specific about how many events occurred. The most popular form of this hypothesis is that there were **2 R**ounds of genome duplication early in the vertebrate lineage, as proposed by Holland *et al.* (1994). This has recently become known as the 2R hypothesis, an abbreviation attributable to Hughes (1999). There is no absolute consensus on the timing of these events, but the majority of references in the literature put one of these events immediately before, and one immediately after the divergence of agnathans from the lineage leading to tetrapods (see Figure 1 in Skrabanek and Wolfe, 1998). These timings are speculative and were probably chosen to coincide with major evolutionary transitions that they were thought to have facilitated. The lower limit on the timing of genome duplication is set by the observation of only a single Hox cluster in the invertebrate chordate amphioxus compared to four clusters in vertebrates (Garcia-Fernandez and Holland, 1994). As an upper limit, it seems unlikely that genome duplications would be viable in the mammalian lineage. Theory predicts that a genome duplication in an organism with a chromosomal basis of sex-determination (such as that of mammals) will result in sterility of the heterogametic sex, and thus inviability (Muller, 1925). Indeed the only known tetraploid mammal, a South American rodent, has duplicated copies of every chromosome except the sex chromosomes (Gallardo *et al.*, 1999).

At the time of writing his book there was little evidence to support Ohno's claim. Very few protein sequences were known, and the hypothesis was based largely on genome size comparisons and matching patterns of cytogenetic bands. Much of the evidence which prompted Ohno to suggest a genome duplication event has lost merit in the light of our current understanding of genetics and genomes (Skrabanek and Wolfe, 1998). For example, differences in genome sizes are largely due to increased amounts of non-coding DNA rather than an increased number of genes; and cytogenetic bands, whose patterns were used to list human chromosomes in pairs (Comings, 1972), are not indicative of the underlying gene content.

The debate on the 2R hypothesis to date has been a war of words (and limited data) between the phylogeneticists and the cartographers. As a general rule, analyses based on phylogenetic methods come out against the genome duplication hypothesis (e.g., Hughes, 1998; Hughes, 1999; Martin, 1999; Hughes *et al.*, 2001; Martin, 2001), whereas map-based studies come out in favour (e.g., Lundin, 1993; Spring, 1997). Two main arguments have been advanced to support the theory of genome duplication in an early vertebrate: that there should be four vertebrate orthologues of each invertebrate gene, the so-called 'one-to-four rule' (Spring, 1997; Meyer and Schartl, 1999; Ohno, 1999); and that paralogous genes are clustered in a similar fashion in different regions of the genome (e.g., Martin *et al.*, 1990; Lundin, 1993).

The one-to-four rule

The one-to-four rule was first proposed by Jürg Spring (1997). He listed human paralogues present on different chromosomes and their *Drosophila* orthologues, and surmised that the maximum ratio of human to *Drosophila* genes was four. These 'tetralogues' seemed to bear the hallmark of a genome-wide event because they were discovered on all 23 female human chromosomes. The observation of some gene families with ratios of 2:6 or 2:5 *Drosophila*:human genes contradicts this hypothesis and Spring suggested that more complete genome sequences would provide data that could split these families into 'tetrapacks'.

The first extensive examination of the one-to-four rule using almost complete proteomes from *D. melanogaster*, *C. elegans*, and human, showed no excess of four-membered vertebrate gene families (see Fig. 49 of International Human Genome Sequencing Consortium [2001] and Fig. 12 of Venter *et al.* [2001]). Furthermore, the observation of gene families with five or more members directly contradicts the expectations of Spring (1997) that membership would be 'maximally four'. It appears that the one-to-four rule is an over-simplification of the history of the vertebrate genome. These data can of course be explained by hypothesising two genome duplications on a background of independent gene duplication and loss. However, as it is impossible to distinguish genome duplication from gene duplication on the basis of gene family size alone, this measure is simply uninformative.

Paralogous chromosomal segments

The analysis of paralogous regions of the human genome is based on the assumption that, although it is expected that many rearrangements will have occurred in the time since the two duplication events envisaged by the 2R hypothesis, there should still be detectable remnants of the 4-way paralogy between some chromosomes, *i.e.*, some portions of some chromosomes should remain almost intact in four copies. This principle seems reasonable, though these studies have suffered for want of extensive genomic data. Finding as few as two genes in several linked clusters in a genome of over 30,000 is hardly overwhelming evidence for a genome duplication event (e.g., Martin *et al.*, 1990). Objections that these observations can easily be explained by regional duplications of segments of chromosomes must be entertained.

HSA 1, 6, 9, and 19

The observation of paralogous regions (around the MHC locus) on human chromosomes 1, 6, 9, and 19, led to the suggestion that these were duplicated by whole genome duplication events at the base of the vertebrate lineage (Kasahara *et al.*, 1996; Katsanis *et al.*, 1996; Kasahara, 1997). This was further supported by the finding of only a single related cluster in amphioxus (Flajnik and Kasahara, 2001). Ten members of particular gene families are present on chromosomes 6 and 9, and four of these are also represented on chromosome 1. The claim that this arrangement resulted from several rounds of polyploidy was refuted by Hughes (1998) using phylogenetic analysis of the nine families with sufficient data (Retinoid X receptor (RXR); α pro-collagen (COL); ATP-binding cassette (ABC) transporter; Proteasome component β (PSMB); Notch; Pre-B-cell-leukemia transcription factor (PBX); Tenascin (TEN); C3/C4/C5 complement components; Heat shock protein 70 (HSP70)). However, Hughes' analysis did indicate that this arrangement could be partly due to block duplication. Trees of these families showed that five (RXR, COL, PBX, TEN, C3/4/5) of the nine families with sufficient phylogenetic information could have duplicated simultaneously, and that this timing was consistent with a duplication in early vertebrate history 550–700 Mya. The phylogenetic analysis indicated that the four genes on chromosome 1 probably duplicated as a block. Similarly, a phylogenetic analysis by Endo *et al.* (1997) rejected the hypothesis that the 11 gene pairs on chromosomes 6 and 9 were duplicated in a single event, but did support the simultaneous duplication of six of the pairs. However, analysis of the remaining genes showed that the ABC transporter genes diverged before the origin of eukaryotes, the PSMB and the HSP70 gene families both originated before the divergence of animals and fungi, and the Notch genes diverged before the origin of deuterostomes (Hughes, 1998). Obviously these gene families did not arise as part of a block duplication event at the base of the vertebrate lineage. However, it can still be argued that these results are consistent with block duplication of this region if one assumes that there was an ancient tandem duplication of some of these genes, and after block duplication there was differential loss of one of the tandems, so that the divergence date of paralogues on two different chromosomes is that of the tandem duplication event rather than of the block duplication event (Kasahara *et al.*, 1996; Smith *et al.*, 1999).

HSA 4, 5, 8, and 10

Pébusque *et al.* (1998) reported the presence of paralogous genes on human chromosomes 4, 5, 8, and 10. In contrast to the analysis of the genes around the MHC discussed above, this study was based on a combination of phylogenetic and map-based methods. These genes are linked on the human chromosomes, with the exception that there is one family member on each of chromosomes 2 and 20, which require genome rearrangements to be reconciled with a block duplication event. The phylogenetic analyses consistently showed that these gene family members diverged in the vertebrate lineage and so are consistent with the 2R hypothesis of genome duplication. This conclusion was criticised by Martin (1999) who pointed out that the gene trees of the ankyrin family and the EGR (early growth response) family indicated different histories for their host chromosomes. The ankyrin gene tree groups chromosome 4 and 10 to the exclusion of chromosome 8, whereas the EGR gene tree groups 8 and 10 to the exclusion of all others. This contradicts the expectation that the family members on each chromosome have had a shared history since the block duplication event.

HSA 2, 7, 12, and 17

The quadruplication of the Hox cluster is the touchstone of the 2R hypothesis. There are four colinear Hox clusters in the vertebrate genome (Kappen *et al.*, 1989), but only one in the invertebrate chordate amphioxus (Garcia-Fernandez and Holland, 1994). Phylogenetic analysis of the clusters showed that they duplicated early in vertebrate history. It seems certain that these clusters duplicated *en bloc*. The question is whether they arose by genome duplication events, or by sub-genomic duplication events, or a mixture of both. In the analysis of Zhang and Nei (1996) Hox clusters C and D were grouped with a high bootstrap, but there is not enough information in the alignments of the 61 amino acids of the homeodomain to resolve the phylogeny further. Instead, Bailey *et al.* (1997) analysed the relationship of the linked fibrillar-type collagen genes, which presumably shared the same duplication history. Assuming the collagen genes have a shared history with the Hox clusters, then the results can be interpreted as a topology (outgroup(HoxD(HoxA(HoxB,HoxC)))), which contradicts the grouping of HoxC and HoxD found by Zhang and Nei (1996). Furthermore, this is contrary to the expectations of the 2R hypothesis, which predicts a symmetric topology, but may be explained by three rounds of genome duplication with loss of 4 clusters, or by independent cluster duplications (Bailey *et al.*, 1997).

In a phylogenetic analysis of the human Hox-bearing chromosomes (2, 7, 12, 17) Hughes *et al.* (2001) examined 35 gene families with members on at least two of the Hox chromosomes. 15 of these families could be classified as either pre-vertebrate, or post-mammalian duplicates and so are inconsistent with the 2R hypothesis. For the remaining 17 gene families the tree topologies did not exclude duplication at the same time as the Hox clusters. There were 15 of these for which the molecular clock was not rejected and estimates for the divergence dates of these gene families were calculated. Six of the gene families were dated to within the time of divergence of the Hox clusters, 528–750 Mya (as defined by lineage divergences), and two others had divergence estimates that were not significantly different from the time of Hox duplication. Phylogenies of gene families with members on at least three of the four Hox bearing chromosomes did not reveal a common topology for the relationship of these chromosomes.

Other regions

Some of the supposed paralogous regions of the vertebrate genome that can be found listed in the literature are based on rather sparse evidence. For example Gibson and Spring (2000) list human chromosomes X, 4, 5, and 11 as a possible paralogous quartet based only on the presence of members of two gene families (alpha-amino–3-hydroxy–5-methyl–4-isoxazole-propionic acid (AMPA) and androgen / mineralocorticoid / glucocorticoid / progesterone nuclear receptors) on all of these chromosomes.

Testing the (AB)(CD) topology prediction

In its simplest form, the hypothesis of two rounds of genome duplication predicts a symmetric (A,B)(C,D) phylogenetic tree topology (where A, B, C, D, represent any four-membered gene family), with the age of the AB split the same as the age of the CD split, thus displaying the history of successive genome duplications. The alternative hypothesis, that of sequential gene duplication, will not always predict a symmetric topology. Under a sequential duplication model a four-membered family must arise from the duplication of one member of a three membered family. There is only one possible topology for three sequences, namely (A(C,D)) (Figure 1). Duplication of gene A will result in a symmetric topology, and duplication of either C or D will result in an asymmetric topology. Assuming that all three genes are equally likely to be duplicated, sequential gene duplication will give rise to a symmetric (A,B)(C,D) topology 1/3 of the time, and an asymmetric topology (A((B,C)D)) or (A(C(B,D))) the remaining 2/3 of the time. Thus, the null hypothesis is that the symmetric topology should be found in 1/3 of trees, and not 1/5 as proposed by Gibson and Spring (2000).

Hughes (1999) and Martin (2001) employed similar methodologies to test the phylogenies of gene families listed as exemplars of the one-to-four rule (Sidow, 1996; Spring, 1997) for congruence with the 2R hypothesis (*i.e.*, whether or not they displayed a symmetric topology, and if they duplicated in the vertebrate lineage). The symmetric topology was only observed in a small minority of the cases (one out of nine trees in Hughes [1999]; and two out of ten trees in Martin [2001]), although in Martin's analysis seven of the eight minimum-length trees that were not symmetric were not significantly shorter than a symmetric tree.

Variations on the 2R hypothesis result in different predictions for the phylogenies of vertebrate gene families. For example, if vertebrate genome doubling occurred by allopolyploidy (*i.e.*, hybridisation of two species, as has been suggested; (Spring, 1997)) or by segmental allopolyploidy (*i.e.*, behaving as an autopolyploid at some loci, and as an allopolyploid at others) then a single genome doubling event will produce paralogues with two different coalescence dates (Gaut and Doebley, 1997; Wolfe, 2001). Alternative models hypothesise that the two rounds of genome duplication may have occurred in short succession and thus not allowing the diploidisation procedure time to complete before the second genome duplication event. This would result in some tetrasomic loci, and some octasomic loci, in the quadruplicated genome (Gibson and Spring, 2000).

Diploidisation

Diploidisation is a natural consequence of polyploidy. With some rare exceptions (*e.g.*, some loci of recent salmonid tetraploids; Allendorf and Thorgaard, 1984) all hypothesised paleopolyploid genomes have reverted to disomic inheritance at all loci. There is an increased incidence of non-disjunction of chromosomes when they form multivalents rather than divalents, so selection for increased fertility probably causes the reinstatement of disomic inheritance (Allendorf and Thorgaard, 1984).

Immediately after autotetraploidy all loci in the genome will be tetrasomic. These duplicated genes will not separate into two independently diverging loci until disomic inheritance is established (Ohno, 1970). This is important for our interpretation of what a paleopolyploid genome should look like because one of the properties we test in assessing genome duplication is the synchronicity of divergence of duplicated loci. Depending on the manner and speed of diploidisation this may or may not be an appropriate test for a paleopolyploid genome. In a diploid organism, chromosomes are arranged in pairs at meiosis (*i.e.*, chromosomes are bivalent). These pairs can exchange segments of DNA by recombination, and drift and gene conversion maintain a high degree of similarity between most alleles. In a tetraploid genome, chromosomes are arranged in tetravalents,

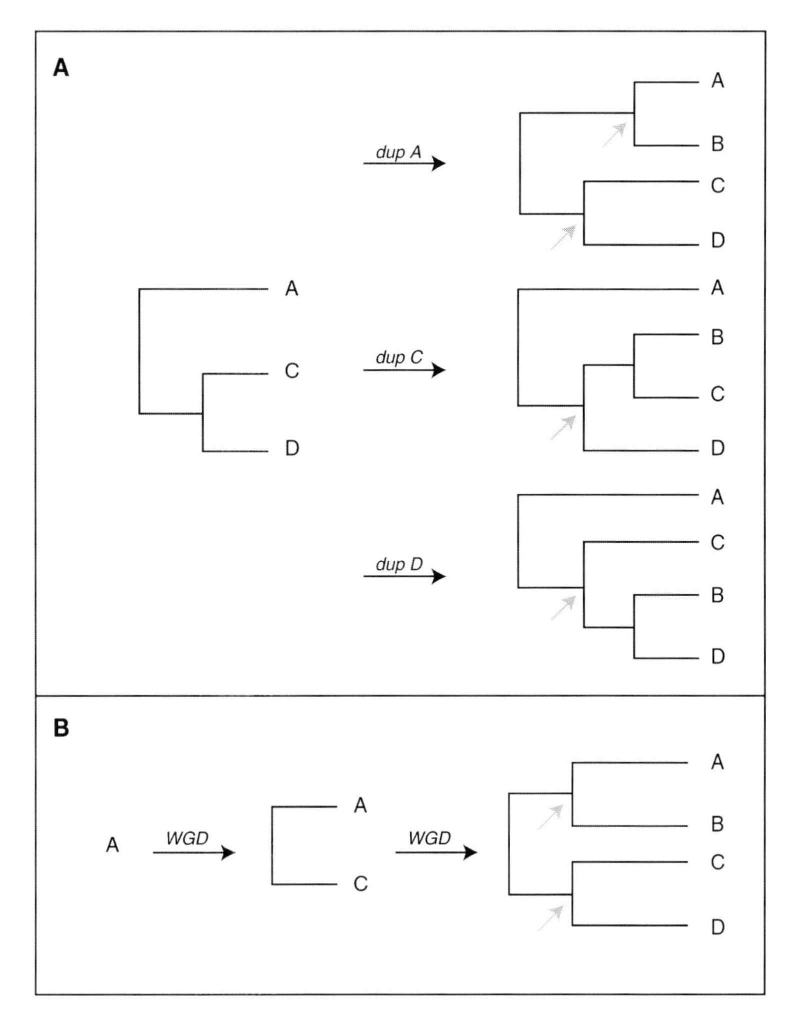

Figure 1. Alternative phylogenetic tree topologies of four-membered families resulting from sequential gene duplication or genome duplication. Grey arrows indicate the nodes that are critical to define the symmetry or asymmetry of the topology. (A) Phylogenetic tree topologies resulting from duplication of one member of a three-membered gene family. Three different trees result. The tree from the duplication of gene C and that from the duplication of gene D have asymmetric topologies. (B) Phylogenetic tree topologies resulting from two whole genome duplication (WGD) events. All genes are duplicated at each step, resulting in a symmetric tree topology.

rather than pairs, at meiosis. Diploidisation can be reduced to a problem of chromosome association. By what mechanism does a genome convert from forming chromosome quartets to forming chromosome pairs, *i.e.*, from tetraploid to diploid behaviour?

The answer to this question probably lies in a deeper understanding of the mechanisms of chromosome association. Is chromosome sequence divergence a cause or a consequence of diploidisation? If chromosome association occurs by homologous sequence attraction, then sequence divergence (by chromosome rearrangements) will cause diploidisation of chromosomes. On the other hand, if chromosome association is controlled by some other mechanism, such as attraction of homologous centromeres or telomeres, then chromosomal rearrangements may allow the independent evolution of the relocated loci and their previous partners in a tetrasomic locus, as separate loci without actually causing the diploidisation of the chromosomes in question.

The mammalian Y chromosome may serve as a model for this process. It is an unusual chromosome because it is partially diploid (at the pseudoautosomal region), and the rest is haploid. Lahn and Page (1999) identified homologous genes on the human X and Y chromosomes, which would have been part of the same locus when these chromosomes behaved autosomally (the sex chromosomes are thought to have evolved from autosomes; (Graves, 1996)). They measured the amount of divergence at synonymous sites (K_s) between homologous gene pairs. From this they found that the homologues were in four age classes arranged sequentially along the X chromosome. They interpreted this as the result of inversions of large sections of the Y chromosome, leaving the X intact, which had the effect of suppressing recombination between these portions of the chromosomes. These chromosomes have diverged substantially, and most of the Y chromosome loci are haploid. The X and Y chromosome still pair at meiosis (at the pseudoautosomal region), and thus behave like diploid chromosomes, yet most of the loci are haploid. It may be the case that chromosomal tetravalency and locus tetrasomy can be separated in the same way.

The wheat genome (*Triticum aestivum*) is hexaploid, with three contributory genomes (*A*, *B*, and *D*). There is evidence for genetic control of chromosome association in wheat through the *Ph1* locus on chromosome V of the B genome (Riley and Kempanna, 1963). In the presence, but not the absence, of a particular allele of this locus, non-homologous associated centromeres separate at the beginning of meiosis (Martinez-Perez *et al.*, 2001). The *Ph1* locus probably acts to amplify the differences between non-homologous chromosomes.

The most widely accepted hypothesis is that diploidisation proceeds by structural divergence of chromosomes. Allendorf and Thorgaard (1984) discuss a model whereby some loci may appear disomic while others apparently segregate tetrasomically. In their model they assumed that chromosome pairing occurred at the telomeres, but it can easily be modified to assume centromere association as indicated from the wheat study (Martinez-Perez *et al.*, 2001). The model of residual tetrasomic inheritance hypothesises that there are two stages of chromosome pairing. The first stage will allow pairing between homoeologous chromosomes (partially similar chromosomes), thereby allowing recombination events between paralogous loci on different chromosomes. The second stage of pairing in this hypothesis resolves non-homologous chromosome pairing, and ensures that each gamete receives one copy of each chromosome in the normal manner. Evidence in support of this model comes from the observation of Martinez-Perez *et al.* (2001) that some non-homologous centromeres are associated just before the beginning of meiosis. This model predicts that loci closer to the point of association of the chromosomes (*i.e.*, closer to the centromere) will retain residual tetrasomic inheritance longer than others. For any locus, the likelihood that it behaves disomically rather than tetrasomically in a particular meiosis will be correlated with its distance from the centromere.

Paralogon searches in the human genome sequence

In our laboratory we have begun to analyse the draft sequence of the human genome for evidence of ancient large-scale duplications, such as might be expected under the 2R hypothesis, and describe our approach here. The detection of paralogous chromosomal blocks (termed 'paralogons' by Popovici *et al.* (2001)) essentially involves the search for closely grouped sets of genes with homologues that occur in close vicinity in one or more other locations within the genome. Thus, the basic requirements for paralogon detection consist of the position and the sequence information of all genes. For a thorough and precise analysis the most complete and accurate set of human genes together with their map position is desirable. Further annotational data such as gene descriptions can add valuable background information, particularly about gene functions. A final version of the human genome sequence will probably not be available until 2003 (International Human Genome Sequencing Consortium, 2001), but several groups are striving to annotate the genome sequence in its current state. We have chosen to use data releases from Ensembl (Hubbard *et al.*, 2002), a joint project between EMBL-EBI and the Sanger Institute which aims to develop a software system for automatic annotation of eukaryotic genomes, because it stands out in several respects:

- Ensembl employs an 'open source' philosophy, which allows public insight into every detail of the data assembly procedure. This can be very valuable to trace back steps that lead to decisions in the gene prediction process.
- Strategies and plans are discussed openly through a mailing list. This offers the chance to receive early information on data-related issues and also allows for interaction with the developers.
- The data are provided in the common SQL database format, which facilitates their integration into a local computer system.
- Frequent version releases provide a constant update of information.

- Most importantly, all data and programs are freely available and can be used without any restrictions.

In the genome annotation process, Ensembl integrates information of known proteins from SP-TREMBL (Bairoch and Apweiler, 2000). The GeneWise program (Birney and Durbin, 2000) converts these to exon structures on the human sequence. Additionally, new genes are detected *ab initio* through the GenScan program (Burge and Karlin, 1997). Functional annotation is derived from the InterPro, OMIM and SAGE databases. This results in a set of predicted and confirmed genes, the latter of which numbered 27,615 in Ensembl data release version 1.0. Since sequencing of the human genome is still in progress, this also affects the annotation process, which leads to regular data updates. Ensembl provides chromosomal locations for most of the confirmed genes through integration of mapping data from the 'Golden Path', an arrangement of BAC-cloned human sequences assembled and maintained by the University of California at Santa Cruz (Kent and Haussler, 2001). The underlying data comprise approximately 830 Mb of finished sequences and 2,300 Mb of draft sequences. A further 100 Mb had not been sequenced at the time of the data freeze. Data in the draft stage have only been sequenced once or twice and contain basepair ambiguities, gaps and segments with unknown order or orientation. High quality sequence data are expected from tenfold coverage.

Sequence similarity search

Establishing homology relationships among genes in an automated fashion is based on sequence similarity searches. The first step in the analysis therefore requires the comparison of all human proteins with each other. We included invertebrate proteomes (nematode and fly) in the search database to act as an approximate natural orthology threshold (any human proteins that are less similar than an invertebrate protein to the human query protein probably duplicated before the invertebrate-vertebrate lineage divergence, and so are not relevant to the 2R hypothesis). We carried out BLASTP searches, running on a 20-node Beowulf cluster, with the SEG filter to exclude low complexity regions. For organising the large volume of resulting query/hit pairings, we found the freely available MySQL database system very useful.

ORF collapsing

Tandem gene duplications are usually relatively recent evolutionary events and occur quite frequently. They can artificially inflate gaps between pairs of paralogues, and can inflate the number of hits reported between different chromosomes (or chromosomal regions). We therefore attempted to detect and collapse tandem duplicates. Similar to the method applied by Vision *et al.* (2000) in an analysis of the *A. thaliana* genome, all genes within a close neighbourhood that show strong sequence similarity to each other were removed from the map and replaced by a single representative, *i.e.*, the longest peptide of each group was retained. Any BLASTP hits to a gene that is part of a tandem array were 'redirected' to the single remaining gene representing the array.

Parameter optimisation

The paralogon detection process has to take into account evolutionary events like inversions, rearrangements, deletions and mutations, which are likely to obfuscate traces of genome duplication. A program was developed that is controlled by parameters to deal with gaps between pairs of duplicates, high-copy protein families, and the distinction between spurious similarities and true homologies. The values for these parameters greatly determine the outcome of the program. Too strict a set of values will mostly show highly conserved or recent blocks, where only closely grouped genes with very strong similarity are detected. By contrast, a relaxed definition of paralogons can lead to inflated block sizes and numbers as a result of inclusion of insignificant pairings. In the end a trade-off between selectivity and sensitivity must be chosen. We carried out extensive tests with different combinations of parameters to determine suitable parameter values.

Paralogon detection results

A paralogon was 'built' starting from an anchor: a pair of homologous genes at different chromosomal locations. This was extended by including protein pairs on these chromosomes that were positioned no further than 30 genes distance from another protein included in the paralogon. Hits with a BLASTP expectation threshold higher than 1e–7 where excluded, as well as proteins with more than 20 hits.

The resulting paralogons range in sizes from only two pairs of duplicated genes to up to 29. Some of these paralogons, particularly the smaller ones, might have arisen from chance constellation of similar genes. To determine statistical significance, the paralogon detection method was applied to an artificial genome in which chromosome number and size has been retained but where genes have been assigned random locations. Using the same sets of parameters and repeating the shuffling and detection 1000 times allowed an estimation of significant block sizes. The results indicate that paralogons occur much more frequently in the real genome than expected by chance. Paralogons defined by at least six duplicated genes were in excess of 50 standard deviations more frequent in the real genome than expected from the simulations. The only alternative hypothesis that could fit these data is selection for clustering of these genes on a chromosome, as has been suggested for the mammalian MHC gene complex and the Surfeit locus (Hughes, 1999).

A web-based, interactive user interface was developed to allow navigation of duplicated regions and zooming between chromosomal and gene level. An example of the graphical presentation of a paralogon is shown in Figure 2. The web graphic contains integrated links that lead to more detailed information for genes and their similarity search results, as well as to external databases like Ensembl or GenBank. Only the paired duplicates are shown but intermittent genes can be switched on for closer inspection. The number of paralogues within a block lies typically between 9 and 23 percent of the genes that are covered by the region. This gives an indication of the large amount of evolutionary changes that happened since the occurrence of the duplication event.

Blocks with sizes of six or greater cover more than 44% of the human genome. Some of the largest regions are found on the four Hox chromosomes 2/7/12/17. This is also one of the few examples where duplications among four locations were found. The graphical overview of these chromosomes shown in Figure 3 proves the effectiveness of the block detection algorithm: the paralogons exactly cover the position of the Hox clusters, which are often used as a prime example of duplications within the human genome. Additionally, the covered areas expand previously reported regions and indicate more extensive duplications than formerly estimated.

Hughes *et al.* (2001) recently reported widely differing duplication dates for 42 gene families having members on the Hox-bearing chromosomes. Comparison of their results to ours shows that 139 of the 175 genes used in their study were present in our genome dataset, but only 31 of these genes (and a further 9 associated tandem repeats) form links that make up our paralogons; the other possible pairs were removed by our chordate-specific filters. Our paralogons (containing three or more duplicated genes) on the Hox chromosomes include a total of 426 duplicated genes (*i.e.*, 395 genes not included in Hughes *et al.*). There is no disagreement between the two sets of observations; chromosomes 2/7/12/17 contain some large paralogons formed by chordate-specific duplications, as well as many members of gene families formed by older duplications.

Beyond known examples, our analysis also uncovered interesting new duplications such as a region shared between chromosomes 8/14/16/20 in which copines and matrix metalloproteinases are found in close vicinity. Figure 4 provides a detailed view of the core areas of the duplicated regions. Copines form a recently discovered family of calcium-dependent, phospholipid-binding proteins that are suggested to be involved in membrane trafficking (Tomsig and Creutz, 2000). The metalloproteinases found in their vicinity are classified as the transmembrane subtype of the membrane-type MMPs (MT-MMPs) (Sato *et al.*, 1997; Kojima *et al.*, 2000). A literature search produced no results that indicate a connection between these two groups, which is not surprising, considering that research on copines is in its early stages. Their colocation in the same blocks, together with their membrane-association, suggests some kind of relationship, in particular because the copine and transmembrane MT-MMPs gene families are both small. These highly significant results provide an interesting base for a separate research project.

Comparison with Celera data

The only equally comprehensive report on paralogous blocks in human so far can be found in a study which is part of the private-sector human genome project led by Celera (Venter *et al.*, 2001). A version of the program MUMmer (Delcher *et al.*, 1999), modified to align protein sequences, was used to carry out intragenomic comparison based on the Celera sequence data. Results are presented as one large graphic (Fig. 13 in Venter *et al.*, 2001), which shows paralogous regions for each chromosome. Blocks were defined by at least three linked genes. Due to

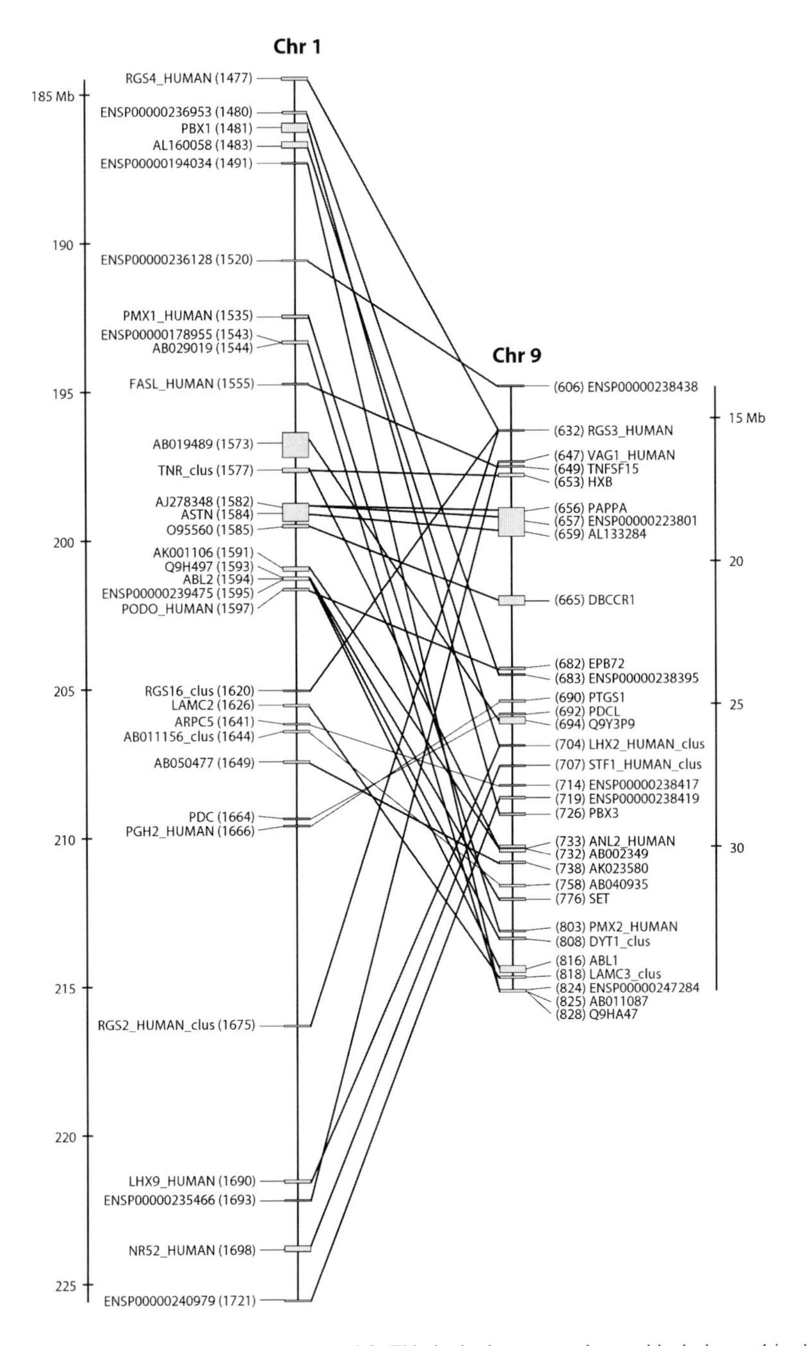

Figure 2. Paralogous block between human chromosomes 1 and 9. This is the largest paralogous block detected in the human genome including 29 duplicated genes. Blocks can be viewed at wolfe.gen.tcd.ie/dup.

lack of details a comprehensive comparison with our paralogons is not possible. The only feasible method consists of counting the presence or absence of links between each pair of chromosomes for both data sets: Of the 276 possible chromosome pairs, our method detected 151. We detected 55 regions that were not found in the analysis by Venter *et al.*, and we did not detect any relationship between 21 pairs of chromosomes for which they illustrated pairings. Pairings between chromosomes 18 and 20 were provided in more detail and are shown in Figure 5 together with the corresponding blocks detected by our method. The overall appearance of cross-links seems to be the same except for a region near the centre of chromo-

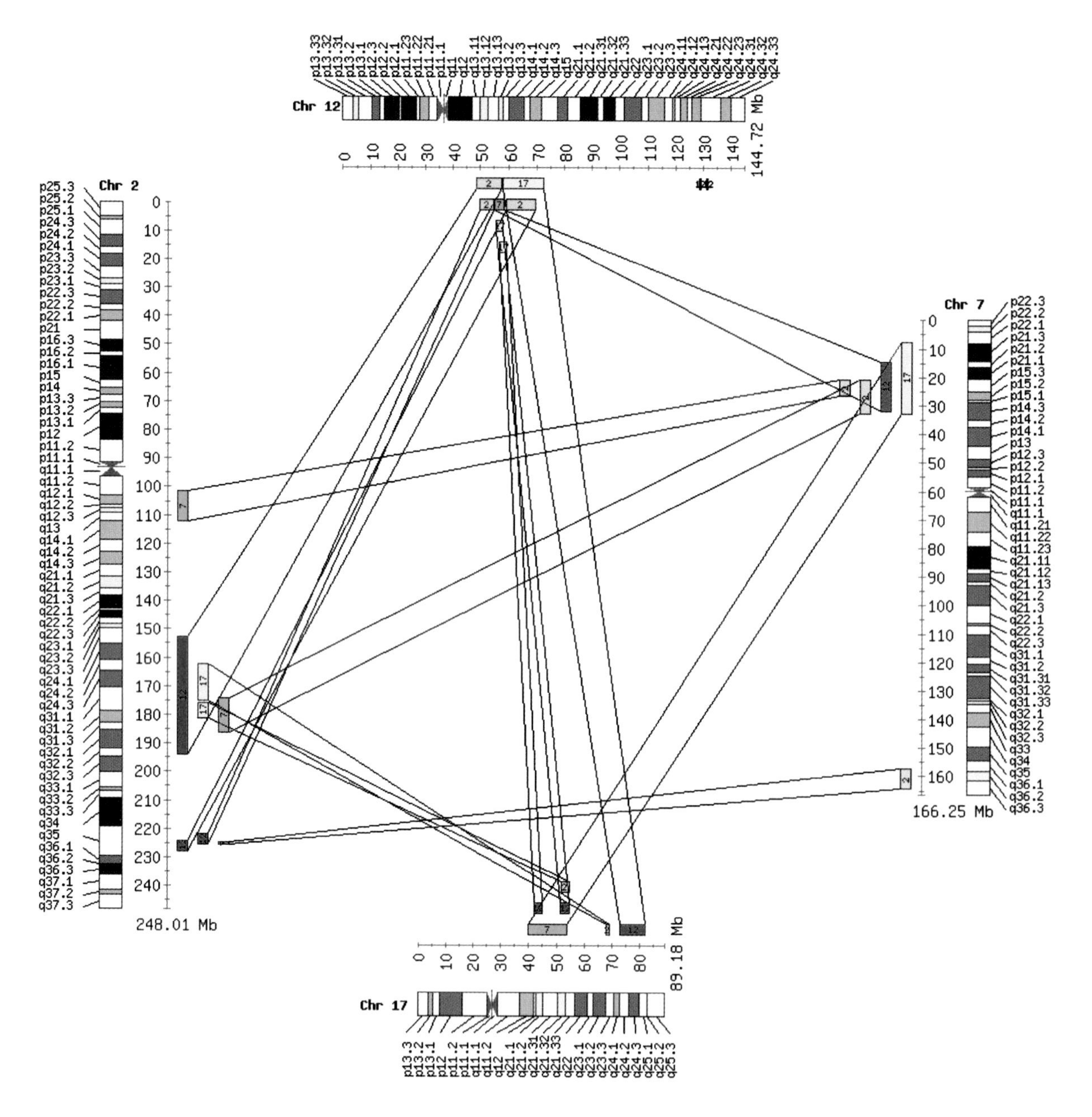

Figure 3. Blocks between Hox chromosomes. All detected paralogous blocks containing 6 or more duplicated genes between human chromosomes 2, 7, 12 and 17 are shown. Blocks can be viewed at wolfe.gen.ted.ie/dup.

some 20. The segment from gene 'GATA rel' to 'Krup rel' on chromosome 20 in the MUMmer graph might correspond to the far end of chromosome 20 (> 64 Mb) in our graph, because both seem to be connected to roughly the same region on chromosome 18. A translocation such as this could occur from differences in the assembly process. Large discrepancies exist in the underlying data: Celera reports 217 protein assignments on chromosome 18 and 322 on chromosome 20. This corresponds to 388 and 748 proteins in the Ensembl data. Venter *et al.* state that their analysis found 64 protein pairs in the blocks between chromosome 18 and 20, and that these blocks have a duplicate gene density of 20–30%. In our case four blocks of sizes 6, 7, 7 and 8 are detected which link a total of 29 and 28 genes on chromosome 18 and 20, respectively. The density of involved genes ranges from 12–39% with a median at 19.7%. Unfortunately,

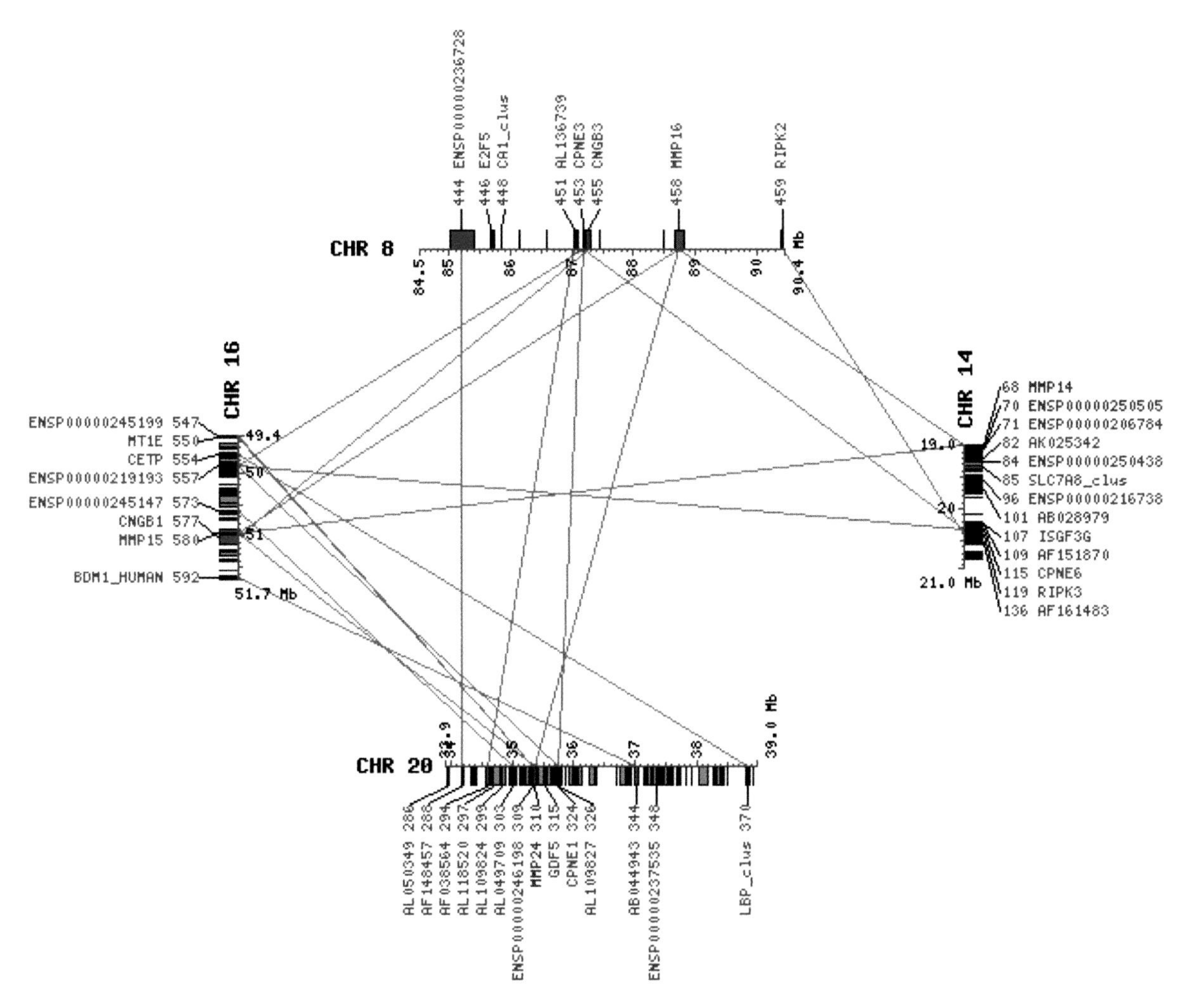

Figure 4. Blocks with copines and MMPs. The core areas of duplicated regions between human chromosomes 8, 14, 16 and 20 are shown. A copine (CPNE) and a matrix metalloproteinase (MMP) can be found in each of them in close vicinity (protein ENSP00000219193 on chromosome 16 is predicted to be a copine by the Ensembl automated annotation system). Linked genes are drawn and labelled in blue. Genes with links to two or three other blocks are highlighted in red and green, respectively. Intermediate genes are filled with grey.

only 7 of the reported gene names (TGIF, VAPA, VAPB, NFATC1, NFATC2, KCNG2, KCNB1) match between the two data sets so a more detailed comparison was not possible. A possible explanation for differences between both graphs can be found in the recent finishing of chromosome 20 by Deloukas *et al.* (2001), where discrepancies between the private and the public sequence affecting order of genes and chromosomal blocks were discussed.

A question of parsimony

One way in which the genome duplication hypothesis is more parsimonious than alternative hypotheses that explain the distribution of paralogues in the genome is in the number of words it takes to describe it, a fact which may be related to its popularity as a hypothesis. Austin Hughes has challenged the assumption that block duplication is the most parsimonious way to generate paralogous regions within a genome using a parsimony statistic. The statistic considers the relative parsimony of the hypothesis that paralogous regions were made by a block duplication event (perhaps as part of a whole genome duplication event), or the alternative hypothesis that they are the result of tandem duplication of genes followed by translocation (Hughes, 1998; Hughes *et al.*, 2001). Following Gu and Huang (2002) we refer to these as the 'BD' (block duplication) model and 'TD' (tandem duplication)

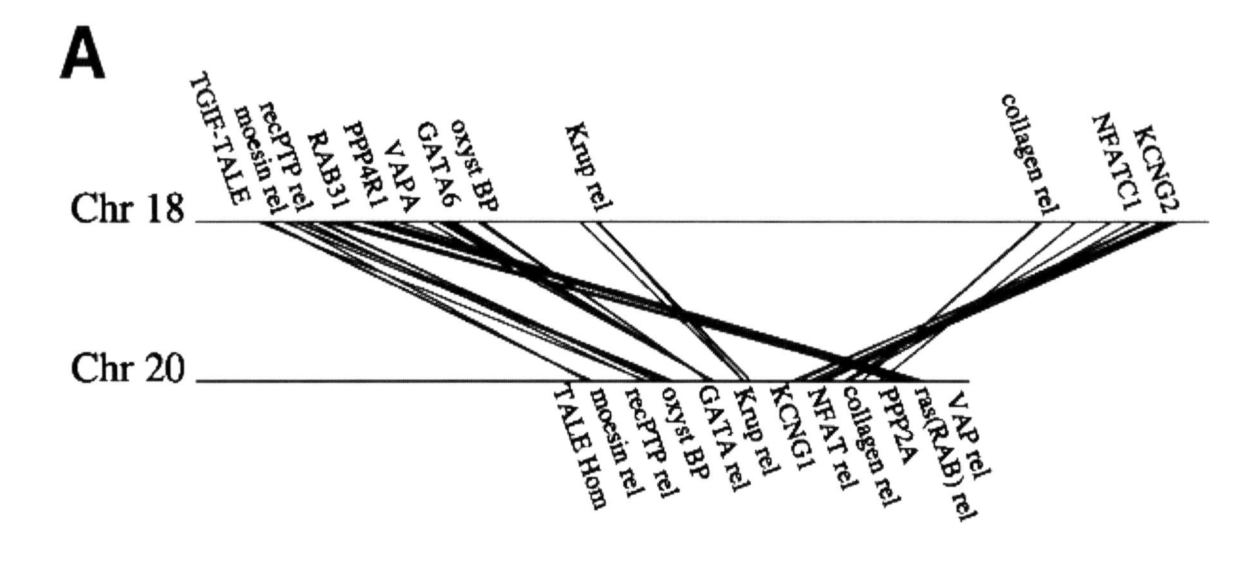

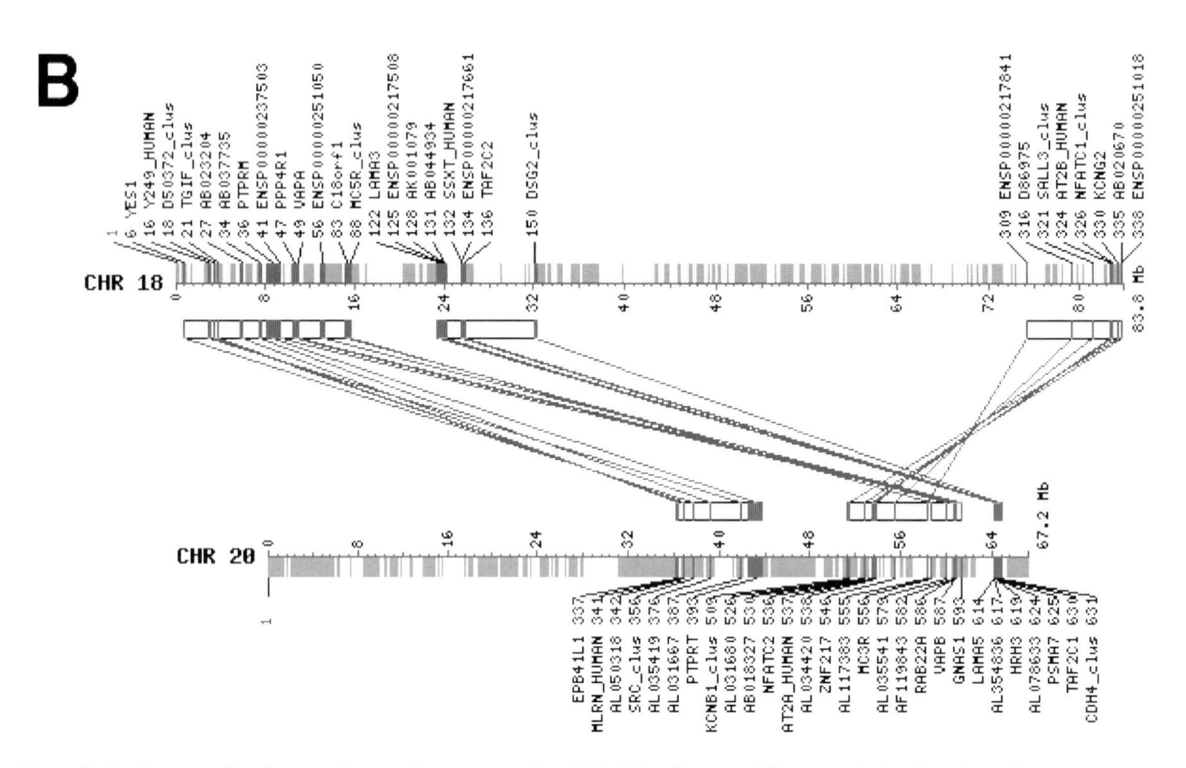

Figure 5. Paralogous regions between human chromosomes 18 and 20 defined by two different methods. Lines drawn between the chromosomes connect paralogues. (A) From Venter *et al.* (2001), showing labels for a selection of genes. (B) From our analyses with all duplicated genes labelled.

model, respectively. Hughes found for both the Hox cluster regions and the chromosome 1/6/9/19 region that the TD model was more parsimonious than the BD model as an explanation for the observed gene orders and phylogenetic trees. However, his reasoning may have been flawed as shown below. The applicability of the parsimony statistic to the paleopolyploid *Arabidopsis* genome has also been challenged by Gu and Huang (2002).

Here we consider the simple case of a single genome duplication. The BD model has an inbuilt disadvantage in the parsimony count method of Hughes (1998; 2001). Each gene that is no longer present in duplicate is counted as an individual deletion event (or equally, as a single translocation event removing it from the scope of detection as part of a paralogous region). By contrast, the method is very generous to the TD model, assuming that a single translocation event brings each gene to its current position within a paralogous region. In fact, as shown below, Hughes' TD model will always require fewer events than a BD model, so long as fewer than 1/3 of genes are retained in duplicate.

Let G be the number of genes in the pre-duplication genome. Let q be the proportion of the pre-duplication genome retained in duplicate in the modern genome, and p be the proportion in single copy.

Then $q + p = 1$ (1)

and $Gq + Gp = G$ (2)

Gq is the number of genes retained in duplicate.

The TD model requires Gq tandem duplication events and a further Gq translocation events, totalling $2Gq$ events. The BD model requires one large duplication event (in this example, a genome duplication) followed by Gp gene deletion events (the number of genes seen in single copy). For these two hypotheses to have an equal number of events (*i.e.*, to be equally parsimonious) then:

$$2Gq = 1 + Gp \quad (3)$$

Replace p with $1-q$ from Equation 1:

$$\rightarrow 2Gq = 1 + G(1 - q) \quad (4)$$

$$\rightarrow q = 1/3G + 1/3 \quad (5)$$

For genomes with a large number of genes (*e.g.*, $G = 6000$ for yeast):

$$\rightarrow q \approx 1/3 \quad (6)$$

The TD model will be more parsimonious than the genome duplication (BD) model if $q < 1/3$, *i.e.*, whenever the retention of genes in duplicate is less than 1/3 of the pre-duplication genome. This result is also apparent from the work of Gu and Huang (2002) who separately analysed 103 duplicated blocks in the *Arabidopsis* genome. Their Figure 2 shows empirically that the TD model is more parsimonious in the 94 blocks having $q < 1/3$, whereas the BD model is more parsimonious only in the remaining 9 blocks with $q > 1/3$. A retention frequency (q) of 1/3 corresponds to a duplication level of 50% in the post-duplication genome (*i.e.*, 50% of genes in the modern genome will have polyploidy-derived paralogues).

Table 1. Crossover values for q (proportion of genes retained in duplicate) and d (average number of genes deleted in a single event) at which the genome duplication and TD models are equally parsimonious (calculated from equation 9).

q	d
0.33	1.0
0.20	2.0
0.10	4.5
0.08	5.8
0.05	9.5
0.01	49.5

The above calculations were based on Hughes' assumption that genes are deleted individually, *i.e.*, that only one gene is deleted per deletion event. It may be more biologically realistic to allow for several neighbouring genes to be duplicated in a single event. If d is the average number of genes deleted in a gene deletion event, then Equation 4 can be rephrased as:

$$2q = 1/G + (1 - q)/d \quad (7)$$

$$\rightarrow 2q \approx (1 - q)/d \quad (8)$$

$$\rightarrow d \approx 1/2q - 1/2 \quad (9)$$

Solving Equation 9 for different values of q shows the average size of a deletion event that is required for the two hypotheses to have equal probability for different frequencies of duplicate gene retention (Table 1).

One of the observations of the well-documented case of paleopolyploidy in yeast was that only 8% of the pre-duplication genome was retained in duplicate (Seoighe and Wolfe, 1998). For $q = 0.08$ the average size of a deletion event (d) needs to be 6 genes or larger (Table 1) to favour the genome duplication hypothesis by the simple parsimony statistic. Intuitively this seems like a biologically feasible size. Indeed it seems more acceptable than another assumption built-in to the alternative tandem-duplication and translocation model, *i.e.*, that selection can create genomic regions with similar gene contents by favouring particular translocations (Hughes, 1999).

We consider an example from the yeast genome. Within a block first described by Pohlmann and Philippsen (1996), and later numbered block 39 by Wolfe and Shields (1997) there are six duplicated genes and 20 unduplicated genes (eight on chromosome XIV and 12 on chromosome IX). Under Hughes' TD model the formation of this block would require 12 steps (six tandem duplications, and six translocations). Under a whole genome duplication model, with each gene deleted individually, the formation of this block would require 21 events (one whole genome duplication, and 20 gene deletions) and would thus be less parsimonious by this statistic. However, if each deletion event included on average three genes then only seven deletion events would be required to explain the current state of this paralogous region, and the block duplication model would be more parsimonious. Thus it appears that, even in the well-documented case of yeast, which Hughes and colleagues agree is a likely polyploid (Friedman and Hughes, 2001), this simple parsimony statistic is not appropriate to determine the relative probability of paralogous region formation by block duplication versus tandem duplication and translocation.

Discussion

The approach described here acknowledges two important aspects of the genome duplication hypothesis. Ones is that it proposes a whole genome event, and anecdotal evidence for individual gene families is unlikely to result in a firm conclusion – we have therefore analysed the whole human genome. The other important aspect is that genome duplication is not proposed as the only mechanism of gene family expansion – we have therefore removed obvious tandem duplicates from the analysis, and deliberately limited this study to look at the mechanisms of gene-family expansion in the vertebrate lineage. Nobody realistically expects that all gene families that exist in the vertebrate genome were singletons before vertebrate origins. What is detectable as a gene family, i.e., a group of paralogous genes, will often include members representing a long evolutionary history, only some of which is vertebrate specific. Studies where gene families are included indiscriminately will doubtless include diverse evolutionary histories, and closer inspection will reveal a straw man easy to knock down.

Our map-based method found paralogons covering over 44% of the human genome. These are most probably vertebrate specific, and are distributed throughout the genome. This can be explained by either extensive sub-genomic duplications, or by polyploidy. An additional phylogenetic analysis carried out recently on the same data set indicates a significant accumulation of duplication activity during a relatively short period between 350 and 650 Mya (McLysaght *et al.*, 2002). Both findings together lend support to the hypothesis of one polyploidy event early in the vertebrate lineage. There is no specific evidence for two as opposed to one tetraploidy event, or for auto- rather than allotetraploidy. Neither do current findings from investigations of the one-to-four rule or tree topologies give clear signals for any of these scenarios. Further refinement of the human genome sequence and complete gene identification will hopefully enable more precise analyses in the future. It is also likely that the genome sequencing project for the tunicate *Ciona* will contribute useful data in the near future. However, in view of the immense time span under consideration and the huge complexity of genomic changes that might have occurred, it remains unsure if the complete history of ancient duplication events that led to the current shape of the human genome will ever be revealed.

References

Allendorf, F.W. and Thorgaard, G.H. (1984) Tetraploidy and the evolution of salmonid fishes. In *Evolutionary Genetics of Fishes* (Ed. Turner, B.), Plenum Press, New York, NY, pp. 1–46.

Bailey, W.J., Kim, J., Wagner, G.P. and Ruddle, F.H. (1997) Phylogenetic reconstruction of vertebrate Hox cluster duplications. *Mol. Biol. Evol.*, **14,** 843–853.

Bairoch, A. and Apweiler, R. (2000) The SWISS-PROT protein sequence database and its supplement TrEMBL in 2000. *Nucleic Acids Res.*, **28,** 45–48.

Birney, E. and Durbin, R. (2000) Using GeneWise in the Drosophila annotation experiment. *Genome Res.*, **10,** 547–548.

Burge, C. and Karlin, S. (1997) Prediction of complete gene structures in human genomic DNA. *J. Mol. Biol.*, **268,** 78–94.

Comings, D.E. (1972) Evidence for ancient tetraploidy and conservation of linkage groups in mammalian chromosomes. *Nature*, **238,** 455–457.

Delcher, A.L., Kasif, S., Fleischmann, R.D., Peterson, J., White, O. and Salzberg, S.L. (1999) Alignment of whole genomes. *Nucleic Acids Res.*, **27,** 2369–2376.

Deloukas, P., *et al.* (2001) The DNA sequence and comparative analysis of human chromosome 20. *Nature*, **414,** 865–871.

.Endo, T., Imanishi, T., Gojobori, T. and Inoko, H. (1997) Evolutionary significance of intra-genome duplications on human chromosomes. *Gene*, **205,** 19–27.

Flajnik, M.F. and Kasahara, M. (2001) Comparative genomics of the MHC: glimpses into the evolution of the adaptive immune system. *Immunity*, **15,** 351–362.

Friedman, R. and Hughes, A.L. (2001) Gene duplication and the structure of eukaryotic genomes. *Genome Res.*, **11,** 373–381.

Gallardo, M.H., Bickham, J.W., Honeycutt, R.L., Ojeda, R.A. and Kohler, N. (1999) Discovery of tetraploidy in a mammal. *Nature*, **401,** 341.

Garcia-Fernandez, J. and Holland, P.W. (1994) Archetypal organization of the amphioxus Hox gene cluster. *Nature*, **370,** 563–566.

Gaut, B.S. and Doebley, J.F. (1997) DNA sequence evidence for the segmental allotetraploid origin of maize. *Proc. Natl. Acad. Sci. USA*, **94,** 6809–6814.

Gibson, T.J. and Spring, J. (2000) Evidence in favour of ancient octaploidy in the vertebrate genome. *Biochem. Soc. Trans.*, **28,** 259–264.

Graves, J.A. (1996) Mammals that break the rules: genetics of marsupials and monotremes. *Annu. Rev. Genet.*, **30,** 233–260.

Gu, X. and Huang, W. (2002) Testing the parsimony test of genome duplications: a counterexample. *Genome Res.*, **12,** 1–2.

Holland, P.W.H., Garcia-Fernandez, J., Williams, N.A. and Sidow, A. (1994) Gene duplications and the origins of vertebrate development. *Development*, **Suppl. 1994,** 125–133.

Hubbard, T., *et al.* (2002) The Ensembl genome database project. *Nucleic Acids Res.*, **30,** 38–41.

Hughes, A.L. (1998) Phylogenetic tests of the hypothesis of block duplication of homologous genes on human chromosomes 6, 9, and 1. *Mol. Biol. Evol.*, **15,** 854–870.

Hughes, A.L. (1999) Phylogenies of developmentally important proteins do not support the hypothesis of two rounds of genome duplication early in vertebrate history. *J. Mol. Evol.*, **48,** 565–576.

Hughes, A.L., da Silva, J. and Friedman, R. (2001) Ancient genome duplications did not structure the human Hox-bearing chromosomes. *Genome Res.*, **11,** 771–780.

International Human Genome Sequencing Consortium (2001) Initial sequencing and analysis of the human genome. *Nature* **409,** 860–921.

Kappen, C., Schughart, K. and Ruddle, F.H. (1989) Two steps in the evolution of Antennapedia-class vertebrate homeobox genes. *Proc. Natl. Acad. Sci. USA*, **86,** 5459–5463.

Kasahara, M. (1997) New insights into the genomic organization and origin of the major histocompatibility complex: role of chromosomal (genome) duplication in the emergence of the adaptive immune system. *Hereditas*, **127,** 59–65.

Kasahara, M., Hayashi, M., Tanaka, K., Inoko, H., Sugaya, K., Ikemura, T. and Ishibashi, T. (1996) Chromosomal localization of the proteasome Z subunit gene reveals an ancient chromosomal duplication involving the major histocompatibility complex. *Proc. Natl. Acad. Sci. USA*, **93,** 9096–9101.

Katsanis, N., Fitzgibbon, J. and Fisher, E.M.C. (1996) Paralogy mapping: identification of a region in the human MHC triplicated onto human chromosomes 1 and 9 allows the prediction and isolation of novel PBX and NOTCH loci. *Genomics*, **35,** 101–108.

Kent, W.J. and Haussler, D. (2001) Assembly of the working draft of the human genome with GigAssembler. *Genome Res.*, **11,** 1541–1548.

Kojima, S., Itoh, Y., Matsumoto, S., Masuho, Y. and Seiki, M. (2000) Membrane-type 6 matrix metalloproteinase (MT6-MMP, MMP–25) is the second glycosyl-phosphatidyl inositol (GPI)-anchored MMP. *FEBS Lett.*, **480,** 142–146.

Lahn, B.T. and Page, D.C. (1999) Four evolutionary strata on the human X chromosome. *Science*, **286,** 964–967.

Lundin, L.G. (1993) Evolution of the vertebrate genome as reflected in paralogous chromosomal regions in man and the house mouse. *Genomics*, **16,** 1–19.

Martin, A. (2001) Is tetralogy true? Lack of support for the 'one-to-four' rule. *Mol. Biol. Evol.*, **18,** 89–93.

Martin, A.P. (1999) Increasing genomic complexity by gene duplication and the origin of vertebrates. *Amer. Nat.*, **154,** 111–128.

Martin, G.R., Richman, M., Reinsch, S., Nadeau, J.H. and Joyner, A. (1990) Mapping of the two mouse engrailed-like genes: close linkage of En–1 to dominant hemimelia (Dh) on chromosome 1 and of En–2 to hemimelic extratoes (Hx) on chromosome 5. *Genomics*, **6,** 302–308.

Martinez-Perez, E., Shaw, P. and Moore, G. (2001) The Ph1 locus is needed to ensure specific somatic and meiotic centromere association. *Nature*, **411,** 204–207.

McLysaght, A., Hokamp, K. and Wolfe, K.H. (2002) Extensive genomic duplication during early chordate evolution. *Nature Genet.*, **31**, 200–204.

Meyer, A. and Schartl, M. (1999) Gene and genome duplications in vertebrates: the one-to-four (-to-eight in fish) rule and the evolution of novel gene functions. *Curr. Opin. Cell Biol.*, **11,** 699–704.

Muller, H.J. (1925) Why polyploidy is rarer in animals than in plants. *Amer. Nat.*, **9,** 346–353.

Ohno, S. (1970) *Evolution by Gene Duplication*, George Allen and Unwin, London, UK.

Ohno, S. (1999) Gene duplication and the uniqueness of vertebrate genomes circa 1970–1999. *Semin. Cell Devel. Biol.*, **10,** 517–522.

Pébusque, M.-J., Coulier, F., Birnbaum, D. and Pontarotti, P. (1998) Ancient large scale genome duplications: phylogenetic and linkage analyses shed light on chordate genome evolution. *Mol. Biol. Evol.*, **15,** 1145–1159.

Pohlmann, R. and Philippsen, P. (1996) Sequencing a cosmid clone of *Saccharomyces cerevisiae* chromosome XIV reveals 12 new open reading frames (ORFs) and an ancient duplication of six ORFs. *Yeast*, **12,** 391–402.

Popovici, C., Leveugle, M., Birnbaum, D. and Coulier, F. (2001) Coparalogy: Physical and functional clusterings in the human genome. *Biochem. Biophys. Res. Commun.*, **288,** 362–370.

Riley, R. and Kempanna, C. (1963) The homeologous nature of the non-homologous meiotic pairing in *Triticum aestivum* deficient for chromosome V (5B) *Heredity*, **18,** 287–306.

Sato, H., Tanaka, M., Takino, T., Inoue, M. and Seiki, M. (1997) Assignment of the human genes for membrane-type–1, –2, and –3 matrix metalloproteinases (MMP14, MMP15, and MMP16) to 14q12.2, 16q12.2-q21, and 8q21, respectively, by in situ hybridization. *Genomics*, **39,** 412–413.

Seoighe, C. and Wolfe, K.H. (1998) Extent of genomic rearrangement after genome duplication in yeast. *Proc. Natl. Acad. Sci. USA*, **95,** 4447–4452.

Sidow, A. (1996) Gen(om)e duplications in the evolution of early vertebrates. *Curr. Opin. Genet. Devel.*, **6,** 715–722.

Skrabanek, L. and Wolfe, K.H. (1998) Eukaryote genome duplication – where's the evidence? *Curr. Opin. Genet. Devel.*, **8,** 694–700.

Smith, N.G.C., Knight, R. and Hurst, L.D. (1999) Vertebrate genome evolution: a slow shuffle or a big bang? *BioEssays*, **21,** 697–703.

Spring, J. (1997) Vertebrate evolution by interspecific hybridisation – are we polyploid? *FEBS Lett.*, **400,** 2–8.

Tomsig, J.L. and Creutz, C.E. (2000) Biochemical characterization of copine: a ubiquitous Ca^{2+}-dependent, phospholipid-binding protein. *Biochemistry*, **39,** 16163–16175.

Venter, J.C. *et al.* (2001) The sequence of the human genome. *Science*, **291,** 1304–1351.

Vision, T.J., Brown, D.G. and Tanksley, S.D. (2000) The origins of genomic duplications in *Arabidopsis*. *Science*, **290,** 2114–2117.

Wolfe, K.H. (2001) Yesterday's polyploids and the mystery of diploidization. *Nature Reviews Genet.*, **2,** 333–341.

Wolfe, K.H. and Shields, D.C. (1997) Molecular evidence for an ancient duplication of the entire yeast genome. *Nature*, **387,** 708–713.

Zhang, J. and Nei, M. (1996) Evolution of Antennapedia-class homeobox genes. *Genetics*, **142,** 295–303.

A. Meyer, Y. Van de Peer (eds.), Genome Evolution, 111-116.

Introns in, introns out in plant gene families: a genomic approach of the dynamics of gene structure

Alain Lecharny*, Nathalie Boudet, Isabelle Gy, Sébastien Aubourg[1] & Martin Kreis
Institut de Biotechnologie des Plantes, Unité Mixte de Recherche-Centre National de la Recherche Scientifique 8618, Université de Paris-Sud, Bât. 630, F–91405 Orsay Cedex, France; [1]Present address: URGV, Unité Mixte de Recherche INRA-CNRS, 2 rue Gaston Crémieux, CP 5708, F–91057 Evry, France.
**Author for correspondence: E-mail: lecharny@ibp.u-psud.fr*

Received 05.02.2002; accepted in final form 29.08.2002

Key words: *Arabidopsis*, gene families, intron, structure

Abstract

Gene duplication is considered to be a source of genetic information for the creation of new functions. The *Arabidopsis thaliana* genome sequence revealed that a majority of plant genes belong to gene families. Regarding the problem of genes involved in the genesis of novel organs or functions during evolution, the reconstitution of the evolutionary history of gene families is of critical importance. A comparison of the intron/exon gene structure may provide clues for the understanding of the evolutionary mechanisms underlying the genesis of gene families. An extensive study of *A. thaliana* genome showed that families of duplicated genes may be organized according to the number and/or density of intron and the diversity in gene structure. In this paper, we propose a genomic classification of several *A. thaliana* gene families based on introns in an evolutionary perspective.

Abbreviations: BGAL, β-galactosidases; PCMP, plant combinatorial and modular protein

Gene duplication in plants

Duplicated genes have been identified in all the eukaryotic genomes sequenced today. Duplications provide the potential for the creation of metabolic and developmental complexities. In recent years, the rates of genome duplications, and the importance of the mechanisms underlying functional divergence of genome or segment duplications have received a renewed interest (Sankoff, 2001). Even if all eukaryotic genomes share a main common mechanims leading to families of duplicated genes, extant genomes exhibit some peculiarities between them, notably the duplication rates, the fixation of both copies or the rates at which genes are lost (Lynch and Conery, 2000), that may be of interest to consider in an evolutionary perspective. The genome of the model plant *Arabidopsis thaliana* is one of the smallest plant genomes, containing about 120 Mb. Many data indicate that, similar to the relatively small animal genomes (Petrov *et al.*, 2000), the *A. thaliana* genome is under a high constraint against an increase in its size as manifested by comparatively small introns and intergenic regions, few repeats and transposons. It nevertheless contains 26,000 genes, similar to the number of genes identified in the human genome, i.e. 32,000 (AGI, 2000; IHGSC, 2001), but higher than in fly (Adams *et al.*, 2000) or worm (The *C. elegans* Sequencing Consortium, 1998). This situation, unexpected some years ago, is at least partly explained both by a higher number of families of duplicated genes as well as more members per family in *A. thaliana* than in other sequenced eukaryotic genomes (Aubourg *et al.*, 1997, 1998, 1999; Gy *et al.*, 1998). For instance, there are more than 1,700 genes coding for transcriptional factors in *A. thaliana*, many of them specific to plants, and only 700 and 500 in *Drosophila melanogaster* and *Caenorhabditis elegans*, respectively (Riechmann and Ratcliffe, 2000). The amplification and dispersal of genes resulting in the formation of families, appears to have been even more important in conifers than in angiosperms (Kinlaw and Neale, 1997). Therefore, it seems that in plants gene duplication has been the privileged way to increase the diversity in transcripts and gene products rather than other mechanisms, like alternative splicing (1–2%, The European Union Chromosome 3 ASC *et al.*, 2000) and multiple transcription starts. The latter two mechanisms are more frequently observed in animals (Mironov *et al.*, 1999), particularly in human. In the genome of *A. thaliana*, and depending on the criteria retained, about 15 (chromosome 4, The European Union Arabidopsis Genome Sequencing Consortium *et al.*, 2000) to 27% (chromosome 2, Lin *et al.*, 2000) of the predicted genes are without recognizable homologues and may represent plant specific genes. Among these potential plant-specific genes, many are members of families of more than 5 paralogues (Ride *et al.*, 1999; Aubourg *et al.*, 2000; Tavares *et al.*, 2000; Aubourg *et al.*, 2002). Therefore, these families have been formed relatively recently by gene duplication. Even in *A. thaliana*, a plant with an extant genome essentially diploid, the presence of large segment duplications has been shown and diversely interpreted as vestiges of a unique event of an entire genome duplication (Blanc *et al.*, 2000) or of independent smaller duplications (Vision *et al.*, 2000). Any one of these scenarios may simply explain the existence of large families of tens of dispersed genes relatively recently formed in plants. Hence, mechanisms of gene translocation following tandem duplications by recombination are likely to be important. As a consequence, *A. thaliana* gene families that have homologues outside the plant kingdom, may be suspected to contain duplicated genes formed by both ancient duplications, predating the plant/animal separation, and recent duplications. The proportion of the two components, i.e. ancient and recent duplicated genes in representative families associated with genomic functional data should inform us on the evolutionary or adaptative value of the expansion or regression of paralogous gene families in a given species.

Evolution of gene structures vs protein sequences

The reconstitution of the history of the families of duplicated genes is of critical importance when we tackle the problem of genes or pathways involved in the apparition of novel organs or functions during evolution. The genesis of a gene family can be studied using only exons, hence only the structure of the protein, or by taking into consideration both introns and exons. The evolution of exon sequences is under a high selection pressure, at least in protein regions involved in the definition of the protein activity. Introns are potentially interesting structures for evolutionary studies since they are both rapidly diverging in size and sequence and slowly diverging if their position in the genes is considered. Thus, changes in sizes and sequences may be used to compare recently duplicated genes (Liss *et al.*, 1997) or geographical isolates of a single species, while gene structures may be used to compare more ancient duplications or inter species relationships. Effectively, on one hand, the conservation of gene structure in homologues genes is usually high enough to recognize the lineage of introns in evolution (Hardison, 1996). Indeed, it has recently been shown that exon structure may even be conserved despite low sequence conservation (Betts *et al.*, 2001). On the other hand, departures from the common family structure are due to relatively rare events of individual intron deletions, insertions or both, after duplications. Differences in intron/exon structures, created by the above events, between paralogues may often be used as guides to reconstruct the history of gene families (Brown *et al.*, 1995; Robertson, 1998). Thus, classifications of paralogous genes into subfamilies based on shared variations in gene structures have been proposed recently (Wattler *et al.*, 1998; Tognolli *et al.*, 2000; Trapp and Croteau, 2001).

Introns should always be considered for both their presence or absence and for the possible information their sequences may contain. Introns are not all neutral elements, a number of them have a regulatory function that has been demonstrated experimentally (Fu *et al.*, 1995a, 1995b; Bolle *et al.*, 1996). Introns may thus participate in the creation of divergence between paralogues by subfunctionalization, similarly to promoters or other regulatory regions (Lynch and Force, 2000). After duplication, and during a neutral phase of evolution, the two copies diverge by a distribution, that may only be partial, of the elements defining the functional specialization of their ancestor. The insertion or deletion of introns between codons (*i.e.* phase 0 introns) has no effect on the nature of the gene product, but may promote some subfunctionalization either directly by introducing a difference in regulatory elements or by helping exon shuffling.

A genomic classification of *A. thaliana* gene families based on introns

Families of duplicated genes may be organized according to the number or density of introns and the diversity observed in the gene structures. An overview of plant gene families shows the organization of three extreme families.

Families of genes with a conserved structure

In many duplicated gene families, the gene structure is highly conserved between the paralogues as well as between homologues. For instance, most of the intron positions within genes coding for invertases (Haouazine-Takvorian *et al.*, 1997), shaggy-related protein kinases (Dornelas *et al.*, 1998, 1999), terpene synthases (Aubourg *et al.*, 2002) or the MYST, the homeodomain-GLABRA2 and the like cinnamyl alcohol dehydrogenase proteins (Tavares *et al.*, 2000) can unambiguously be attributed to an ancestral gene. Intron distribution and parsimony groups using the alignments of the protein sequences, defined between the different paralogues allowed the reconstruction of the same history.

Families of genes with almost no introns

In *A. thaliana* the number of genes identified without introns is close to 20% (The European Union Arabidopsis Genome Sequencing Consortium *et al.*, 2000). Nevertheless, we discovered a family of genes, named *AtPCMP*, in *A. thaliana* containing more than 200 paralogues, i.e. about 1% of the genome (Aubourg *et al.*, 2000). The data indicate that the latter gene family is specific to plants. Genes belonging to this unexpectedly large family are orphan of function, but the members present a highly characteristic combinatorial arrangement of 8 motifs. The structure of the *PCMP* genes exhibits an unusual characteristic since out of 111 full-length predicted genes, only 13 have an intron. Furthermore, all the predicted or verified introns interrupt the coding sequence at a specific position for each of the 13 genes. We showed that most of the *PCMP* genes are transcribed. The genome of *A. thaliana* contains an other family of genes with all the same characteristics as *PCMP*, including a high number of members, a comparable protein structure and the same paucity of introns (Small and Peeters, 2000).

Families of genes with an unconserved structure

In some large families, subgroups of paralogous genes are clearly identified by their different structures. This is the case for the *A. thaliana* DEAD box RNA helicases (Aubourg *et al.*, 1999; Boudet *et al.*, 2001), syntaxins (Sanderfoot *et al.*, 2000), polygalacturonases (Torki *et al.*, 2000) and cytochrome P450 (Paquette *et al.*, 2000). For the DEAD helicases, the level of similarity between products from both paralogous and homologous genes exclude the possibility that the unrelated gene structures may indicate non homologous ancestor genes because similarities due to convergence are generally considered to be limited to small regions of the genes. Families of genes that evolved by duplications from a single ancestral gene but that exhibit diverse exon/intron structures are also found in animals (Rzhetsky *et al.*, 1997; Gotoh, 1998; Boudet *et al.*, 2001).

Reverse transcription and plant gene evolution

Without entering the debate concerning the intron early or late theories, we may with Trotman (1998) postulate that duplications of ancestral mosaic genes have been followed by more recent gains and losses of introns. Generally, these two processes are slow and gene structures are often well recognizable between evolutionarily distant homologues. Nevertheless, the genesis of gene families with an unconserved structure (see above), are not easily explained by this mechanism. We studied in great detail the DEAD helicase family in *A. thaliana* (Boudet *et al.*, 2001) and showed that it exhibits an apparent continuity from genes without introns (in the conserved catalytic domain) to genes with up to 18 introns. The genesis of genes without introns in a family of paralogues with introns (Liaud *et al.*, 1992; Frugoli *et al.*, 1998; Charlesworth *et al.*, 1998; Aubourg *et al.* 1999; Tavares *et al.*, 2000; Boudet *et al.*, 2001) is consistantly explained by events of recombinations with reverse transcribed pre-mRNAs as formerly proposed by Lewin (1983), Fink (1987) and Martinez *et al.* (1989). This mechanism of gene structure evolution has probably been frequently used in plants, even recently.

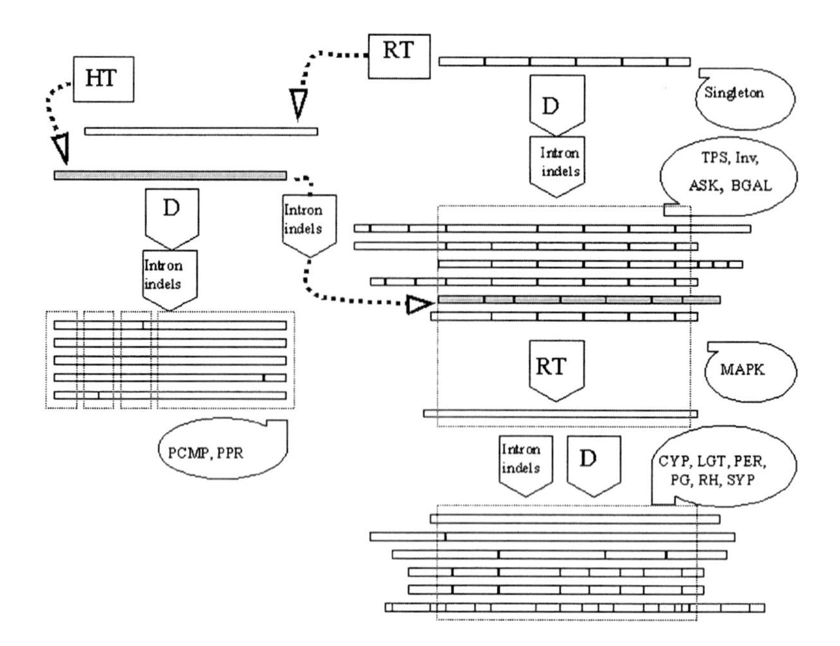

Figure 1. Model for the genesis and evolution of gene families in *A. thaliana*. The three extreme family organizations, described in the text, are represented with the events that are thought to have been involved during evolution. Abbreviations: D, duplication; Intron indels, deletions and insertion of introns; RT, reverse transcription of a gene with introns that create a copy without introns; HT, horizontal transfer. Gene family names: ASK, SHAGGY-related protein kinases (Dornelas, 1998, 1999); BGAL, β-galactokinases (Gy, personnal communication); CYP, cytochrome P450 (Paquette *et al.*, 2000); Inv, invertases (Haouazine-Takvorian *et al.*, 1997); LGT, like glycosyl transferase family (Tavares *et al.*, 2000); MAPK, elements of MAPK modules (Henry, personnal communication); PCMP, plant combinatorial and modular proteins (Aubourg *et al.*, 2000); PER, peroxidase (Tognolli *et al.*, 2000); PG, polygalacturonase (Torki *et al.*, 2000); PPR, pentatricopeptide repeat family of proteins (Small and Peeters, 2000); Singletons represent 35.0% of all genes (AGI, 2000); RH, DEAD box RNA helicases (Boudet *et al.*, 2001); SYP, syntaxins (Sanderfoot *et al.*, 2000); TPS, terpene syntases (Aubourg *et al.*, 2002).

Evolution of gene structure after horizontal gene transfer

The β-galactosidase gene family in *A. thaliana* contains 17 genes (Gy I, personnal communication). For all these genes, but two, the intron distribution and sequence relation trees strongly indicate a common ancestral gene with 18 introns. The paralogues have between 10 to 18 introns that are all located at conserved positions, *i.e.* found at one of the intron positions proposed for the ancestral gene. Thus, most of the AtBGAL genes belong to a family with a conserved gene structure. Based on sequence comparisons we showed that one of the AtBGAL genes, namely AtBGAL17, has a procaryotic origin and thus has been acquired by the plant either through a direct transfer from a bacterium or, more probably, through the endosymbiotic pathway that lead to the formation of mitochondria from a α-proteobacterium or chloroplasts from a cyanobacterium. Analysis of the whole sequence of the *A. thaliana* genome indicates that more than 800 nuclear encoded proteins are of possibly plastid descent (AGI, 2000). Interestingly enough, AtBGAL17 has 16 introns, a much higher number than the average in *A. thaliana* genes, and they are all at an original position as compared to intron positions in the other AtBGAL. This strongly suggests the possibility that after the transfer of the AtBGAL17 ancestral prokaryotic gene without introns, selection has probably acted on the acquisition of a highly mosaic gene structure for an unknown functional reason.

A general model for gene structure evolution in plants

Figure 1 represents a mechanistic model for the genesis and evolution of gene families in *A. thaliana*. Our model points on some important features. (i) In plants, events of reverse transcription made an important contribution to evolution of gene structure. (ii) Insertions/deletions of introns have been frequent. (iii) Gene structures seem to be under a selective pressure, that differs from one family to an other. It probably applies to all plants, because a number of the characteristics shown are also found in animals.

References

Adams, M.D., Celniker, S.E., Holt, R.A., Evans, C.A., Gocayne, J.D., Amanatides, P.G., Scherer, S.E., Li, P.W., Hoskins, R.A., Galle, R.F. *et al.* (2000) The genome sequence of *Drosophila melanogaster. Science*, **287**, 2185–2195.

AGI, Arabidopsis Genome Initiative. (2000) Analysis of the genome sequence of the flowering plant *Arabidopsis thaliana. Nature*, **408**, 796–815.

Aubourg, S., Takvorian, A., Chéron, A., Kreis, M. and Lecharny, A. (1997) Structure, organization and putative function of the genes identified within a 23-kb fragment from *Arabidopsis thaliana* chromosome IV. *Gene*, **199,** 241–253.

Aubourg, S., Chéron, A., Kreis, M. and Lecharny, A. (1998) Structure and expression of an asparaginyl-tRNA synthetase gene located on chromosome IV of *Arabidopsis thaliana* and adjacent to a novel gene of 15 exons. *Biochim. Biophys. Acta*, **1398,** 225–231.

Aubourg, S., Picaud, A., Kreis, M. and Lecharny, A. (1999) Structure and expression of three *src2* homologues and a novel subfamily of flavoprotein monooxygenase genes revealed by the analysis of a 25 kb fragment from *Arabidopsis thaliana* chromosome IV. *Gene*, **230,** 197–205.

Aubourg, S., Kreis, M. and Lecharny, A. (1999) The DEAD box RNA helicase family in *Arabidopsis thaliana. Nucleic Acids Res.*, **27,** 628–636.

Aubourg, S., Lecharny, A. and Bohlmann, J. (2002) Genomic analysis of the terpenoid synthase (AtTPS) gene of *Arabidopsis thaliana. Mol. Genet. Genomics*, **267**, 730–745.

Betts, M.J., Guigo, R., Agarwal, P. and Russel, R.B. (2001) Exon structure conservation despite low sequence similarity: a relic of dramatic events in evolution. *EMBO J.*, **20,** 5354–5360.

Blanc, G., Barakat, A., Guyot, R., Cooke, R. and Delseny, M. (2000) Extensive duplication and reshuffling in the *Arabidopsis* genome. *Plant Cell*, **12,** 1093–1102.

Bolle, C., Herrmann, R.G., and Oelmuller, R. (1996) Intron sequences are involved in the plastid- and light-dependent expression of the spinach PsaD gene. *Plant J.*, **10,** 919–924.

Boudet, N., Aubourg, S., Toffano-Nioche, C., Kreis, M. and Lecharny, A. (2001) Evolution of intron/exon structure of DEAD helicase family genes in *Arabidopsis*, *Caenorhabditis*, and *Drosophila. Genome Res.*, **11,** 2101–2114.

Brown, N.P., Whittaker, A.J., Newell, W.R., Pawlings, C.J. and Beck, S. (1995) Identification and analysis of multigene families by comparison of exon fingerprints. *J. Mol. Biol.*, **249,** 342–359.

Dornelas, M.C., Lejeune, B., Dron, M. and Kreis, M. (1998) The *Arabidopsis shaggy*-related protein kinase (ASK) gene family: structure, organisation and evolution. *Gene*, **212,** 249–257.

Dornelas, M.C., Wittich, P., von Recklinghausen, I., vanLammeren, A. and Kreis, M. (1999) Characterization of three novel members of the *Arabidopsis SHAGGY*-related protein kinase (*ASK*) multigene family. *Plant Mol. Biol.*, **39,** 137–147.

European Union Chromosome 3 Arabidopsis Sequencing Consortium; The Institute for Genomic Research and Kazusa DNA Research Institute. (2000) Sequence and analysis of chromosome 3 of the plant *Arabidopsis thaliana. Nature*, **408,** 820–821.

Frugoli, J.A., McPeek, M.A., Thomas, T.L. and McLung, C.R. (1998) Intron loss and gain during evolution of the catalase gene family in angiosperms. *Genetics*, **149,** 355–365.

Fu, H., Kim, S.Y. and Park, W.D. (1995a) High-level tuber expression and sucrose inducibility of a potato Sus4 sucrose synthase gene require 5' and 3' flanking sequences and the leader intron. *Plant Cell*, **7,** 1387–1394.

Fu, H., Kim, S.Y. and Park, W.D. (1995b) A potato Sus3 sucrose synthase gene contains a context-dependent 3' element and a leader intron with both positive and negative tissue-specific effects. *Plant Cell*, **7,** 1395–1403.

Gotoh, O. (1998) Divergent structures of *Caenorhabditis elegans* cytochrome P450 genes suggest the frequent loss and gain of introns during the evolution of nematodes. *Mol. Biol. Evol.*, **15,** 1447–1459.

Gy, I., Aubourg, S., Sherson, S., Cobett, C.S., Chéron, A., Kreis, M. and Lecharny, A. (1998) Analysis of a 14-kb fragment containing a putative cell wall gene and a candidate for the *ARA1*, arabinose kinase, gene from chromosome IV of *Arabidopsis thaliana. Gene*, **209,** 201–210.

Haouazine-Takvorian, N., Tymowska-Lalanne, Z., Takvorian, A., Tregear, J., Lejeune, B., Lecharny, A. and Kreis, M. (1997) Characterization of two members of the *Arabidopsis thaliana* gene family, *AtBfruct3* and *AtBfruct4*, coding for vacuolar invertases. *Gene*, **197,** 239–251.

International Human Genome Sequencing Consortium. (2001) Initial sequencing and analysis of the human genome. *Nature*, **409,** 860–921.

Kinlaw, C.S. and Neale, D.B. (1997) Complex gene families in pine genomes. *Trends Plant Sci.*, **2,** 356–359.

Liss, M., Kirk, D.L., Beyser, K. and Fabry, S. (1997) Intron sequences provide a tool for high-resolution phylogenetic analysis of volvocine algae. *Curr. Genet.*, **31,** 214–227.

Lynch, M. and Conery, J.S. (2000) The evolutionary fate and consequences of duplicate genes. *Science*, **290,** 1151–1155.

Mironov, A.A., Fickett, J.W. and Gelfand, M.S. (1999) Frequent alternative splicing of human genes. *Genome Res.*, **9,** 1288–1293.

Paquette, S.M., Bak, S. and Feyerreisen, R. (2000) Intron-exon organisation and phylogeny in a large superfamily, the paralogous cytochrome P450 genes of *Arabidopsis thaliana. DNA Cell Biol.*, **19,** 307–317.

Petrov, D.A., Sangster, T.A., Johnston, J.S., Hartl, D.L. and Shaw, K.L. (2000) Evidence for DNA loss as a determinant of genome size. *Science*, **287,** 1060–1062.

Ride, J.P., Davies, E.M., Franklin, F.C.H. and Marshall, D.F. (1999) Analysis of *Arabidopsis* genome sequence reveals a large new gene family in plants. *Plant Mol. Biol.*, **39,** 927–932.

Riechmann, J.L. and Ratcliffe, O.J. (2000) A genomic perspective on plant transcription factors. *Curr. Opin. Plant Biol.*, **3,** 423–434.

Robertson, H.M. (1998) Two large families of chemoreceptor genes in the Nematodes *Caenorhabditis elegans* and *Caenorhabditis briggsae* reveal extensive gene duplication, diversification, movement, and intron loss. *Genome Res.*, **8,** 449–463.

Rzhetsky, A., Ayala, F.J., Hsu, L.C., Chang, C. and Yoshida, A. (1997) Exon/intron structure of aldehyde dehydrogenase genes supports the 'introns-late' theory. *Proc. Natl. Acad. Sci. USA*, **94,** 6820–6825.

Sanderfoot, A.A., Assaad, F.F. and Raikhel, N.V. (2000) The *Arabidopsis* genome. An abundance of soluble N-ethylmaleimide-sensitive factor adaptor protein receptors. *Plant Physiol.*, **124,** 1558–1569.

Sankoff, D. (2001) Gene and genome duplication. *Curr. Opin. Genet. Dev.*, **11,** 681–684.

Small, I.D. and Peeters, N. (2000) The PPR motif - a TPR-related motif prevalent in plant organellar proteins. *Trends Biochem. Sci.*, **25,** 46–47.

The *C. elegans* Sequencing Consortium. (1998) Genome Sequence of the Nematode *C. elegans*. A Platform for Investigating Biology. *Science*, **282,** 2012–2027.

Theologis A. *et al*. (2000) Sequence and analysis of chromosome 1 of the plant *Arabidopsis thaliana*. *Nature*, **408,** 816–820.

Tavares, R., Aubourg, S., Lecharny, A. and Kreis, M. (2000) Organization and structural evolution of four multigene families in *Arabidopsis thaliana*: AtLCAD, AtLGT, AtMYST and AtHD-GL2. *Plant Mol. Biol.*, **42,** 703–717.

Tognolli, M., Overney, S., Penel, C., Greppin, H. and Simon, P. (2000) A genetic and enzymatic survey of *Arabidopsis thaliana* peroxidases. *Plant Perox. Newslett.*, **14,** 3–12.

Torki, M., Mandaron, P., Mache, R. and Falconet, D. (2000) Characterization of a ubiquitous expressed gene family encoding polygalacturonase in *Arabidopsis thaliana*. *Gene*, **242,** 427–436.

Trapp, S.C., and Croteau, R.B. (2001) Genomic organization of plant terpene synthases and molecular evolutionary implications. *Genetics*, **158,** 811–832.

Trotman, C.N.A. (1998) Introns-early: slipping lately? *Trends Genet.*, **14,** 132–134.

Vision, T.J., Brown, D.G. and Tanksley, S.D. (2000) The origin of genomic duplications in *Arabidopsis*. *Science*, **290,** 2114–2117.

Wattler, S., Russ, A., Evans, M. and Nehls, M. (1998) A combined analysis of genomic and primary protein structure defines the phylogenetic relationship of new members of the T-box family. *Genomics*, **48,** 24–33.

A. Meyer, Y. Van de Peer (eds.), Genome Evolution, 117-129.

Investigating ancient duplication events in the *Arabidopsis* genome

Jeroen Raes[†], Klaas Vandepoele[†], Cedric Simillion, Yvan Saeys & Yves Van de Peer*
Department of Plant Systems Biology, Flanders Interuniversity Institute for Biotechnology, Ghent University, K.L. Ledeganckstraat 35, B–9000 Gent, Belgium;
**Author for correspondence: E-mail yves.vandepeer@gengenp.rug.ac.be*
[†] The two authors contributed equally to this work.

Received 26.03.2002; accepted in final form 29.08.2002

Key words: large-scale gene duplications, plant genome evolution, polyploidy, synonymous substitution rate

Abstract

The complete genomic analysis of *Arabidopsis thaliana* has shown that a major fraction of the genome consists of paralogous genes that probably originated through one or more ancient large-scale gene or genome duplication events. However, the number and timing of these duplications still remains unclear, and several different hypotheses have been put forward recently. Here, we reanalyzed duplicated blocks found in the *Arabidopsis* genome described previously and determined their date of divergence based on silent substitution estimations between the paralogous genes and, where possible, by phylogenetic reconstruction. We show that methods based on averaging protein distances of heterogeneous classes of duplicated genes lead to unreliable conclusions and that a large fraction of blocks duplicated much more recently than assumed previously. We found clear evidence for one large-scale gene or even complete genome duplication event somewhere between 70 to 90 million years ago. Traces pointing to a much older (probably more than 200 million years) large-scale gene duplication event could be detected. However, for now it is impossible to conclude whether these old duplicates are the result of one or more large-scale gene duplication events.

Abbreviations: dA, fraction of amino acid substitutions; Kn, number of nonsynonymous substitutions per non-synonymous site; Ks, number of synonymous substitutions per synonymous site; MYA, million years ago

Introduction

For over 30 years, geneticists, evolutionists and, more recently, developmental biologists have been debating on the number of genome duplications in the evolution of animal lineages and its impact on major evolutionary transitions and morphological novelties. Thanks to the recent progress made in gene mapping studies and large-scale genomic sequencing, the debate has been livelier than ever before. Indeed, huge amounts of sequence data have become available, amongst which the complete genome sequences of invertebrates, such as *Drosophila melanogaster*, *Caenorhabditis elegans*, and vertebrates, such as pufferfish and human, while others are being finalized. With these data at our disposition, we expect to address the ancient questions and hypotheses regarding genome duplications, as formulated by pioneers like J.B.S. Haldane (who already contemplated the benefits and evolutionary impact of polyploidy events in 1933) and S. Ohno. However, a great deal of controversy still exists on the prevalence of genome duplications in certain lineages. For example, the classic hypothesis of Ohno (1970) that at least one genome duplication occurred in the evolution of the vertebrates has not been evidenced yet. Several theories, which differ in the proposed number of duplications as well as in their timing, have been proposed, but without confirmation (Skrabanek and Wolfe, 1998; Hughes, 1999; Wolfe, 2001). More recently, a putatively ancient fish-specific genome duplication before the teleost radiation has been the subject of lively debate (Robinson-Rechavi *et al.*, 2001; Taylor *et al.*, 2001a, 2001b; Van de Peer *et al.*, this issue). Given the already controversial nature of the occurrence and date of these genome duplications in vertebrates, their precise role in the evolution of new body plans (Holland, 1992) or in speciation (Lynch and Conery, 2001; Taylor *et al.*, 2001c) remains even more speculative.

For plants, controversy about ancient genome duplications has long been nearly nonexisting. Polyploidy seems to have occurred frequently in plants. Up to 80% of angiosperms are estimated to be polyploid, with variation from tetraploidy (maize) and hexaploidy (wheat) to 80-ploidy (*Sedum suaveolens*) (for a review, see Leitch *et al.*, 1997). Because of the complexity of many plant genomes and lack of sequence data, research on plant genome evolution was basically restricted to experimental techniques (Wendel, 2000) and, until very recently, few computational analyses had been performed to investigate the prevalence and timing of older large-scale duplications and their impact on plant evolution.

In 1996, the plant community decided to determine the complete genome sequence of *Arabidopsis thaliana*. This model plant was chosen because it has a small genome with a high gene density and seemed to be an 'innocent' diploid. However, during and even before this huge enterprise, some indications were found that large-scale duplications had occurred (Kowalski *et al.*, 1994; Paterson *et al.*, 1996; Terryn *et al*, 1999; Lin *et al.*, 1999; Mayer *et al.*, 1999). After bacterial artificial chromosome sequences representing approximately 80% of the genome had been analyzed, almost 60% of the genome was found to contain duplicated genes and regions (Blanc *et al.*, 2000). This phenomenon could only be explained by a complete genome duplication event, an opinion shared by the Arabidopsis Genome Initiative (2000). Previously, comparative studies of bacterial artificial chromosomes between *Arabidopsis* and soybean (Grant *et al.*, 2000) and between *Arabidopsis* and tomato (Ku *et al.*, 2000) had led to similar notions. In the latter study, two complete genome duplications were proposed: one 112 and another 180×10^6 years ago (MYA). Vision *et al.* (2000) rejected the single-genome duplication hypothesis by dating duplicated blocks through a molecular clock analysis. Several different age classes among the duplicated blocks were found, ranging from 50 to 220 MYA and at least four rounds of large-scale duplications were postulated. One of these classes, dated approximately 100 MYA, grouped nearly 50% of all the duplicated blocks, suggesting a complete genome duplication at that time (Vision *et al.*, 2000). However, the dating methods used for these gene duplications were based on averaging evolutionary rates of different proteins, which was later criticized because of their high sensitivity to rate differences (Sankoff, 2001; Wolfe, 2001). Because the same methodology was also used by Ku *et al.* (2000), their results should also be considered with caution. On the other hand, Vision *et al.* (2000) discovered overlapping blocks, a phenomenon that can be explained only by multiple duplication events. Neither Blanc *et al.* (2000) nor the Arabidopsis Genome Initiative (2000) detected these overlapping blocks.

Using a different method of dating based on the substitution rate of silent substitutions, Lynch and Conery (2000) discovered that most *Arabidopsis* genes had duplicated approximately 65 MYA, which brings us back to a single polyploidy event. However, no duplicated blocks of genes, but only paralogous gene pairs were taken into account.

Apparently, the evolutionary history of the first fully sequenced plant seems a lot more complex than originally expected. There is no clear answer on whether one single or multiple polyploidy events took place nor when they occurred. The results of the different analyses seem to be highly dependent of the methods used. For this reason, we reinvestigated the ancient large-scale gene duplications described by Vision *et al.* (2000) by applying two alternative dating methodologies on several of the more anciently duplicated blocks found in their study. Furthermore, we compared the results obtained to pinpoint the strengths and weaknesses of the methodology used in the two studies.

Materials and methods

Strategy

The original goal was to reinvestigate whether one or several ancient large-scale gene duplication(s) had occurred in the evolution of *Arabidopsis thaliana*. Furthermore, because Vision *et al.* (2000) dated one of the large-scale duplication events as approximately 200×10^6 years old, we were curious to see whether this event pre- or postdated the monocot-dicot split, which is estimated to have occurred at about that time: 170–235 MYA (Yang *et al.*, 1999) and

143–161 MYA (Wikström *et al.*, 2001). We focused on the blocks that according to Vision *et al.* (2000), originated during this ancient round of duplication and consisted of six regions in the genome (class F). We mapped these regions to a more up-to-date data set (see below) and subjected them to two dating methodologies: dating based on synonymous substitution rates and molecular phylogeny. The former was done with three different approaches to estimate synonymous substitution rates, namely those of Li (1993), of Nei and Gojobori (1986) and of Yang and Nielsen (2000). Molecular phylogeny-based dating was performed through the construction of evolutionary trees by the Neighbor-joining method (Saitou and Nei, 1987). By using these different approaches, the possibility of drawing wrong conclusions caused by weaknesses of one particular method is minimized.

However, during the course of this study, it became clear that the most ancient blocks described by Vision *et al.* (2000) contained genes that had duplicated much more recently. Because the dating methodology of Vision *et al.* (2000) had been criticized before (Sankoff, 2001; Wolfe, 2001), we subsequently focused on two sets of 10 blocks of two younger age classes, D and E, estimated to be 140 and 170×10^6 years old, respectively. These data sets were chosen in such a way that they represented a wide distribution in block size (number of anchor points) as well as amino acid substitution rate (dA) within each age class.

Data set of duplicated genes

From the complete set of segmentally duplicated blocks defined by Vision *et al.* (2000) that consisted of 103 regions with seven or more duplicated genes, we analyzed selected blocks covering the three oldest classes. This selection consisted of all six blocks from class F (200×10^6 years old), 10 from class E (170×10^6 years old) and 10 from class D (140×10^6 years old). Because the original data set (i.e., the chromosomal DNA sequences) represented a preliminary version of the *Arabidopsis* genome sequence (incomplete and not always correctly assembled), the positions of these duplicated blocks were transferred to a data set that had been built recently. This new data set consisted of a genome-wide non-redundant collection of *Arabidopsis* protein-encoding genes, which were predicted with GeneMark.hmm (Lukashin and Borodvsky, 1998; genome version of January 18th, 2000 (v180101), downloaded from the Institute for Protein Sequences center [Martiensried, Germany; ftp://ftpmips.gsf.de/cress/]). In addition to the protein sequence, the position and orientation of the genes within the *Arabidopsis* genome were determined.

Within this protein set, all pairs of homologous gene products between two chromosomes were determined and the result stored in a matrix of (m, n) elements (m and n being the total number of genes on a certain chromosome). Two proteins were considered as homologous if they had an E-value $< 1^{e-50}$ within a BLASTP (Altschul *et al.*, 1997) sequence similarity search (Friedman and Hughes, 2001).

The synchronization of our data set with the blocks detected by Vision *et al.* (2000) was done using their supplementary data (website: http://www.igd.cornell.edu/~tvision/arab/science_supplement.html). Initially, for a set of anchor points (i.e. pairs of duplicated genes), defining a duplicated block (Vision *et al.*, 2000), the corresponding protein couples were detected in our data set and then these protein couples were localized in the matrix. To check whether these proteins were indeed part of a segmentally duplicated block, an automatic and manual detection was performed. The automatic detection was done with a new tool (Vandepoele *et al.*, 2002), primarily based on discovering clusters of diagonally organized elements (representing duplicated blocks) within the matrix of homologous gene products. Similar to the strategy of Vision *et al.* (2000), tandem repeats were remapped before defining a duplicated block. An overview of blocks analyzed in this study, together with the number of anchor points per block, is presented in Table 1.

Dating based on Ks

Blocks of duplicated genes were dated using the NTALIGN program in the NTDIFFS software package (Conery and Lynch, 2001). This package first aligns the DNA sequence of two mRNAs based on their corresponding protein alignment and then calculates Ks by the method of Li (1993). We calculated Ks also with two alternative dating methodologies (Nei and Gojobori, 1986; Yang and Nielsen, 2000) based on the same alignments. These two methods are implemented in the PAML phylogenetic analysis package (Yang, 1997). The time since duplication was calculated as $T = Ks/2\lambda$, with λ being the mean rate of synonymous substitution; in *Arabidopsis* the estimation is $\lambda = 6.1$ synonymous subsitutions per 10^9 years (Lynch and Conery, 2000). The mean Ks value (average of the estimates obtained by the three

Table 1. Re-analysis of the duplicated blocks as described by Vision *et al.* (2000)

Vision *et al.* (2000)							This study					
Block number	Chr1[a]	Chr2[a]	Anchors	dA	Age class	Age in MY	Anchors[b]	Ks[c]	Ks[d]	Ks[e]	Mean age[f]	StdDev
15	1	3	7	0.8975	F	200	7	1.8641	2.5378	2.1679	213	92
25	1	5	7	0.8012	F	200	6	1.6757	1.7008	2.5515	160	27
37	1	5	11	0.8146	F	200	17	0.8386	0.8138	0.9698	72	19
39	1	3	8	0.8375	F	200	7	1.6053	1.9744	1.8768	170	62
57	2	3	7	0.8521	F	200	7	2.9251	3.2702	2.4395	269	64
59	2	5	15	0.8473	F	200	18	1.8078	2.3744	2.0642	191	70
34	1	5	23	0.7165	E	170	27	0.8723	0.8308	0.8900	71	18
71	3	5	31	0.6814	E	170	70	0.7933	0.8262	0.8312	67	19
100	4	5	20	0.6899	E	170	15	1.8656	1.9727	2.1682	170	45
78	3	5	26	0.701	E	170	35	0.7382	0.7551	0.8475	64	11
47	2	5	8	0.7397	E	170	8	1.8475	3.0169	2.1072	218	87
16	1	3	8	0.6562	E	170	7	0.8390	0.8536	1.0224	74	19
55	2	5	14	0.685	E	170	9	1.7585	2.0966	1.8341	162	32
9	1	3	24	0.6947	E	170	20	0.9098	0.9966	1.1350	83	20
87	3	4	11	0.7231	E	170	8	1.6049	1.8936	2.1889	164	67
48	2	3	11	0.7045	E	170	8	1.7175	1.9716	2.0465	162	56
6	1	5	30	0.6106	D	140	30	0.7754	0.8138	0.9228	69	17
30	1	3	92	0.5262	D	140	106	0.8047	0.8325	0.9668	71	20
95	4	5	88	0.5592	D	140	61	0.7337	0.7884	0.8707	65	10
17	1	1	153	0.5684	D	140	167	0.8110	0.8175	0.8983	69	18
92	4	5	97	0.6064	D	140	107	0.8741	0.8849	1.0507	77	25
33	1	4	18	0.5381	D	140	11	1.6283	1.6707	1.5669	133	26
5	1	4	13	0.5631	D	140	6	1.5232	1.5657	1.5324	126	16
73	3	5	26	0.5855	D	140	25	0.7965	0.8187	0.9105	69	15
93	4	5	42	0.6263	D	140	28	0.7719	0.8174	0.9010	68	16
26	1	4	35	0.5273	D	140	42	0.8719	0.8946	1.0867	78	23

[a] Chromosome numbers on which the two duplicated blocks are found.
[b] Number of anchor points in blocks detected in this study.
[c] Ks values calculated according to Li (1993).
[d] Ks values calculated according to Nei and Gojobori (1986).
[e] Ks values calculated according to Yang and Nielsen (2000).
[f] Mean age of the block was derived from the mean Ks, excluding outliers (see Materials and Methods).

methods) for each block was derived for each duplicated pair. These values were then used to calculate the mean Ks for each block, excluding outliers using the Grubbs test (Grubbs, 1969; Stefansky, 1972) with a 99% confidence interval.

Phylogenetic analysis

The public databases (PIR, GenBank/EMBL/DDBJ, Swiss-PROT) were scanned for homologues of the anchor points using BLASTP (Altschul *et al.*, 1997). When homologues were found in other species next to the *Arabidopsis* paralogues, the gene family was selected for phylogenetic analysis. Protein sequences were subsequently aligned with ClustalW (Thompson *et al.*, 1994). Duplicates or sequences that were too short were removed from the data set. After manual optimization of the alignment and reformatting using BioEdit (Hall, 1999) and ForCon (Raes and Van de Peer, 1999), the more conserved positions of the alignment were subjected to phylogenetic analysis. Trees were constructed based on Poisson or Kimura

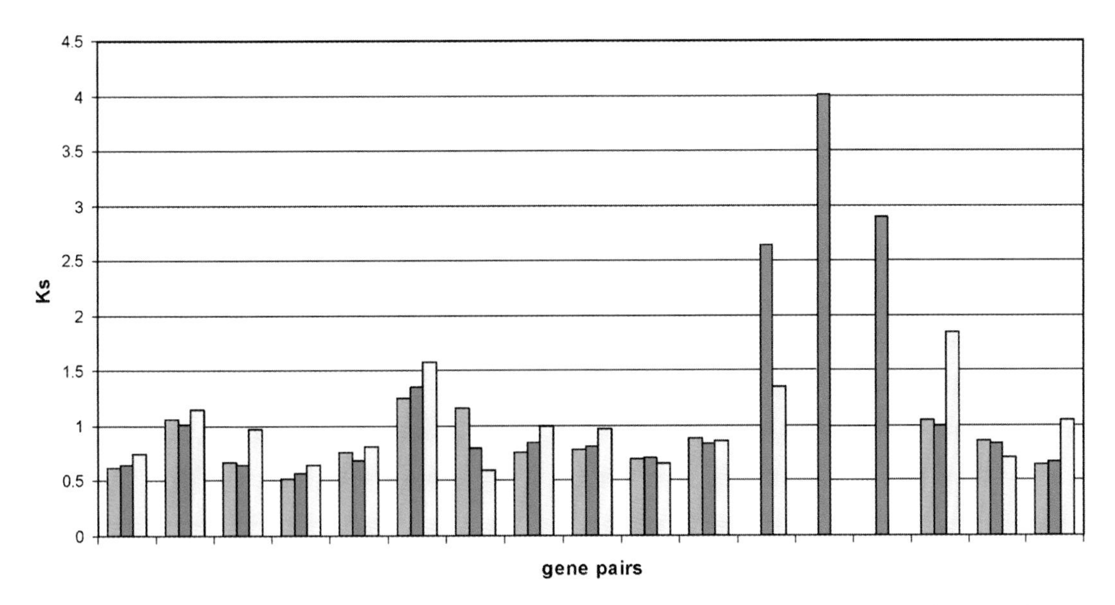

Figure 1. Distribution of Ks values for duplicated genes as found in block 37, and calculated with the methods of Li (green bars), Nei and Gojobori (blue bars) and Yang and Nielsen (yellow bars).

distances using the Neighbor-joining algorithm as implemented in the TREECON package (Van de Peer and De Wachter, 1997).

Supplementary data such as sequences, accession numbers, alignments, and trees can be obtained from the authors upon request.

Results

Dating based on Ks

In contrast to mutations that result in amino acid changes (nonsynonymous substitutions), silent or synonymous substitutions do not affect the biochemical properties of the protein. As such they are generally believed not to be subjected to natural selection and, consequently, to evolve in a (nearly) neutral, clock-like way (Li, 1997). Absolute dating based on synonymous substitution rates (Ks) should be more accurate than dating based on the estimation of genetic distances between duplicated protein sequences. However, because of rapid saturation of synonymous sites, dates of older (Ks > 1) divergences/ duplications will become unreliable (Li, 1997).

We calculated Ks values with three different methods for all pairs of duplicated genes in 26 old blocks (classes D, E, and F, estimated to have originated between 140 and 200 MYA; Vision *et al.*, 2000). From these values we calculated the duplication date of each block. The results of this analysis are given in Table 1.

Interestingly, several block duplications were dated to be much younger than what was found by Vision *et al.* (2000). For example, a duplication between chromosome 1 and 5, denoted as block 37 and based on 11 gene pairs (17 in our study; Table 1), was found to have occurred 72 MYA, and not 200 MYA. The distribution of the Ks values of the duplicated pairs in this block, calculated with the three different methods, confirmed our hypothesis that this is a younger block. With only a few exceptions, almost all duplicated pairs seemed to have Ks values between 0.5 and 1 synonymous substitutions per synonymous site, and this for the three methods used (Fig. 1). For three pairs of genes within the duplicated block, the situation is less clear (Fig. 1). No results were obtained with the method of Li (1993), probably because the duplicated gene sequences are too divergent to calculate a Ks value using this method, whereas the two other methods gave extremely high or no Ks values. One possible explanation is a higher synonymous mutation rate specific for these genes, because fluctuations in Ks have been reported before (Li, 1997; Zeng *et al.*, 1997). Another possible explanation could be that these genes originated earlier than the other genes in that block and that the situation observed is due to differential deletions of alter-

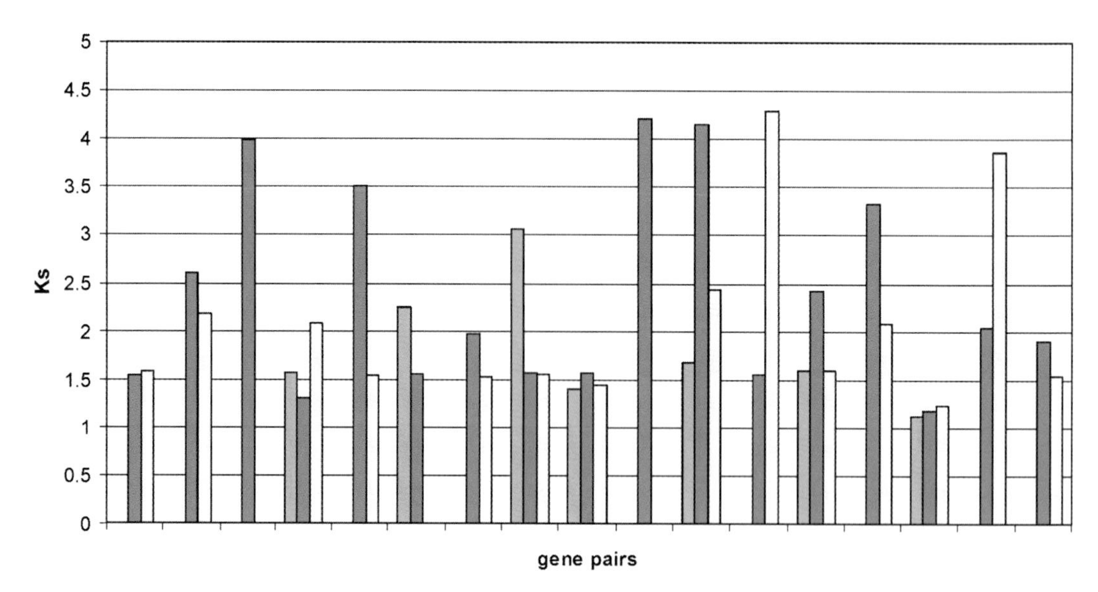

Figure 2. Distribution of Ks values for duplicated genes found in block 59, and calculated with the methods of Li (green bars), Nei and Gojobori (blue bars) and Yang and Nielsen (yellow bars).

nate members of duplicated tandem pairs (Friedman and Hughes, 2001). For this reason, these gene pairs were not included in the calculation of the duplication date of the whole block (see Materials and Methods).

However, most blocks of age class F had significantly higher Ks values and consequently older divergence dates, which indeed points to a more ancient large-scale duplication event. This observation was strengthened by the fact that, with a few exceptions, duplicated blocks of this age class had less anchor points (Table 1) and Ks values seemed to fluctuate more between members of the same block (see, for example, the distribution of block 59, estimated to have duplicated approximately 190 MYA; Fig. 2). The latter is probably due to saturation of synonymous substitutions, by which larger errors in Ks estimation are introduced, causing values of Ks > 1 to be unreliable.

In our evaluation of class E blocks (170 MYA; Vision *et al.*, 2000), the situation is even more peculiar. From the 10 blocks we selected, a large part again seemed to be much younger than what was derived based on dA values. Five out of 10 blocks seemingly originated only approximately 70 MYA, less than half the age calculated by Vision *et al.* (2000). Here also, the distribution of Ks values clearly showed that a large majority of duplicated pairs in these blocks belonged to the same, much younger, age class, with only a few exceptions (data not shown). However, the other half of the 10 selected blocks seem to be older.

In the class D sample, dated 140×10^6 years old by Vision *et al.* (2000), 8 out of 10 blocks seemed to have duplicated approximately 70 MYA. The distribution of Ks values within one block again gave similar results as above: most pairs had Ks values between 0.5 and 1, with a minor fraction of exceptions (data not shown).

Although only a subset of the complete set of duplicated blocks of age classes D and E were analyzed, many blocks appeared to be much younger than proposed by Vision and *et al.* (2000). Preliminary results of a more rigorous analysis seem to confirm our findings (unpublished results).

Dating by phylogenetic analysis

Absolute dating methods based on substitution numbers per site are very useful in high-throughput analyses, such as those by Lynch and Conery (2000) and Vision *et al.* (2000), but they have some serious drawbacks. Inferred divergence dates based on amino acid substitutions are not as quickly underestimated due to saturation, although saturation at the amino acid level has been demonstrated (Van de Peer *et al.*, 2002). However, when using this technique, there is a serious risk of overestimating the age of more rapidly

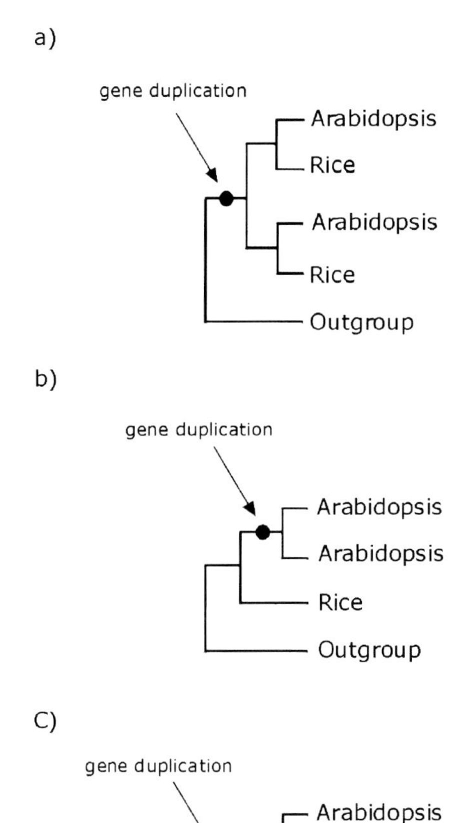

Figure 3. a) Expected tree topology for genes formed by a gene/genome duplication event prior to the split of monocots and dicots. b) Expected tree topology for genes formed by a gene/genome duplication event that occurred after the split of monocots and dicots and specific to *Arabidopsis*. c) Even if only one of the paralogues is known, due to gene loss or absence in the databases, the gene duplication can be inferred.

evolving blocks, or underestimating the age of blocks containing more slowly evolving proteins. The use of synonymous mutation rates is probably favorable because these positions evolve at nearly neutral rates and, so, give a more reliable estimate in the case of fast or slowly evolving genes. Unfortunately, these analyses are compromised for older duplications because of the rapid saturation of these sites.

To validate the results, an alternative technique was applied, namely relative dating using phylogenetic methods. If a duplication occurred before the monocot-dicot split, this could be proven by a tree topology (Fig. 3a), in which the two dicot members of a gene family each group with a monocot sequence. If, however, the two *Arabidopsis* duplicates originated more recently, i.e. after the dicot-monocot split, the two dicot branches should be sister sequences, outgrouped by their monocot orthologue (Fig. 3b). Even if certain sequences are still missing from the databases (because of gene loss or nondetection), conclusions can be drawn. For example, the tree topology presented in Figure 3c could only be explained by a duplication that occurred before the monocot-dicot split.

For all the anchor points of the oldest blocks (F), we searched the protein databases for homologues in other plant species to construct evolutionary trees. Unfortunately, it was impossible to construct trees for many of the duplicated genes, the main reason being the absence of homologues from plant species other than *Arabidopsis* in the databases. Furthermore, the sequences often contained too few conserved positions to get statistically significant results (i.e. high bootstrap values).

An overview of constructed trees and conclusions is presented in Table 2. Gene families for which no homologues from other species than *Arabidopsis thaliana* could be found in the databases are not shown.

Although we could not draw conclusions on many of the genes/blocks, we would like to consider some of the constructed trees. A first interesting result was obtained from the analysis of the gluthatione synthase gene family; it has two members on chromosomes 1 and 5 that are part of block 37, which is a duplicated block of class F (200 MYA; Vision *et al.*, 2000); but, according to our estimation, it had duplicated approximately 72 MYA. The tree topology (Fig. 4) for this family clearly showed that the duplication that yielded the two duplicates occurred before the divergence of *Arabidopsis* and *Brassica*, but after the split between Asteridae and Rosidae. In consequence, the duplication between these two genes must have happened between 15–20 (Yang *et al.*, 1999; Koch *et al*, 2001) and 135 MYA (the latter value being the mean of two estimations, 112–156 MYA [Yang *et al.*, 1999]) and 114–125 MYA [Wikström *et al.*, 2001]), which is in accordance with our findings for this block.

A second tree of interest is that of the GATA transcription factor family with a pair of duplicates on chromosomes 2 and 3 that belong to block 57, also of age class F. It was very hard to date this block with our dating methods, because the sequences were

Table 2. Gene families selected for phylogenetic analysis for each paralogous block, belonging to age class F (Vision *et al.*, 2000; 200 MYA)

Block[a]	Family[b]	Sites[c]	Conclusion	Reason
15	Unknown	279	None	No statistical support
25	-	None	No trees possible due to the absence of sequences from other species	
37	Calmodulin	105	None	No statistical support
	Calmodulin-like	112	Probably younger than the split between eurosids I and eurosids II	Genetic distance
	Glutamine synthase	314	Younger than the split with asteridae and older than the *Arabidopsis-Brassica* divergence (see Fig. 3)	Topology with statistical support
39	Unknown	287	None	Too few monocot sequences for this family
57	DOF Zinc-finger	85	None	Highly inequal rates of evolution between duplicates
	GATA transcription factor	148	Older than the monocot-dicot split (see Fig. 4)	Topology with statistical support
	Apetala 2	81	None	No statistical support
	Expansin	180	None	No statistical support
59	Protein phosphatase 2C	174	None	Too few monocot sequences available
	Putative Rab5 interacting protein	100	Probably younger than the monocot-dicot split	Genetic distance
	Cyclophilin	141	None	No statistical support
	Phosphoprotein phosphatase 1	305	None	No statistical support
	Apetala 2 (see also B57)	81	None	No statistical support

[a] Block number as defined by Vision *et al.* (2000).
[b] Name of the family analyzed, as far as could be deduced from the description line of the entries.
[c] Length of sequence alignment used for tree construction.

apparently saturated for synonymous substitutions. However, all Ks values calculated for pairs in this block were above 2.2 synonymous substitutions per synonymous site (see Table 1), suggesting that this block is genuinely old. When we investigated the topology of the GATA family (Fig. 5), we observed a topology similar to that described in Figure 3c: although there is only one monocot sequence, this topology could be only explained if the duplication that gave rise to the two *Arabidopsis* genes occurred before the monocot-dicot split. This would mean that this block occurred at least 190 MYA (Yang *et al.*, 1999; Wilkström *et al.*, 2001).

In some cases, evolutionary distances can be informative of duplication dates. As illustration, an example from the age class D (140 MYA; Vision *et al.*, 2000) is given. Figure 6 shows the topology of the casein kinase gene family that has two members on both chromosomes 1 and 5, all four of them belonging to the same duplicated block 6.

Using Ks-based dating, we determined that this block had duplicated approximately 70 MYA, with approximately 80% of the Ks values in this block being smaller than 1. As can be seen from the tree topology, the two members of block 6 first originated (probably) through tandem duplication (arrow 1) and then through a larger-scale duplication including the other members of that block (arrow 2). Both these events happened after the monocot-dicot split, as can be derived from the fact that the group containing these four proteins is outgrouped by a rice sequence. The evolutionary distance from each of the duplicates to the block duplication point is approximately 0.025 amino acid substitutions per site, whereas the evolutionary distance between the genes originating by tandem duplication is approximately 0.158 amino acid substitutions per site. The average evolutionary distance between the sequences of rice and *Arabidopsis* is approximately 0.206 amino acid substitutions per site, meaning that, if a divergence date for monocots and dicots of 190 MYA (Yang *et al.*, 1999; Wilkström *et al.*, 2001) and a molecular clock-like evolution of this protein were assumed, the block duplication would have happened somewhere

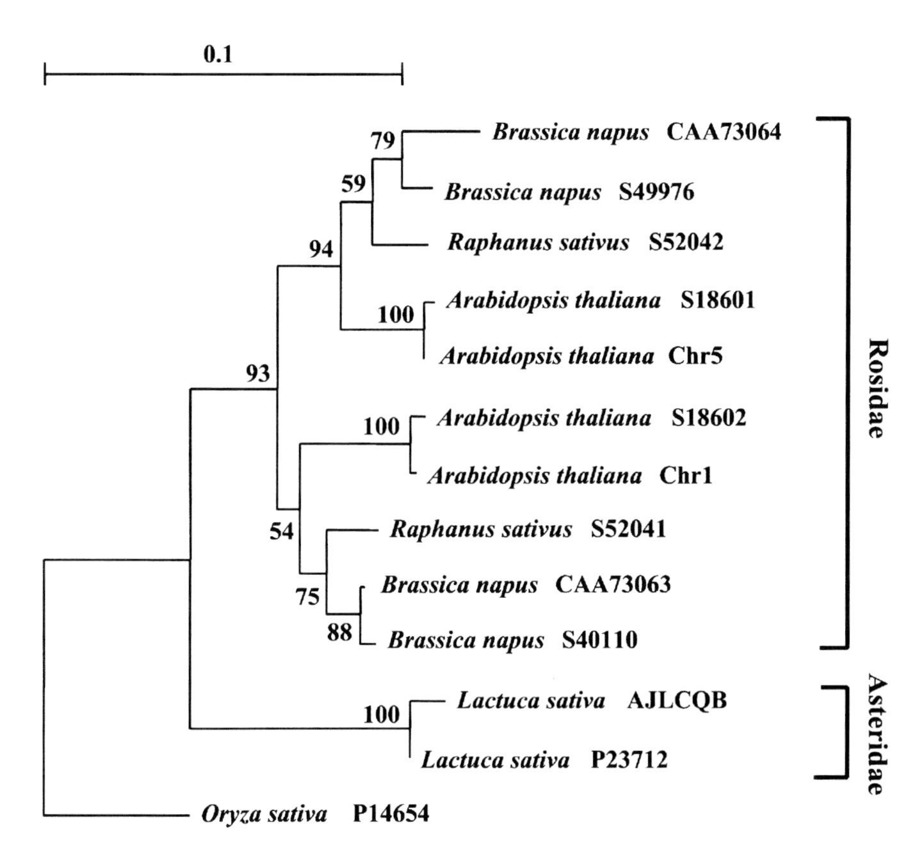

Figure 4. Neighbor-joining tree of the glutamine synthase family, inferred from Poisson-corrected evolutionary distances. Sequences that belong to the analyzed duplicated blocks are indicated with their chromosome number. Bootstrap values (above 50%) are shown in percentages at the internodes. Scale = evolutionary distance in substitutions per amino acid.

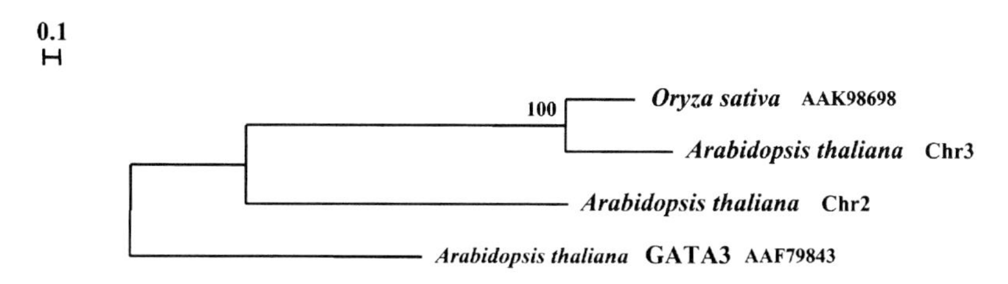

Figure 5. Neighbor-joining tree of the GATA family of transcription factors, inferred from Poisson corrected evolutionary distances. Sequences that belong to the analyzed duplicated blocks are indicated by their chromosome number. Bootstrap values (above 50%) are shown in percentages at the internodes. Scale = evolutionary distance in substitutions per amino acid.

46 MYA (with $\lambda = K/2T = 0.206$ substitutions per site/380 MY $= 5.42 \times 10^{-4}$ substitutions per site/MY). This value is much closer to our estimation based on Ks than that of 140 MYA obtained by Vision *et al.* (2000).

Discussion

Currently, three different methods to date gene duplication events are generally used: absolute dating based on synonymous substitution rates, absolute dating based on nonsynonymous substitution rates or protein-based distances, and relative dating through the construction of phylogenetic trees. Here, we provide some evidence that protein distances are not very reliable for large-scale dating of heterogeneous classes of proteins. For example, classes containing blocks of the same age based on mean protein distance (classes D, E, and F; Vision *et al.*, 2000) seem to be very heterogeneous in age when dating is based on synonymous substitution rates. Protein-based dis-

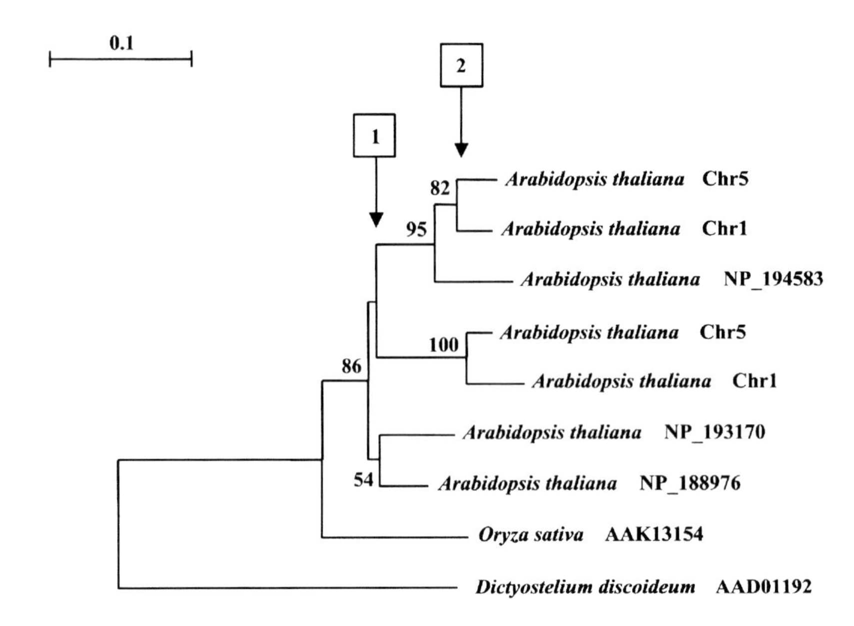

Figure 6. Neighbor-joining tree of the casein kinase family, using Poisson correction for evolutionary distance calculation. Sequences that belong to the analyzed duplicated blocks are indicated by their chromosome number. Arrows indicate (1) a tandem duplication and (2) the block duplication. Bootstrap values (above 50%) are shown in percentages at the internodes. Scale = evolutionary distance in substitutions per amino acid.

tances are known to vary considerably among proteins (e.g. Easteal and Collet, 1994); therefore, duplicated blocks that contain a larger fraction of fast-evolving genes will have a relatively high mean protein distance between the paralogous regions and appear older than they actually are. In our opinion, the use of synonymous and, consequently, neutral substitutions for evolutionary distance calculations is more reliable. However, there is one important caveat: dating based on silent substitutions can only be applied when $Ks < 1$. A $Ks > 1$ points to saturation of synonymous sites and can no longer be used to draw any reliable conclusions regarding the origin of duplicated genes or blocks. In this case, a solution could be relative dating with phylogenetic means. Although the dating is rather crude, it offers a way of determining duplication dates relative to known divergences. The main problem here, however, is the availability of plant sequence data. Only a few duplicated pairs had enough orthologues in the public databases to allow any conclusions to be drawn. Furthermore, if orthologues would be found, the sequences may not be very suitable for phylogenetic analysis. Consequently, it seems that phylogenetic inference cannot yet be as widely applied to plant as to animal genomes (e.g., Wang and Gu, 2000; Friedman and Hughes, 2001; Van de Peer *et al.*, 2001). However, as soon as more sequence data from key species such as mosses, ferns, and monocots, become available, this approach may become more useful.

From the three oldest age classes defined by Vision *et al.* (2000), only one (F) seems to contain many old duplicated blocks, whereas several blocks of the two other age classes have seemingly been duplicated approximately 70–90 MYA. In our opinion, the hypothesis of Vision *et al.* (2000) that at least four large-scale duplications have occurred is far from being proven. In contrast with the multimodal distribution of large-scale gene duplication, our results show that a major fraction of blocks has duplicated approximately at the same time and has probably originated by a complete genome duplication. On the other hand, a fraction of block duplications seems much older than the others. Unfortunately, because synonymous sites were saturated and trees were not reliable enough, these duplications could not be dated more accurately. Although these old duplicated blocks are scattered throughout the genome (Table 1), it is hard to prove that they are the result of a single duplication event.

The question of whether large-scale gene duplications have occurred before the divergence of monocots and dicots still remains to be answered. Some of these events are probably anterior to the monocotyldicotyl split, as suggested by the GATA transcription factor topology (Fig. 5). Large-scale gene duplication events prior to the monocot-dicot split may have led to the origin of flowering or even of seed plants: Duplications of (sets of) developmentally important genes could have given the opportunity to develop new reproductive organs and strategies and consequently cause reproductive isolation, which may have resulted in speciation. The ongoing accumulation of sequence data delivered by several plant expressed sequence tags and genome sequencing projects will provide the means to answer the questions regarding the prevalence and timing of gen(om)e duplications in the evolution of plants and will hopefully help elucidating the role of these events in the diversification and evolution of plant species.

Acknowledgments

The authors would like to thank Eric Bonnet, Sven Degroeve, and John S. Taylor for helpful discussions and Martine De Cock for help with the manuscript. K.V. and C.S. are indebted to the Vlaams Instituut voor de Bevordering van het Wetenschappelijk-Technologisch Onderzoek in de Industrie for a predoctoral fellowship. Y.V.d.P. was a Research Fellow of the Fund for Scientific Research (Flanders).

Note added in proof

Since acceptance of this paper, novel tools to identify heavily degenerated block duplications allowed us to find evidence for the recent genome duplication described in this study. The occurrence of two additional, but probably no more, ancient genome duplicatons in *Arabidopsis* was also demonstrated [Simillian, C., Vandepoele, K., Van Montagu, M.C.E., Zabeau, M., and Van de Peer, Y. (2002). The hidden-duplication past of *Arabidopsis thaliana. Proc. Natl. Acad. Sci. USA* **99**, 13627–13632].

References

Altschul, S.F., Madden, T.L., Schäffer, A.A., Zhang, J., Zhang, Z., Miller, W., and Lipman, D.J. (1997) Gapped BLAST and PSI-BLAST: a new generation of protein database search programs. *Nucleic Acids Res.*, **25,** 3389–3402.

The Arabidopsis Genome Initiative (2000) Analysis of the genome sequence of the flowering plant *Arabidopsis thaliana. Nature*, **408,** 796–815.

Blanc, G., Barakat, A., Guyot, R., Cooke, R., and Delseny, M. (2000) Extensive duplication and reshuffling in the Arabidopsis genome. *Plant Cell*, **12,** 1093–1101.

Conery, J.S., and Lynch, M. (2001) Nucleotide substitutions and the evolution of duplicate genes. In *Pacific Symposium on Biocomputing 2001* (Eds., Altman, R.B., Dunker, A.K., Hunter, L., Lauderdale, K. and Klein, T.E.), World Scientific, Singapore, pp. 167–178.

Easteal, S., and Collet, C. (1994) Consistent variation in amino-acid substitution rate, despite uniformity of mutation rate: protein evolution in mammals is not neutral. *Mol. Biol. Evol.*, **11,** 643–647.

Friedman, R., and Hughes, A.L. (2001) Pattern and timing of gene duplication in animal genomes. *Genome Res.*, **11,** 1842–1847.

Grant, D., Cregan, P., and Shoemaker, R.C. (2000) Genome organization in dicots: genome duplication in *Arabidopsis* and synteny between soybean and *Arabidopsis. Proc. Natl. Acad. Sci. USA*, **97,** 4168–4173.

Grubbs, F. (1969) Procedures for detecting outlying observations in samples. *Technometrics*, **11,** 1–21.

Haldane, J.B.S. (1933) The part played by recurrent mutation in evolution. *Am. Nat.*, **67,** 5–19.

Hall, T.A. (1999) BioEdit: a user-friendly biological sequence alignment editor and analysis program for Windows 95/98/NT. *Nucleic Acids Symp. Ser.*, **41,** 95–98.

Holland, P. (1992) Homeobox genes in vertebrate evolution. *BioEssays*, **14,** 267–273.

Hughes, A.L. (1999) Phylogenies of developmentally important proteins do not support the hypothesis of two rounds of genome duplication early in vertebrate history. *J. Mol. Evol.*, **48,** 565–576.

Koch, M., Haubold, B., and Mitchell-Olds, R. (2001) Molecular systematics of the Brassicaceae: evidence from coding plastidic *matK* and nuclear *Chs* sequences. *Am. J. Bot.*, **88,** 534–544.

Kowalski, S.P., Lan, T.-H., Feldmann, K.A., and Paterson, A.H. (1994) Comparative mapping of *Arabidopsis thaliana* and *Brassica oleracea* chromosomes reveals islands of conserved organization. *Genetics*, **138,** 499–510.

Ku, H.-M., Vision, T., Liu, J., and Tanksley, S.D. (2000) Comparing sequenced segments of the tomato and *Arabidopsis* genomes: large-scale duplication followed by selective gene loss creates a network of synteny. *Proc. Natl. Acad. Sci. USA*, **97,** 9121–9126.

Leitch, I.J., and Bennett, M.D. (1997) Polyploidy in angiosperms. *Trends Plant. Sci.*, **2,** 470–476.

Li, W.-H. (1993) Unbiased estimation of the rates of synonymous and nonsynonymous substitution. *J. Mol. Evol.*, **36,** 96–99.

.Li, W.-H. (1997) *Molecular Evolution*, Sinauer Associates, Sunderland, MA.

Lin, X., Kaul, S., Rounsley, S., Shea, T.P., Benito, M.-I., Town, C.D., Fujii, C.Y., Mason, T., Bowman, C.L., Barnstead, M., Feldblyum, T.V., Buell, C.R., Ketchum, K.A., Lee, J., Ronning, C.M., Koo, H.L., Moffat, K.S., Cronin, L.A., Shen, M., Pai, G., Van Aken, S., Umayam, L., Tallon, L.J., Gill, J.E., Adams, M.D., Carrera, A.J., Creasy, T.H., Goodman, H.M., Somerville, C.R., Copenhaver, G.P., Preuss, D., Nierman, W.C., White, O., Eisen, J.A., Salzberg, S.L., Fraser, C.M., and Venter, J.C.

(1999) Sequence and analysis of chromosome 2 of the plant *Arabidopsis thaliana*. *Nature*, **402,** 761–768.

Lukashin, A.V., and Borodovsky, M. (1998) GeneMark.hmm: new solutions for gene finding. *Nucleic Acids Res.*, **26,** 1107–1115.

Lynch, M., and Conery, J.S. (2000) The evolutionary fate and consequences of duplicate genes. *Science*, **290,** 1151–1155.

Mayer, K., Schüller, C., Wambutt, R., Murphy, G., Volckaert, G., Pohl, T., Düsterhöft, A., Stiekema, W., Entian, K.-D., Terryn, N., Harris, B., Ansorge, W., Brandt, P., Grivell, L., Rieger, M., Weichselgartner, M., de Simone, V., Obermaier, B., Mache, R., Müller, M., Kreis, M., Delseny, M., Puigdomenech, P., Watson, M., Schmidtheini, T., Reichert, B., Portatelle, D., Perez-Alonso,, M., Boutry, M., Bancroft, I., Vos, P., Hoheisel, J., Zimmermann, W., Wedler, H., Ridley, P., Langham, S.-A., McCullagh, B., Bilham, L., Robben, J., Van der Schueren, J., Grymonprez, B., Chuang, Y.-J., Vandenbussche, F., Braeken, M., Weltjens, I., Voet, M., Bastiaens, I., Aert, R., Defoor, E., Weitzenegger, T., Bothe, G., Ramsperger, U., Hilbert, H., Braun, M., Holzer, E., Brandt, A., Peters, S., van Staveren, M., Dirkse, W., Mooijman, P., Klein Lankhorst, R., Rose, M., Hauf, J., Kötter, P., Berneiser, S., Hempel, S., Feldpausch, M., Lamberth, S., Van den Daele, H., De Keyser, A., Buysschaert, C., Gielen, J., Villarroel, R., De Clercq, R., Van Montagu, M., Rogers, J., Cronin, A., Quail, M., Bray-Allen, S., Clark, L., Foggett, J., Hall, S., Kay, M., Lennard, N., McLay, K., Mayes, R., Pettett, A., Rajandream, M.-A., Lyne, M., Benes, V., Rechmann, S., Borkova, D., Blöcker, H., Scharfe, M., Grimm, M., Löhnert, T.-H., Dose, S., de Haan, M., Maarse, A., Schäfer, M., Müller-Auer, S., Gabel, C., Fuchs, M., Fartmann, B., Granderath, K., Dauner, D., Herzl, A., Neumann, S., Argiriou, A., Vitale, D., Liguori, R., Piravandi, E., Massenet, O., Quigley, F., Clabauld, G., Mündlein, A., Felber, R., Schnabl, S., Hiller, R., Schmidt, W., Lecharny, A., Aubourg, S., Chefdor, F., Cooke, R., Berger, C., Montfort, A., Casacuberta, E., Gibbons, T., Weber, N., Vandenbol, M., Bargues, M., Terol, J., Torres, A., Perez-Perez, A., Purnelle, B., Bent, E., Johnson, S., Tacon, D., Jesse, T., Heijnen, L., Schwarz, S., Scholler, P., Heber, S., Francs, P., Bielke, C., Frishman, D., Haase, D., Lemcke, K., Mewes, H.W., Stocker, S., Zaccaria, P., Bevan, M., Wilson, R.K., de la Bastide, M., Habermann, K., Parnell, L., Dedhia, N., Gnoj, L., Schutz, K., Huang, E., Spiegel, L., Sehkon, M., Murray, J., Sheet, P., Cordes, M., Abu-Threideh, J., Stoneking, T., Kalicki, J., Graves, T., Harmon, G., Edwards, J., Latreille, P., Courtney, L., Cloud, J., Abbott, A., Scott, K., Johnson, D., Minx, P., Bentley, D., Fulton, B., Miller, N., Greco, T., Kemp, K., Kramer, J., Fulton, L., Mardis, E., Dante, M., Pepin, K., Hillier, L., Nelson, J., Spieth, J., Ryan, E., Andrews, S., Geisel, C., Layman, D., Du, H., Ali, J., Berghoff, A., Jones, K., Drone, K., Cotton, M., Joshu, C., Antonoiu, B., Zidanic, M., Strong, C., Sun, H., Lamar, B., Yordan, C., Ma, P., Zhong, J., Preston, R., Vil, D., Shekher, M., Matero, A., Shah, R., Swaby, I'K., O'Shaughnessy, A., Rodriguez, M., Hoffman, J., Till, S., Granat, S., Shohdy, N., Hasegawa, A., Hameed, A., Lodhi, M., Johnson, A., Chen, E., Marra, M., Martienssen, R., and McCombie, W.R. (1999) Sequence and analysis of chromosome 4 of the plant *Arabidopsis thaliana*. *Nature*, **402,** 769–777.

Nei, M., and Gojobori, T. (1986) Simple methods for estimating the numbers of synonymous and nonsynonymous nucleotide substitutions. *Mol. Biol. Evol.*, **3,** 418–426.

Ohno, S. (1970) *Evolution by Gene Duplication*, Springer-Verlag, Berlin.

Paterson, A.H., Lan, T.-H., Reischmann, K.P., Chang, C., Lin, Y.-R., Liu, S.-C., Burow, M.D., Kowalski, S.P., Katsar, C.S., DelMonte, T.A., Feldmann, K.A., Schertz, K.F., and Wendel, J.F. (1996) Toward a unified genetic map of higher plants, transcending the monocot—dicot divergence. *Nature Genet.*, **14,** 380–382.

Raes, J., and Van de Peer, Y. (1999) ForCon: a software tool for the conversion of sequence alignments. *EMBnet.news*, 6 (http://www.ebi.ac.uk/embnet.news/vol6_1).

Robinson-Rechavi, M., Marchand, O., Escriva, H., and Laudet, V. (2001) An ancestral whole-genome duplication may not have been responsible for the abundance of duplicated fish genes. *Curr. Biol.*, **11,** R458-R459.

Saitou, N., and Nei, M. (1987) The neighbor-joining method: a new method for reconstructing phylogenetic trees. *Mol. Biol. Evol.*, **4,** 406–425.

Sankoff, D. (2001) Gene and genome duplication. *Curr. Opin. Genet. Dev.*, **11,** 681–684.

Skrabanek, L., and Wolfe, K.H. (1998) Eukaryote genome duplication - where's the evidence? *Curr. Opin. Genet. Dev.*, **8,** 694–700.

Stefansky, W. (1972) Rejecting outliers in factorial designs. *Technometrics*, **14,** 469–479.

Taylor, J.S., Van De Peer, Y., Braasch, I., and Meyer, A. (2001a) Comparative genomics provides evidence for an ancient genome duplication event in fish. *Phil. Trans. R. Soc. Lond.*, **B 356,** 1661–1679.

Taylor, J.S., Van de Peer, Y., and Meyer, A. (2001b) Genome duplication, divergent resolution and speciation. *Trends Genet.*, **17,** 299–301.

Taylor, J.S., Van de Peer, Y., and Meyer, A. (2001c) Revisiting recent challenges to the ancient fish-specific genome duplication hypothesis. *Curr. Biol.*, **11,** R1005-R1007.

Terryn, N., Heijnen, L., De Keyser, A., Van Asseldonck, M., De Clercq, R., Verbakel, H., Gielen, J., Zabeau, M., Villarroel, R., Jesse, T., Neyt, P., Hogers, R., Van den Daele, H., Ardiles, W., Schueller, C., Mayer, K., Déhais, P., Rombauts, S., Van Montagu, M., Rouzé, P., and Vos, P. (1999) Evidence for an ancient chromosomal duplication in *Arabidopsis thaliana* by sequencing and analyzing a 400-kb contig of the *APETALA2* locus on chromosome 4. *FEBS Lett.*, **445,** 237–245.

Thompson, J.D., Higgins, D.G., and Gibson, T.J. (1994) CLUSTAL W: improving the sensitivity of progressive multiple sequence alignment through sequence weighting, position-specific gap penalties and weight matrix choice. *Nucleic Acids Res.*, **22,** 4673–4680.

Van de Peer, Y., and De Wachter, R. (1994) TREECON for Windows: a software package for the construction and drawing of evolutionary trees for the Microsoft Windows environment. *Comput. Appl. Biosci.*, **10,** 569–570.

Van de Peer, Y., Taylor, J.S., Braasch, I., and Meyer, A. (2001) The ghost of selection past: rates of evolution and functional divergence of anciently duplicated genes. *J. Mol. Evol.*, **53,** 436–446.

Van de Peer, Y., Frickey, T., Taylor, J.S., and Meyer, A. (2002) Dealing with saturation at the amino acid level: a case study based on anciently duplicated zebrafish genes. *Gene*, **295**, 205–211.

Vandepoele, K., Saeys, Y., Simillion, C., Raes, J., and Van de Peer, Y. (2002) A new tool for the Automatic Detection of Homologous Regons (ADHoRe) and its application to microcolinearity between *Arabidopsis* and rice. *Genome Res.*, in press.

40. Vision, T.J., Brown, D.G., and Tanksley, S.D. (2000) The origins of genomic duplications in *Arabidopsis*. *Science*, **290,** 2114–2117.

Wang, Y., and Gu, X. (2000) Evolutionary patterns of gene families generated in the early stage of vertebrates. *J. Mol. Evol.*, **51,** 88–96.

Wendel, J.F. (2000) Genome evolution in polyploids. *Plant Mol. Biol.*, **42,** 225–249.

Wikström, N., Savolainen, V., and Chase, M.W. (2001) Evolution of the angiosperms: calibrating the family tree. *Proc. R. Soc. Lond.*, **B 268,** 2211–2220.

Wolfe, K.H. (2001) Yesterday's polyploids and the mystery of diploidization. *Nature Rev. Genet.*, **2,** 333–341.

Yang, Y.-W., Lai, K.-N., Tai, P.-Y., and Li, W.-H. (1999) Rates of nucleotide substitution in angiosperm mitochondrial DNA sequences and dates of divergence between *Brassica* and other angiosperm lineages. *J. Mol. Evol.*, **48,** 597–604.

Yang, Z. (1997) PAML: a program package for phylogenetic analysis by maximum likelihood. *Comput. Appl. Biosci.*, **13,** 555–556.

Yang, Z., and Nielsen, R. (2000) Estimating synonymous and nonsynonymous substitution rates under realistic evolutionary models. *Mol. Biol. Evol.*, **17,** 32–43.

Zeng, L.-W., Comeron, J.M., Chen, B., and Kreitman, M. (1998) The molecular clock revisited: the rate of synonymous vs. replacement change in *Drosophila*. *Genetica*, **103,** 369–382.

A. Meyer, Y. Van de Peer (eds.), Genome Evolution, 131-137.

Crystallin genes: specialization by changes in gene regulation may precede gene duplication

Joram Piatigorsky
Laboratory of Molecular and Developmental Biology, National Eye Institute, NIH, Bethesda, Maryland 20892–2730, USA
e-mail: joramp@nei.nih.gov

Received 29.01.2002; accepted in final form 29.08.2002

Key words: crystallins, enzyme-crystallins, gene regulation, gene sharing, lens evolution, Pax–6, small heat shock proteins

Abstract

The crystallins account for 80–90% of the water-soluble proteins of the transparent lens. These diverse proteins are responsible for the optical properties of the lens and have been recruited from metabolic enzymes and stress proteins. They often differ among species (i.e. are taxon-specific) and may be expressed outside of the lens where they have non-refractive roles (a situation we call gene sharing). Crystallin recruitment has occurred by changes in gene regulation resulting in high lens expression. Duck lactate dehydrogenase/ϵ-crystallin and α-enolase/τ-crystallin are each encoded in single-copy genes, consistent with these enzymes acquiring a crystallin role, without loss of their nonlens metabolic function, by a change in gene regulation in the absence of gene duplication. The small heat shock protein/α-crystallins and avian argininosuccinate lyase/δ-crystallins were also recruited by a change in gene regulation leading to high lens expression, except this was followed by a gene duplication with further lens specialization of the αA and the δ1 (in chickens) crystallin genes. Cephalopod (squid and octopus) S-crystallins were recruited from glutathione S-transferase apparently after duplication of the original gene encoding the enzyme, although this remains uncertain. We speculate that one of the new genes (glutathione S-transferase/S11-crystallin) specialized for lens expression by a change in gene regulation and subsequently duplicated many times to form the lens-specialized, multiple S-crystallins that lack enzymatic activity. That similar transcription factors (e.g. Pax–6, retinoic acid receptors, maf, Sox, AP–1, CREB) regulate different crystallin genes suggest that common features of lens-specific expression have played a pivotal role for recruiting the diverse, multifunctional proteins as crystallins.

Introduction

Complex eyes with cellular lenses exist from jellyfish to man (Tomarev and Piatigorsky, 1996). In most cases the primary job of the transparent lens is to focus an image on the photoreceptor cells of the eye, although simply gathering light may also be important, especially in lenses of some invertebrates. Lenses from different animals share many properties, such as loss of organelles and, especially, accumulation of water-soluble proteins called crystallins, which account for approximately 90% of the water-soluble proteins of eye lenses. The crystallins contribute to transparency and affect refraction by forming a uniform concentration gradient, with the highest protein concentration being present at the center of the lens. Since the structure and function of eye lenses are surprisingly similar among animals, it was assumed that their crystallins would be correspondingly similar. Comparative studies, however, have revealed unexpected heterogeneity and diversity among lens crystallins (Wistow and Piatigorsky, 1988). Indeed, many species use different proteins in their lenses as crystallins, a situation called taxon-specificity. Crystallin diversity differs markedly from the divergent evolution of similar proteins for specialized tasks, such as oxygen transport and hemoglobin, or phototransduction and rhodopsin. While the reasons for crystallin diversity are not established, the mechanisms by which such diversity occurred are of evolutionary interest.

Crystallins are borrowed proteins: the concept of gene sharing

Surprisingly, α-crystallins, present in all vertebrate lenses, are homologous to the small heat shock proteins of *Drosophila* (Ingolia and Craig, 1982) and can act as molecular chaperones to protect against physiological stress (Horwitz, 1992). Subsequently it was found that the taxon-specific crystallins are similar or identical to metabolic enzymes. For example, ϵ-crystallin in ducks and crocodiles is lactate dehydrogenase (Wistow *et al.*, 1987), τ-crystallin in lampreys and turtles is α-enolase, and δ-crystallin in birds and reptiles is argininosuccinate lyase (Wistow and Piatigorsky, 1987). More comprehensive lists and reviews of lens crystallins can be found elsewhere (Wistow and Piatigorsky, 1988; de Jong *et al.*, 1994; Piatigorsky, 1998).

In many cases in vertebrate lenses, the taxon-specific enzyme-crystallins have enzyme activity. Moreover, the ubiquitous (in vertebrates) α-crystallins can act as molecular chaperones and protect against physiological stress (Horwitz, 1992), with αB-crystallin being stress-inducible and an active member of the small heat shock protein family (Klemenz *et al.*, 1991; de Jong *et al.*, 1993; Sax and Piatigorsky, 1994). Consistent with the idea that these lens crystallins have additional functions, they are also expressed in lesser amounts outside of the lens. The dual roles of many lens crystallins – namely, a structural protein affecting refractive index in the lens and a metabolic enzyme or stress protective protein outside of the lens – has been called gene sharing (Piatigorsky *et al.*, 1988; Piatigorsky and Wistow, 1989).

It is believed that the α-crystallins/small heat shock proteins function as protective chaperones as well as refractive proteins in the lens. However, the non-refractive lens role for crystallins that are known to contain enzyme activity is less certain. For example, δ-crystalllin/argininosuccinate lyase purified from duck lenses contains very high enzymatic activity, while δ-crystallin purified from chicken lenses has relatively little enzyme activity (Piatigorsky *et al.*, 1988; Piatigorsky and Horwitz, 1996). At present it is not known whether the duck lens, in contrast to the chicken lens, requires excessive argininosuccinate lyase activity, or whether the high argininosuccinate activity of δ-crystallin is unnecessary for the duck lens. Indeed, the abundance of any enzyme-crystallin in the lens is not consistent with it having a catalytic role. Thus, the extents to which enzyme-crystallins have metabolic roles in the lens need further investigation.

Invertebrate crystallins

Invertebrate lenses also contain crystallins that are homologous to metabolic enzymes. S-crystallins of cephalopods are related to glutathione S-transferase (Wistow and Piatigorsky, 1987; Tomarev and Zinovieva, 1988), and Ω-crystallin of cephalopods (Zinovieva *et al.*, 1991; Montogmery and McFall-Ngai, 1991) and scallops (Piatigorsky *et al.*, 2000) is homologous to aldehyde dehydrogenase class 1 proteins. Unlike the vertebrate enzyme-crystallins, glutathione S-transferase/S-crystallins and aldehyde dehydrogenase/Ω-crystallins of invertebrates appear enzymatically inactive. As described below, the multiple S-crystallins have lost enzymatic activity by exon shuffling and amino acid substitutions. Ω-Crystallin in scallops has lost the ability to bind NAD(H) and consequently has been found inactive for all substrates tested (Piatigorsky *et al.*, 2000). Possibly the inactivity of these invertebrate enzyme-crystallins is a consequence of more ancient recruitment for a crystallin role in the lens.

Thus, the principle of accumulating proteins with non-lens functions as lens crystallins began early in eye evolution in invertebrates. Jellyfish are particularly striking in this respect. *Tripedalia cystophora* is a cubomedusan jellyfish with surprisingly complex eyes called ocelli; the ocelli of this species have cellular lenses containing 3 different crystallins (J1, J2 and J3-crystallin) (Piatigorsky *et al.*, 1989). J1-crystallins are a family of 3, extremely similar, novel proteins (Piatigorsky *et al.*, 1993). Recently we have cloned the cDNA and single-copy gene for J3-crystallin and found that the deduced protein is homologous to the saposin protein family (Piatigorsky *et al.*, 2001). Saposins are conserved proteins that act as mediators between membrane lipids and lysosomal hydrolases and activate enzyme activity (Vaccaro *et al.*, 1999). In addition to being expressed abundantly

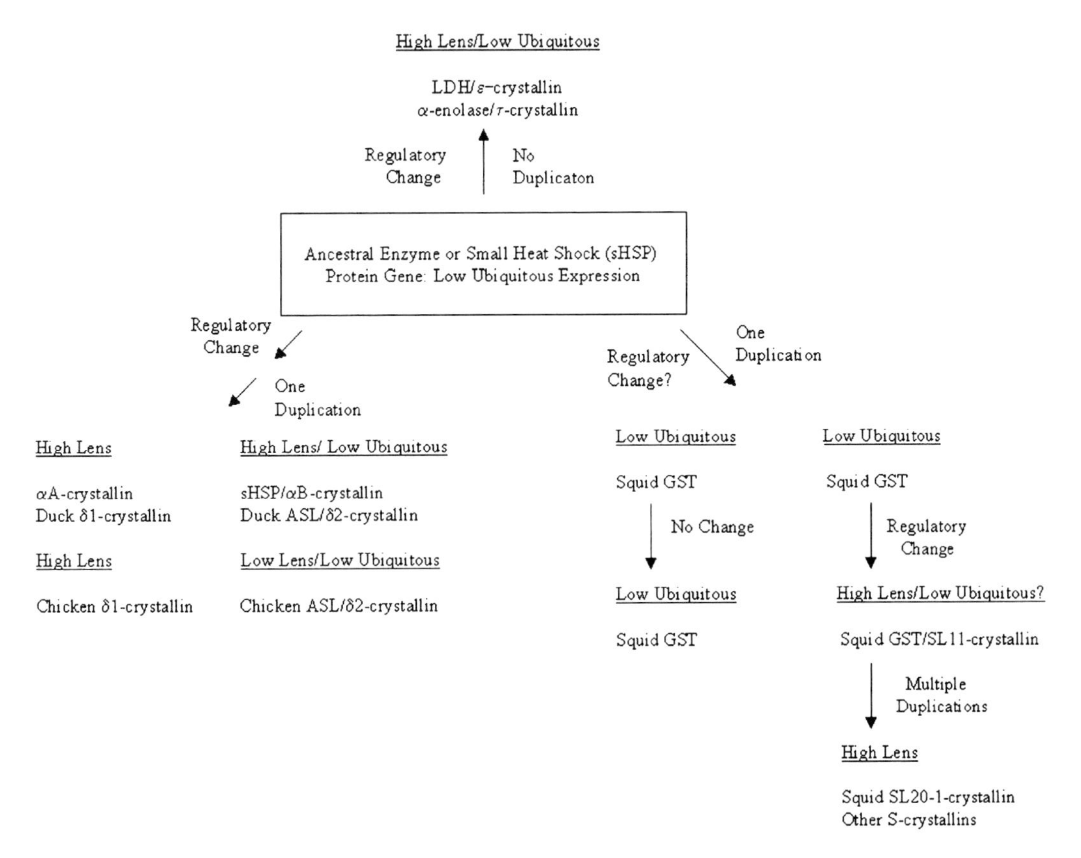

Figure 1. Diagrammatic representation of different evolutionary schemes used for crystallin gene recruitment. The central box represents the gene for a generic enzyme or small heat shock protein that was recruited as a lens crystallin. It is not known whether there was an initial change in gene regulation for squid glutathione S-transferase before gene duplication during its recruitment as a crystallin (lower right scheme). It is written as if gene duplication preceded a change in regulation of one of the new genes. However, it is possible that the regulatory change for high lens expression occurred before duplication, as is depicted in the other schemes, with a subsequent loss of lens specialization in one of the duplicates during further evolution. Such a loss of high lens expression occurred in the chicken argininosuccinate lyase/δ2-crystallin gene. Abbreviations: LDH, lactate dehydrogenase; sHSP, small heat shock protein; ASL, argininosuccinate lyase; GST, glutathione *S*-transferase.

in the jellyfish lens, J3-crystallin mRNA is also associated with the pigmented regions of the jellyfish ocellus as well as being present at the tips of the tentacles and in the cells surrounding the lumen of the statocyst in adult jellyfish. It is not known yet whether J3-crystallin acts as a saposin or has another function in the jellyfish, however the data suggest strongly that it has both a crystallin-like role in the lens and a non-refractive function in other tissues.

The role of gene duplication in crystallin evolution

Crystallin recruitment from ubiquitously expressed genes has occurred both with and without gene duplication (Fig. 1). Tandem Gene duplication has taken place among some of the crystallins, with one of the duplicates having specialized for lens expression. Two examples of this are the ubiquitous α-crystallins and bird δ-crystallins. α-Crystallin gene duplication occurred at least 500 million years ago (de Jong *et al.*, 1993). At present the two α-crystallin genes (αA and αB) are situated on different chromosomes in mice and humans. αB-crystallin gene has remained a functional small heat shock protein, is expressed constitutively in many tissues, including the lens, and is stress-inducible. By contrast, αA-crystallin is highly specialized for lens expression and is not stress-inducible, although it is expressed at low concentrations in a small number of tissues (Sax and Piatigorsky, 1994). Despite their differences, both of the α-crystallin genes are expressed much more highly in the lens than in any other tissue. This suggests strongly that high lens expression for a refractive role was initiated before gene duplication, implying that high lens expression and consequent evolution of a refractive role for the ancestral small heat shock protein preceded duplication of its gene.

The two, tandemly arranged δ-crystallin genes (δ1 and δ2) of ducks and chicken have been derived by duplication of the argininosuccinate lyase gene (Piatigorsky *et al.*, 1988). The expression pattern of δ1 and δ2 differs markedly in these two species (Piatigorsky *et al.*, 1988; Wistow and Piatigorsky, 1990). The enzymatically inactive δ1-crystallin comprises at least 99% of chicken δ-crystallin in the lens. It is likely that loss of enzymatic activity by δ1-crystallin occurred secondarily after gene duplication since argininosuccinate lyase/δ2-crystallin and δ1-crystallin are equally represented in the duck lens. Duck and chicken δ-crystallins have extremely similar sequences and exon/intron gene structures (Piatigorsky *et al.*, 1987; Wistow and Piatigorsky, 1990). Thus, specialization for high lens expression and a crystallin role was probably initiated before gene duplication for δ-crystallin as for α-crystalllin.

The most compelling cases for lens specialization before gene duplication are the enzyme-crystallins that appear to have not duplicated at all. Two examples of this are lactate dehydrogenase/ϵ-crystallin (Hendriks *et al.*, 1988) and α-enolase/τ-crystallin (Wistow *et al.*, 1989) in the duck. These enzymes are encoded in single-copy genes in the duck and thus are playing dual roles in this species. It remains possible, of course, that there are two extremely similar genes for these enzymes in the duck that have not been resolved yet, or that an early duplication took place with subsequent loss of one of the genes.

The cephalopod S-crystallins were derived from glutathione-S-transferase and are encoded in over twenty different genes (Tomarev *et al.*, 1995). Except for one S-crystallin polypeptide, all have lost enzymatic activity. This loss was due to the acquisition of an extra peptide by exon shuffling and numerous amino acid substitutions (Tomarev *et al.*, 1992). The authentic, enzymatically active glutathione S-transferase of octopus has been cloned and is expressed little, if at all, in the lens (Tomarev *et al.*, 1993). We have speculated that the cephalopod glutathione S-transferase gene duplicated and that one of the duplicates was recruited to be a lens crystallin by elevating its lens expression (Tomarev *et al.*, 1995). Additional duplications of the original crystallin gene would have occurred, followed by further changes in the duplicated genes resulting in loss of activity of the encoded lens proteins. All the duplicated S-crystallin genes that have been tested are expressed exclusively in the lens. The first of the recruited S-crystallin genes kept its original gene structure and some enzymatic activity. Another scenario that seems less likely but cannot be excluded, is that the authentic glutathione S-transferase gene became highly expressed in the lens, underwent numerous duplications, and subsequently lost most if not all of its lens expression during further evolution.

Other crystallin genes have undergone multiple duplications. There are at least six to eight β-crystallin and an equivalent number of γ-crystallin genes in vertebrates (Wistow and Piatigorsky, 1988). These are not represented in Fig. 1 since less can be inferred about their ancestral genes. The β/γ-crystallins are structurally related and belong to the same protein superfamily. They appear to have been derived from stress-related genes (Wistow and Piatigorsky, 1988), and some members are expressed outside of the lens (Magabo *et al.*, 2000). Consequently, it is likely that the β/γ-crystallins have a non-refractive function that has not been identified yet. The multiple β- and γ-crystallin genes are expressed predominantly in the lens, although the expression patterns of the different genes within the lens vary with respect to amounts, spatial distribution, and developmental regulation in a species-dependent fashion.

Taken together, there is no one rule that explains the evolutionary strategy of crystallin gene recruitment. Fig. 1 diagrams some of the schemes that have been utilized for recruiting crystallin genes during evolution. The present evidence indicates that gene duplication is not necessary for the innovation of a crystallin role by a ubiquitously expressed protein such as an enzyme or small heat shock protein (Piatigorsky and Wistow, 1991). Rather, crystallin recruitment has taken place by changes in gene regulation of single copy genes (duck lactate dehydrogenase/ϵ-crystallin, α-enolase/τ-crystalllin). In some cases there was subsequent gene duplication, resulting in both genes being highly expressed in the lens (small heat shock proteins/α-crystallins, argininosuccinate lyase/duck δ-crystallins). It also appears as if there are cases where one of the two genes lost much of its high lens expression after duplication (chicken δ2-crystallin). In still other cases, duplication may have preceded specialization for lens expression, and the specialized gene underwent additional duplications and diversification at a later time (cephalopod glutathione S-transferase/S-crystallins). Of particular significance is the fact that there are numerous examples of a lens crystallin gene continuing to be used outside of the lens, where its encoded protein has a non-refractive role (gene sharing). Gene sharing may occur with a single copy gene (ϵ-crystallin, τ-crystallin) or with one member of duplicated genes (αB-crystallin, duck δ2-crystallin). Clearly, nature takes advantage of every opportunity during evolution.

Crystallin gene regulation and crystallin recruitment

Transgenic mouse experiments have indicated that lens expression of crystallin genes is controlled at the transcriptional level (Piatigorsky and Zelenka, 1992). In general, lens specificity is often species-independent, as judged by transgenic mouse experiments. For example, the chicken δ1-crystallin gene (Kondoh *et al.*, 1987) or reporter gene constructs driven by the promoter/intron 3 enhancer of the chicken δ1 or δ2-crystallin/argininosuccinate lyase gene (Li *et al.*, 1997) are expressed principally in the lens of transgenic mice even though δ-crystallin is not used in mice. Other examples are the lens-specific expression of chicken β1- (Duncan *et al.*, 1996) or βA3/A1-(McDermott *et al.*, 1996) crystallin promoters in transgenic mice.

Numerous studies have found that crystallin genes of vertebrates are regulated by a relatively small set of developmentally important transcription factors (Cvekl and Piatigorsky, 1996). One of these, Pax–6, is critical for eye development in all species examined (Gehring and Ikeo, 1999). Retinoic acid receptors and maf, as well as AP–1 and CREB/CREM, are also involved in the regulation of a number of different vertebrate crystallin genes. Although many crystallin genes appear to be regulated by a similar group of transcription factors, the *cis*-elements are often distributed differently in the promoters and enhancers of different crystallin genes (Ilagan *et al.*, 1999). These data suggest that a gene that becomes responsive to transcription factors prevalent in the lens may be tested as a candidate crystallin gene during evolution (Piatigorsky, 1992). This idea implies that the recruitment of crystallin genes involves selective pressures on the transcription factors important for the development and function of the lens and is consistent with the diversity and taxon-specificity of crystallins (Piatigorsky, 1993).

Although very little is known about crystallin gene expression in invertebrates, current experiments suggest that similar ideas may apply and common mechanisms may be used. Invertebrate crystallins, as vertebrate crystallins, are diverse and have been recruited from various enzymes and other multifunctional proteins (Tomarev and Piatigorsky, 1996). Two squid S-crystallin promoters (SI20–1 and SL11) require an AP–1/ARE sequence for activity in transfection experiments (Tomarev *et al.*, 1994). These squid promoter sequences are similar to the functional PL–1 element of the chicken βB1-crystallin promoter and compete with it for complex formation in nuclear extracts, raising the possibility that the non-homologous crystallin genes of chicken and squid evolved a similar *cis*-acting regulatory element for lens expression. We have isolated a retinoic acid X receptor from jellyfish that binds to the jellyfish crystallin promoters, suggesting retinoic acid receptor regulation of invertebrate crystallin genes (Kostrouch *et al.*, 1998)

We are performing current experiments on scallop Ω-crystallin gene expression. Scallop Ω-crystallin is an inactive aldehyde dehydrogenase class 1 protein (Piatigorsky *et al.*, 2000). It is encoded in a gene that is very similar to the aldehyde dehydrogenase genes of vertebrates (Carosa *et al.*, 2001). Interestingly, the Ω-crystallin promoter has overlapping Pax–6 and CREB sites that bind these transcription factors and are functional in transfection tests as judged by site-specific mutagenesis experiments. Other sequences of the scallop Ω-crystallin promoter are also surprisingly analogous to the αA-crystallin promoters of chicken and mice. Although not definitive, these early investigations are consistent with the possibility that the convergent recruitment of crystallin genes have operated by similar mechanisms in invertebrates and in vertebrates. They also indicate that changes in gene regulation have been of paramount importance for the innovation of a crystallin function for an enzyme or other protein, such as a saposin (see above).

Lessons from crystallins

(i) Gene duplication is not necessary for the generation of a new protein phenotype. (ii) Evolutionary tinkering at the level of gene regulation may be as important for the development of new functions as tinkering with the structural gene. Nature is pragmatic and uses the proteins that become available. (iii) Changes in gene regulation may lead to the acquisition of a new role without relinquishing the old one. Some enzyme-crystallins retain enzyme activity. (iv) Multifunctional proteins are under several independent selective pressures. This could give a misleading importance to a conserved trait that is required for a different function than that under investigation. For example, the relative importance of the active site of an enzyme-crystallin is not obvious at the present time. Is it enzyme or refractive protein, or both? Is it always the one or the other, or both, in different species or with different proteins? (v) The functions of diverse proteins may be unified by the regulation of their genes rather than their protein sequences *per se*.

Acknowledgement

I am grateful to Dr. Zbynek Kozmik for critical comments on this manuscript.

References

Carosa, E., Kozmik, Z., Rall, J.E. and Piatigorsky, J. (2002) Structure and expression of the scallop Ω-crystallin gene: evidence for convergent evolution of promoter sequences. *J. Biol. Chem.*, **277**, 656–664.

Cvekl, A. and Piatigorsky, J. (1996) Lens development and crystallin gene expression: many roles for Pax–6. *BioEssays*, **18**, 621–630.

de Jong, W.W. Hendriks, W., Mulders, J.W. and Bloemendal, H. (1989) Evolution of eye lens crystallins: the stress connection. *Trends Biochem. Sci.*, **14**, 365–368.

de Jong, W.W., Leunissen, J.A.M. and Voorter, C.E.M. (1993) Evolution of the α-crystallin/small heat-shock protein family. *Mol. Biol. Evol.*, **10**, 103–126.

de Jong, W.W., Lubsen, N.H. and Kraft, H.J. (1994) Molecular evolution of the eye lens. *Progr. Retinal Eye Res.*, **13**, 391–442.

Duncan, M.K., Li, X., Ogino, H., Yasuda, K. and Piatigorsky, J. (1996) Developmental regulation of the chicken ßB1-crystallin promoter in transgenic mice. *Mech. Dev.*, **57**, 79–89.

Gehring, W.J. and Ikeo, K. (1999) Pax6 mastering eye morphogenesis and eye evolution. *Trends Genet.*, **15**, 371–377.

Hendriks, W., Mulders, J.W.M., Bibby, M.A., Slingsby, C., Bloemendal, H. and de Jong, W.W. (1988) Duck lens e-crystallin and lactate dehydrogenase B4 are identical: a single-copy gene product with two distinct functions. *Proc. Natl. Acad. Sci. USA*, **85**, 7114–7118.

Horwitz, J. (1992) α-crystallin can function as a molecular chaperone. *Proc. Natl. Acad. Sci. USA*, **89**, 10449–10453.

Ilagan, J.G., Cvekl, A., Kantorow, M., Piatigorsky, J. and Sax, C.M. (1999) Regulation of αA-crystallin gene expression. Lens specificity achieved through the differential placement of similar transcriptional control elements in mouse and chicken. *J. Biol. Chem.*, **274**, 19973–19978.

Ingolia, T.D. and Craig, E.A. (1982) Four small *Drosophila* heat shock proteins are related to each other and to mammalian α-crystallin. *Proc. Natl. Acad. Sci. USA*, **79**, 2360–2364.

Klemenz, R. Frohli, E., Steiger, R.H., Schafer, R. and Aoyama, A. (1991) αB-crystallin is a small heat shock protein. *Proc. Natl. Acad. Sci. USA*, **88**, 3652–3656.

Kondoh, H., Katoh, K., Takahashi, Y., Fujisawa, H., Yokoyama, M., Kimura, S., Katsuki, M., Saito, M., Nomura, T., Hiramoto, Y. and Okada, T.S. (1987) Specific expression of the chcken δ-crystallin gene in the lens and the pyramidal neurons of the piriform cortex in transgenic mice. *Dev. Biol.*, **120**, 177–185.

Kostrouch, Z., Kostrouchova, M., Love, W., Jannini, E., Piatigorsky, J. and Rall, J.E. (1998) Retinoic acid X receptor in the diploblast, *Tripedalia cystophora*. *Proc. Natl. Acad. Sci. USA*, **95**, 13442–13447.

Li, X., Cvekl, A., Bassnett, S. and Piatigorsky, J. (1997) Lens-preferred activity of chicken δ1- and δ2-crystallin enhancers in transgenic mice and evidence for retinoic acid-responsive regulation of the δ1-crystallin gene. *Dev. Genet.*, **20**, 258–266.

Magabo, K.S., Horwitz, J., Piatigorsky, J. and Kantorow, M. (2000) Expression of ßB2-crystallin mRNA and protein in retina, brain, and testis. *Invest. Ophthalmol. Vis. Sci.*, **41**, 3056–3060.

McDermott, J.B., Cvekl, A. and Piatigorsky, J. (1996) Lens-specific expression of a chicken ßA3/A1-crystallin promoter fragment in transgenic mice. *Biochem. Biophy. Res. Comm.*, **221**, 559–564.

Montgomery, M.K. and McFall-Ngai, M.J. (1992) The muscle-derived lens of a squid bioluminescent organ is biochemically convergent with the ocular lens. Evidence for recruitment of aldehyde dehydrogenase as a predominant structural protein. *J. Biol. Chem.*, **267**, 20999–21003.

Piatigorsky, J. (1992) Lens crystallins. Innovation associated with changes in gene regulation. *J. Biol. Chem.*, **267**, 4277–4280.

Piatigorsky, J. (1993) Puzzle of crystallin diversity in eye lenses. *Dev. Dyn.*, **196**, 267–272.

Piatigorsky, J. (1998) Gene sharing in lens and cornea: facts and implications. *Progr. Retinal Eye Res.*, **17**, 145–174.

Piatigorsky, J. and Horwitz, J. (1996) Characterization and enzyme activity of argininosuccinate lyase/δ-crystallin of the embryonic duck lens. *Biochim. Biophys. Acta*, **1295**, 158–164.

Piatigorsky, J., Horwitz, J., Kuwabara, T. and Cutress, C.E. (1989) The cellular eye lens and crystallins of cubomedusan jellyfish. *J. Comp. Physiol. A*, **164**, 577–587.

Piatigorsky, J., Horwitz, J. and Norman, B.L. (1993) J1-crystallins of the cubomedusan jellyfish lens constitute a novel family encoded in at least three intronless genes. *J. Biol. Chem.*, **268**, 11894–11901.

Piatigorsky, J., Kozmik, Z., Horwitz, J., Ding. L, Carosa, E., Robison, W.G., Jr., Steinbach, P. and Tamm, E. (2000) Ω-crystallin of the scallop lens. A dimeric aldehyde dehydrogenase class 1/2 enzyme-crystallin. *J. Biol. Chem.*, **275**, 41064–41073.

Piatigorsky, J., Norman, B., Dishaw, L.J., Kos, L., Horwitz, J., Steinbach, P. and Kozmik, Z. (2001) J3-crystallin of the jellyfish lens: similarity to saposins. *Proc. Natl. Acad. Sci. USA*, **98**, 12352–12367.

Piatigorsky, J., Norman, B.L. and Jones, R.E. (1987) Conservation of δ-crystallin gene structure between ducks and chicken. *J. Mol. Evol.*, **25**, 308–317.

Piatigorsky, J., O'Brien, W.E., Norman, B.L., Kalumuck, K., Wistow, G.J., Borras, T., Nickerson, J.M. and Wawrousek, E.F.

(1988) Gene sharing by δ-crystallin and argininosuccinate lyase. *Proc. Natl. Acad. Sci. USA*, **85,**3479–3483.

Piatigorsky, J. and Wistow, G. (1989) Enzyme/crystallins: gene sharing as an evolutionary strategy. *Cell*, **57,** 197–199.

Piatigorsky, J. and Wistow, G. (1991) The recruitment of crystallins: new functions precede gene duplication. Science 252, 1078–1079.

Piatigorsky, J. and Zelenka, P.S. (1992) Transcriptional regulation of crystallin genes: *cis*-elements, *trans*-factors and signal transduction systems in the lens. *Adv. Dev. Biochem.*, **1,** 211–256.

Sax, C.M. and Piatigorsky, J. (1994) Expression of the α-crystallin/small heat shock protein/molecular chaperone genes in the lens and other tissues. *Adv. Enzymol. Related Areas Molec. Biol.*, **69,** 155–201.

Tomarev, S.I., Chung, S. and Piatigorsky, J. (1995) Glutathione S-transferase and S-crystallins of cephalopods: evolution from active enzyme to lens-refractive proteins. *J. Mol. Evol.*, **41,** 1048–1056.

Tomarev, S.I., Duncan, M.K., Roth, H.R., Cvekl, A. and Piatigorsky, J. (1994) Convergent evolution of crystallin gene regulation in squid and chicken: the AP–1/ARE connection. *J. Mol. Evol.*, **39,** 134–143.

Tomarev, S.I. and Piatigorsky, J. (1996) Lens crystallins of invertebrates. Diversity and recruitment from detoxification enzymes and novel proteins. *Eur. J. Biochem.*, **235,** 449–465.

Tomarev, S.I. and Zinovieva, R.D. (1988) Squid major lens polypeptides are homologous to glutathione S-transferease subunits. *Nature*, **336,** 86–88.

Tomarev, S.I., Zinovieva, R.D., Guo, K. and Piatigorsky, J. (1993) Squid glutathione S-transferase. Relationships with other glutathione S-transferases and S-crystallins of cephalopods. *J. Biol. Chem.*, **268,** 4534–4542.

Tomarev, S.I., Zinovieva, R.D. and Piatigorsky, J. (1992) Characterization of squid crystallin genes. Comparison with mammalian glutathione S-transferase genes. *J. Biol. Chem.*, **267,** 8604–8612.

Vaccaro, A.M., Salvioli, R., Tatti, M. and Ciaffoni, F. (1999) Saposins and their interactions with lipids. *Neurochem. Res.*, **24,** 307–314.

Wistow, G.J., Lietman, T., Williams, L.A., Stapels, S.O., de Jong, W.W., Horwitz, J. and Piatigorsky, J. (1988) τ-Crystallin/α-enolase: one gene encodes both an enzyme and a lens structural protein. *J. Cell Biol.*, **107,** 2729–2736.

Wistow, G., Mulders, J.W.M. and de Jong, W.W. (1987) The enzyme lactate dehydrogenases as a structural protein in avian and crocodilian lenses. *Nature*, **326,** 622–624.

Wistow, G.J. and Piatigorsky, J. (1987) Recruitment of enzymes as lens structural proteins. *Science*, **236,** 1554–1556.

Wistow, G. and Piatigorsky, J. (1988) Lens crystallins: the evolution and expression of proteins for a highly specialized tissue. *Ann. Rev. Biochem.*, **57,** 479–504.

Zinovieva, R.D., Tomarev, S.I. and Piatigorsky, J. (1993) Aldehyde dehydrogenase-derived Ω-crystallin of squid and octopus. Specialization for lens expression. *J. Biol. Chem.*, **268,** 11449–11455.

A. Meyer, Y. Van de Peer (eds.), Genome Evolution, 139-150.

Evolution of signal transduction by gene and genome duplication in fish

Jean-Nicolas Volff[1,*] & Manfred Schartl
Physiologische Chemie I, [1]Biofuture Research Group 'Evolutionary Fish Genomics', Biozentrum, University of Würzburg, Am Hubland, D-97074 Würzburg, Germany;
[*]Author for correspondence: E-mail: volff@biozentrum.uni-wuerzburg.de

Received 30.04.2002; Accepted in final form 29.08.2002

Key words: epidermal growth factor receptor, gene duplication, fish, *Xmrk*, Xiphophorus

Abstract

Fishes possess more genes encoding receptor tyrosine kinases from the epidermal growth factor receptor (EGFR) family than other organisms. Three of the four genes present in higher vertebrates have been duplicated early during the evolution of the ray-finned fish lineage possibly as a consequence of an event of whole genome duplication. In the fish Xiphophorus, a much more recent local event of gene duplication of the *egfr* co-orthologue *egfr-b* generated a eighth gene, the *Xmrk* oncogene. This duplicate acquired within a short time a constitutive activity and a pigment cell-specific overexpression responsible for the induction of melanoma in certain interspecific hybrids. Despite its frequent loss during evolution of the genus Xiphophorus, the maintenance of *Xmrk* in numerous species and its evolution under purifying selection suggest a so far unknown function under certain natural conditions. One of the known functions of *Xmrk* in tumor cells is the suppression of differentiation of melanocytes induced by the microphthalmia-associated transcription factor MITF. While only one gene with alternative 5′ exons and promoters is present in higher vertebrates, two *mitf* genes were identified in fish. Subfunctionalization of *mitf* paralogues by differential degeneration of alternative exons and regulatory sequences led particularly to the formation of a *mitf* gene specifically expressed in the melanocyte lineage. These observations validate fish as an outstanding model to study the mechanisms and biological consequences of gene and genome duplication but underline the complexity of the fish model and the caution necessary in transferring knowledge from fish to higher vertebrates and vice versa.

The epidermal growth factor receptor family

Receptor tyrosine kinases (RTKs) are important molecules that are involved in the control of key cellular events like growth, differentiation and apoptosis. They form a large gene family found in all multicellular animals. The encoded proteins share a common gross structure: they consist of an extracellular part, containing the ligand binding domain, a single α-helical transmembrane domain and an intracellular moiety that includes the enzymatic domain, which is a protein tyrosine kinase, and a carboxyterminal tail (Ullrich and Schlessinger, 1990). The kinase domain is highly conserved amongst the members of this gene family and even shares a high degree of similarity to the related cytoplasmic tyrosine kinases. The carboxyterminal tail, which functions predominantly in substrate interactions, is much more divergent.

The extracellular domain can be of different distinct types, which define the subclasses of the RTK family. From the approximately 20 subclasses only subclass I will be considered further. This subclass is also called the epidermal growth factor receptor (EGFR) family because of its prototypic member. The extracellular domain of the subclass I receptors consists of four subdomains. Subdomains II and IV are cysteine-rich and obviously important for the conformation of the molecule. The spacing and number of these cysteines is conserved between paralogous members of the family and also between orthologues. The cysteines are involved in the formation of intramolecular disulfide bonds (Abe *et al.*, 1998).

Subdomains I and III are functioning in ligand binding.

The peptides that bind to the subclass I receptors form a gene family as well, the epidermal growth factor (EGF) family. They are much less conserved, except for several cysteine residues, which define the so-called 'EGF'-motif. These residues also form cystine bridges, which stabilize the structure of the ligand molecules. Because their function in general is to regulate cell growth, they are included in the group of growth factors or cytokines.

Binding of a monomeric ligand to the extracellular domain of the receptor induces dimerization, upon which the intrinsic tyrosine kinase in the intracellular part becomes active. As a first step, the carboxyterminal tail of the other partner of the dimer becomes phosphorylated at tyrosine residues. These phosphotyrosines in their specific amino-acid motif environment act as docking sites for substrate proteins. Some of them are then tyrosine-phosphorylated as well and in that state initiate certain intracellular signaling events. Other substrates that bind are not modified, but as adaptor proteins are also involved in signal transduction (Schlessinger, 2000). Recently, it has been proposed that the dimers are already pre-existing and that ligand-binding merely induces an essential change in conformation (Moriki *et al.*, 2001).

More importantly, the ligands cannot only bind and activate receptor homodimers, they also bind to heterodimers of receptors from the same subclass. Through such mechanisms, the spectrum of specificity of the ligands is greatly enhanced, leading to a complexity of cellular activation and signaling mechanisms that is only beginning to become elucidated.

RTKs of subclass I are crucial for the correct development, for instance mutations in the EGFR of *Caenorhabditis elegans* lead to defects in the development of the reproductive system and in the *Drosophila* EGFR to disturbed eye development. In mammals, the EGFR has been implicated in numerous developmental processes including preimplantation development, implantation, placentation and the development of the skin, hair, kidney, lungs nervous system, mammary glands and teeth (Adamson and Wiley, 1995).

Not surprisingly, RTKs are frequently found to be overexpressed or mutationally altered in tumors. In fact, the first subclass I RTK was characterized as an oncogene: the *v–erb B* oncogene of the avian erythroblastosis virus (Yamamoto *et al.*, 1983). Subsequently, it was recognized that the v-erb B protein is a heavily mutationally altered, tumorigenic version of the EGF-receptor, its corresponding proto-oncogene.

In *C. elegans* and *Drosophila melanogaster* a single member of subclass I RTKs is present, the *C. elegans* EGFR, CER (encoded by the let-23 locus) and the *Drosophila* EGFR, DER. Like for many other gene families, the situation is much more complex in vertebrates. In the human genome four subclass I RTKs have been identified: the EGF-receptor gene (also termed *c-erb B*, *erbB1*, or *HER1*), *HER2* (also termed *HER2/neu*, *neu* or *erbB2*), *HER3* (or *erbB3*) and *HER4* (or *erbB4*). The receptors differ in their expression pattern, their ability to form homodimers or heterodimers with the other family members, their binding affinities to the ligands of the EGF family and the biochemical response they elicit intracellularly through different coupling to the signal transduction molecules. A peculiar situation exists for the erbB3 receptor. Its intracellular domain is enzymatically compromised, i.e. a mutation has led to the loss of the kinase activity. Consequently, erbB3 can only signal in heterodimeric complexes with one of the other family members.

Because of their obvious significance for biology and medicine, the subclass I RTKs have been intensively studied and a large amount of data on their structure and biochemical function is available. However, the specifics of the biological function of the different family members, how they overlap and why the gene family has expanded during evolution, is far from being understood.

Expansion of the EGF receptor family by genome duplication in fish

Gene duplication associated with subsequent functional diversification of duplicates is certainly an important evolutionary process increasing the complexity of living organims (Ohno, 1970; see Aburomania *et al.*, this issue). Even if, in some cases, both redundant duplicates might be maintained because of an advantageous increase of gene expression, one major evolutionary advantage of gene duplication is that one of the versions of the ancestral gene might be free to diverge independently from the selective forces that maintain the original function. Very frequently, this duplicate will be inactivated by mutations and become a pseudogene ('defunctionalization'). More rarely, innovating mutations in regulatory and/or structural sequences might allow one duplicate to encode a product with a new function ('neofunctionalization'). Alternatively, both duplicates might be altered in a way such that the combined activity of the two genes together fulfills the role of the ancestral gene in a complementary fashion ('subfunctionalization', Force *et al.*, 1999).

Genes duplicates can be generated not only by local, segmental events of duplication but also by much larger events of genome duplications. Accumulating evidence supports the involvement of two rounds of genome duplication during the evolution of vertebrates from early deuterostome ancestors (the 'one-to-four' rule or '2R' hypothesis, Ohno, 1999; Hughes, 1999; see also in this issue Aburomania *et al.*; Holland; Hughes and Friedman; Lundin *et al.*; Hokamp *et al.*). The presence in higher vertebrates of four genes encoding EGF receptor-related proteins compared to only one in *D. melanogaster* and *C. elegans* is apparently consistent with this hypothesis.

An additional round of genome duplication might have taken place during the course of evolution of the ray-finned fish lineage (Actinopterygia, will be called for more simplicity 'fish(es)' in this article) after its separation from the sarcopterygian lineage including coelacanth, lungfish and land vertebrates (the one-to-four-to-eight rule, Meyer and Schartl, 1999; Wittbrodt *et al.*, 1998). This might have considerably contributed to the amazing number of duplicated genes present in fish genomes (Amores *et al.*, 1998; Aparicio, 2000; Robinson-Rechavi *et al.*, 2001; Taylor *et al.*, 2001; Van de Peer *et al.*, 2001, also see Van de Peer *et al.* and Josefowicz *et al.*, this issue). The fish-specific genome duplication might have arisen as early as 300–450 million years ago (Taylor *et al.*, 2001; Van de Peer *et al.*, 2001). As many as 20% of the duplicate genes, particularly genes encoding DNA binding proteins, may have been maintained in teleost genomes either by neofunctionalization of a duplicate, and/or by subfunctionalization (Force *et al.*, 1999; Postlethwait *et al.*, 2000; Van de Peer *et al.*, 2001). Fish are therefore an especially suited group of organisms to study the evolution of gene duplicates and the evolutionary consequences of the duplication of genetic information. Phylogenetic and functional analysis of the almost completely sequenced genomes of the pufferfishes *Takifugu rubripes* (Elgar *et al.*, 1999; http://fugu.hgmp.mrc.ac.uk/) and *Tetraodon nigroviridis* (Roest Crollius *et al.*, 2000; http://www.genoscope.cns.fr/) and of data generated by other ongoing fish sequencing projects (for example the zebrafish *Danio rerio* and the medaka *Oryzias latipes* projects) should provide an outstanding opportunity to investigate the evolution of a high number of duplicate pairs over more than 100 million years of evolution.

Evidence for fish-specific duplications of genes encoding members of the EGF receptor family were provided by genetic and functional studies on the platyfish *Xiphophorus maculatus*, a model for the study of hereditary melanoma (Schartl, 1995; Froschauer *et al.*, 2001) and by our present analysis of the data generated by the *Takifugu rubripes* (Japanese pufferfish) genome project. While only one *egfr* gene is present in human and other higher vertebrates, two *egfr* genes (*egfr*-a and *egfr*-b) were identified in the genome of both platyfish (Wittbrodt *et al.*, 1989; Adam *et al.*, 1993; Gómez *et al.*, submitted) and Japanese pufferfish (Figs. 1 and 2). One of them, *egfr-b* (formerly INV-*Xmrk*), has been mapped to the sex-determining region of the sex chromosomes of the platyfish and corrresponds to the precursor of the melanoma-inducing *Xmrk* oncogene (see below; Schartl, 1995; Gutbrod and Schartl, 1999; Froschauer *et al.*, 2001, 2002; Volff and Schartl, 2001). Comparative analysis of Xiphophorus Egfr-a and Egfr-b receptors suggested that they have evolved different biochemical properties (Gómez *et al.*, submitted). In addition, genes orthologous to *erbb2*, *erbb3* and *erbb4* from higher vertebrates could be identified in the genome of the Japanese pufferfish, as well as one *erbB3* gene in a genomic sequence from the zebrafish *Danio rerio* (accession AL591365). Interestingly, as observed for *egfr*, two *erbB3* genes and two *erbB4* genes are present in the genome of *Takifugu rubripes* (Figs. 1 and 2), indicating that at least three different genes encoding receptors from the EGFR family are duplicated in the genome of the Japanese pufferfish. No evidence for a second fish *erbB2* gene could be found.

Analysis of the ratio of synonymous to non-synonymous substitutions strongly suggested that all fish duplicates evolved under purifying selection and therefore probably do not correspond to pseudogenes. The (average) ratios between fish duplicates were 5.7, 9.0 and 14.1 for *egfr*, *erbB3* and *erbB4*, respectively, compared to 7.9, 8.9 and 26.4 between their orthologues in higher vertebrates. Apparently, *erbB4* genes evolved under stronger negative selection than other EGFR-related receptor-encoding genes in vertebrates. Similar ratios indicating purifying selection were also observed in comparison involving fish orthologues, for example platyfish and pufferfish *egfr-a* genes.

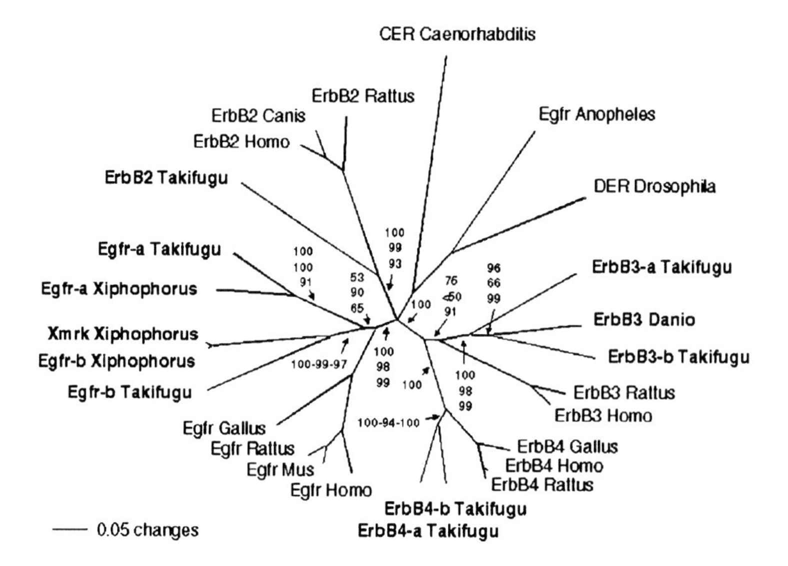

Figure 1. Phylogeny of receptor tyrosine kinases from the EGFR family.
Bootstrap values (%) using neighbour-joining (1,000 replicates, first value) and maximum parsimony analyses (100 replicates, second value), as well as the reliability values (%) for maximum likelihood analysis (quartet puzzling, 10,000 puzzling steps, thirst value) are given. The tree (neighbour-joining) is unrooted. Branches with less than 50% support have been collapsed. A 630 amino-acid alignment of sequences starting in the C-terminal end of the first receptor L domain and ending about 20 amino-acid upstream from the catalytic domain was used for phylogenetic analysis. Sequences were analysed using the GCG Wisconsin package (Version 10.0, Genetics Computer Group, Madison) and PAUP* (D.L. Swofford, Smithsonian Institution, Washington, D.C.) as described (Altschmied *et al.*, 2002). Accession numbers: Egfr *Homo sapiens* XP_044653; Egfr *Mus musculus* AAA17899; Egfr *Rattus norvegicus* AAF14008; Egfr *Gallus gallus* P13387; Egfr-b *Xiphophorus xiphidium* AAD10500; Xmrk *Xiphophorus maculatus* P13388; ErbB2 *Homo sapiens* CAA27060; ErbB2 *Canis familiaris* BAA23127; ErbB2 *Rattus norvegicus* CAA27059; ErbB3 *Homo sapiens* AAA35790; ErbB3 *Rattus norvegicus* AAC28498; ErbB3-a *Takifugu rubripes* AAC34391; ErbB4 *Homo sapiens* NP_005226; ErBb4 *Rattus norvegicus* AAD08899; ErbB4 *Gallus gallus* AAD31764; DER *Drosophila melanogaster* AAD26132; Egfr *Anopheles gambiae* CAC35008; CER *Caenorhabditis elegans* S70712. *Danio rerio* ErbB3 sequence was deduced from the unfinished high throughput genomic sequence AL591365. *Takifugu rubripes* sequences were deduced from data provided by the the Fugu Genome Consortium: Egfr-a (FT:T000306), Egfr-b (FT:T002807), ErbB2 (FT:T001938), ErbB3-b (FT:T012522), ErbB4-a (FT:T000352) and ErB4-b (FT:T001532). Gene structure was determined using programs available on the NIX server (http://menu.hgmp.mrc.ac.uk/menu-bin/Nix/Nix.pl). Deduced sequences and sequence of Xiphophorus Egfr-a receptor are available on request.

Phylogenies based on both amino-acid (Fig. 1) and nucleic sequences (Fig. 2) revealed duplication events having arisen in the (ray-finned) fish lineage after its separation from the sarcopterygian lineage. Particularly, these analyses showed that the formerly called *Xmrk* genes (now *egfr*-b and *Xmrk*) are co-orthologues of the *egfr* gene from higher vertebrates, closing definitively the question of the presence of an *Xmrk* gene in the human genome. The relatively low level of nucleotide identity between fish *egfr* (average 59.6%), *erbB3* (average 65.5%) and *erbB4* paralogues (75.1%) indicated rather ancient events of gene duplications. Accordingly, the duplication that led to the formation of fish *erbB3* paralogues arose before the divergence between zebrafish and pufferfish more than 100 million years ago, and the duplication of *egfr* before the separation between pufferfish and platyfish at least 90 million years ago (Figs. 1 and 2). Taylor *et al.* (2001) estimated the age of several zebrafish paralogues by plotting the number of nucleotide substitutions at third codons positions, corrected for multiple events per site according to Tajima and Nei (1984), against divergence dates for different taxa. Since most third-codon position substitutions do not result in amino-acid replacements, the rate of fixation of these substitutions is expected to be relatively constant even in different protein-coding genes. By this way Taylor *et al.* could estimate that the putative fish-specific genome duplication occurred between 300 and 450 million years ago (ca. 350 million years ago). Using the same method, we could show that the number of substitutions per site at third codon positions between the different fish *egfr* (average 1.36), *erbB3* (average 1.46) and *erbB4* (0.96) paralogues was compatible with the average value (1.02, s.d. = 0.24) obtained by Taylor *et al.* (2001) on the basis of the analysis of 23 pairs of unlinked zebrafish duplicates. As a basis for comparison, the same

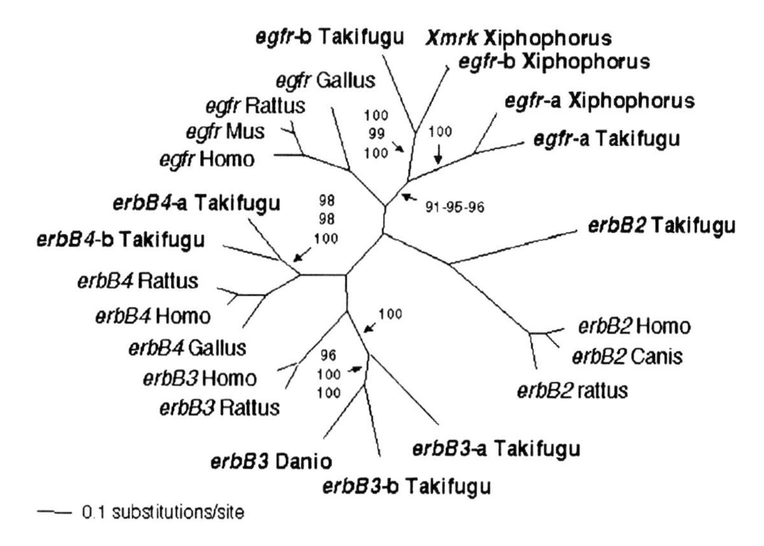

Figure 2. Phylogeny of vertebrate genes encoding receptor tyrosine kinases from the EGFR family.
Bootstrap values (%) using maximum likelihood (100 replicates, first value), maximum parsimony (100 replicates, second value) and neighbour-joining analyses (1,000 replicates, third value) are given. The tree (maximum likelihood) is unrooted. Branches with less than 50% support have been collapsed. A 1767 nt alignment of sequences encoding a receptor region starting in the C-terminal end of the first receptor L domain and ending about 20 amino-acid upstream from the catalytic domain was used for phylogenetic analysis. Accession numbers: *egfr Homo sapiens* XM_044653; *egfr Mus musculus* AK004911; *egfr Rattus norvegicus* M37394; *egfr Gallus gallus* M77637; *egfr-b Xiphophorus xiphidium* U53471; *Xmrk Xiphophorus maculatus* X16891; *erbB2 Homo sapiens* NM_004448; *erbB2 Canis familiaris* AB008451; *erbB2 Rattus norvegicus* NM_017003; *erbB3 Homo sapiens* NM_001982; *erbB3 Rattus norvegicus* NM_017218; *erbB3-a Takifugu rubripes* AF056116; *erbB4 Homo sapiens* NM_005235; *erBb4 Rattus norvegicus* NM_021687; *erbB4 Gallus gallus* NM_021687. The origin of other fish sequences is given in the legend of Figure 1. Invertebrate sequences were not included because of alignment ambiguities.

(average) value between orthologues from mammals and chicken (divergence about 300 million years ago, Kumar and Hedges, 1998) was slightly lower than between the corresponding fish paralogues: 0.97 for *egfr*, and 0.84 for *erbB4*. We conclude that the fish *egfr*, *erbB3* and *erbB4* paralogues have resulted from ancient duplication event(s) and might be remnants of the proposed fish-specific genome duplication.

Compared to other members of the EGFR family, the mammalian ErbB3 protein exhibits an impaired intrinsic tyrosine kinase activity (Guy *et al.*, 1994; Sierke *et al.*, 1997). This has been attributed to substitutions of amino-acid residues that are highly conserved in the catalytic domains of protein kinases. Residues Cys-721, His-740 and Asn-815 are found in human ErbB3 instead of the highly conserved Ala, Glu and Asp in other protein kinases (Fig. 3). In rat and mouse ErbB3, a canonical Asp residue corresponds to Asn-815 in human. Nevertheless, no evidence of catalytic activity could be detected for the rat ErbB3 too (Sierke *et al.*, 1997), suggesting that the change of Asp into Asn-815 was not responsible for the impaired tyrosine kinase activity in the human receptor. Fish ErbB3 receptors exhibit at these three critical positions the canonical Ala (instead of Cys in mammals), His like the mammalian ErbB3 proteins (instead of the canonical Glu) and Asn like the human protein instead of the canonical Asp found in rat and mouse and other RTKs. This suggested that (i) the Ala-Cys substitution occurred in the tetrapod lineage after divergence from the fish lineage, (ii) the Glu-His and Asp-Asn substitutions arose in ErbB3 before the separation between fish and tetrapods, i.e. at least 450 million years ago, and (iii) Asn reverted to Asp (transition AAC to GAC) in the mouse/rat linage after its diverge from the lineage leading to human. Testing the catalytic activity of ErB3 receptors from fish should allow identifying the amino-acid change(s) responsible for the impaired intrinsic tyrosine kinase activity in some ErbB3 receptors.

The *Xmrk* oncogene of Xiphophorus

In addition to the ancient genome duplication event that probably generated the extra members of the subclass I RTK genes in teleosts, a more recent duplication of *egfr*-b is known. This duplicated gene is called X*mrk* (for Xiphophorus melanoma receptor kinase), because it was isolated as the melanoma-inducing gene from fish of the genus *Xiphophorus* (Wittbrodt *et al.*, 1989). During the gene duplication event a second copy of the *egfr*-b gene (which became X*mrk*) was fused in its 5′ region to novel upstream sequences (Adam *et al.*, 1993). This obviously has altered the transcriptional control of the gene. X*mrk* is expressed at low levels in eyes, brain, gills and barely detectable in skin, muscle, and kidney, while transcripts are absent in liver (Schartl *et al.*, 1999). Only in certain hybrids, X*mrk* is overexpressed in a certain type of pigment cells, the macromelanophores, and causes, as a dominant oncogene, the development of melanoma. Together with earlier data on the genetics of melanoma formation in Xiphophorus, this has led to the following model: in non-hybrid parental fish X*mrk* is under transcriptional control in the pigment cell lineage. A low transcript level - like in the other normal organs – does not lead to tumorigenic alterations. The genetic locus that exerts the regulatory control, designated *R* (or *Diff*), resides on another linkage group. If now such fish are crossed to those, which do not have the X*mrk* gene and its corresponding *R* locus, offspring are obtained according to Mendelian laws that have X*mrk* but not *R*. In the absence of regulatory control, X*mrk* is overexpressed. Then, this leads in the pigment cell lineage – and only there – to neoplastic transformation.

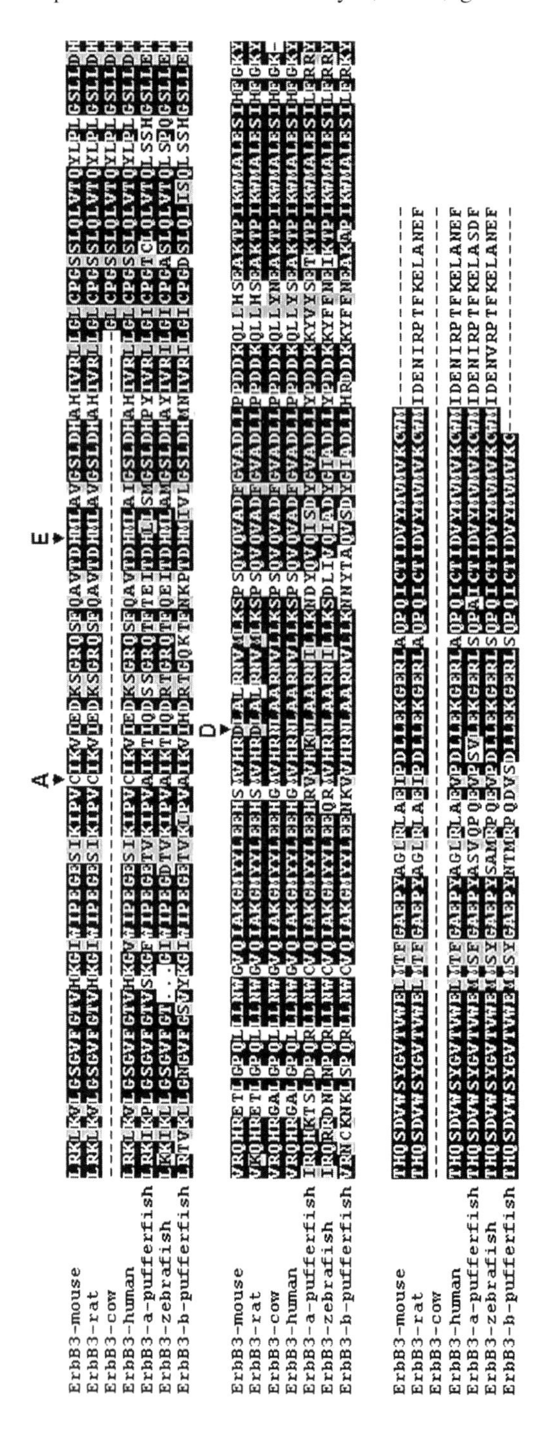

The events that have generated X*mrk* are not easy to reconstruct, because X*mrk* is located in the subtelomeric region of the sex chromosomes (Nanda *et al.*, 2000), which is a region of high genomic plasticity (Froschauer *et al.*, 2001). There is little doubt that X*mrk* arose from the *egfr*-b gene by tandem gene duplication. Both genes locate to the same chromosomal region and their distance is from linkage analysis less than 0,5 centiMorgan (Gutbrod and Schartl, 1999; Volff and Schartl, 2001; Froschauer *et al.*, 2002). The 3′ duplication breakpoint has not been searched so far. The 5′ breakpoint lies somewhere in the promotor region, however is difficult to nail down to the basepair. The X*mrk* upstream region contains parts of the *egfr*-b promoter, which was probably included in the duplication and was rearranged by subsequent deletions and insertions (Froschauer *et al.*, 2001). The X*mrk* promoter harbors – other than the *egfr*-b promotor – an activating GC box element that binds the transcription factor Sp1 (Baudler *et al.*, 1997).

Figure 3. Sequence comparison of the putative tyrosine kinase domains of vertebrate ErbB3 proteins.
A, E, and D indicate canonical residues highly conserved in RTKs substituted in some ErbB3 receptors. Identical residues are in black, conservative substitutions in gray (drawn with MacBoxshade). The mouse sequence corresponds to the translation of a consensus sequence obtained by assembling different *Mus musculus* expressed sequence tags (BG175275, BE376085, BF100702, BF122051, BI151808 and BE912975). The cow sequence is the conceptual translation of an expressed sequence tag from *Bos taurus* (AW336843). The origin of other sequences is given in the legend of Figure 1.

Another difference in the promoter of both genes is that the X*mrk* promotor is highly methylated at CpG dinucleotides in non-transformed cells, but demethylated in melanoma, while the *egfr*-b promoter is highly methylated in melanoma, but less in healthy organs (Altschmied *et al.*, 1997). These data point to the explanation that changes in the promoter region of one of the duplicated *egfr*-b genes were intimately linked to the acquisition of a novel function, namely neoplastic transformation of pigment cells.

Of course, only in rare cases, like for the tumor viruses, acquiring a dominantly acting oncogene is of selective advantage. For a multicellular organism, such a novel function for the *egfr*-b duplicate is difficult to use for explaining why a second copy has been maintained. Before we can address the question why X*mrk* is still found in the genome of these fish, another complication to this issue has to be added from the biochemical data set.

The K_s/K_a ratio of 4.7 for X*mrk* and *egfr-b* from Xiphophorus clearly indicate that X*mrk* does not evolve as a pseudogene. It has not acquired mutations that compromise its function as a subclass I RTK, which is surprising enough considering that it resides in a genomic region of high plasticity, filled with mobile DNA elements that readily are prone to inactivate genes, and being surrounded by genes (or fragments thereof), which are non-functional (Froschauer *et al.*, 2001, 2002). Moreover, two of the amino-acid substitutions in Xmrk have changed the biochemical properties of the receptor in a way that it is now constitutively active and that it signals without a ligand (Gómez *et al.*, 2001). These two mutations, both in the extracellular domain and both inducing ligand independent, covalent dimerization through cystine bridges, are oncogenic. It is interesting to note that Xmrk is the only RTK isolated from the in-vivo tumor situation that carries two tumorigenic aminoacid changes. From the biochemical point of view it can be said, that Xmrk, compared to the Egfr-b, has even become a 'better' RTK: it does not need a growth factor for activating the intracellular tyrosine kinase domain.

For the evolutionary fate of a pair of duplicated genes, three scenarios are generally believed to be possible: (i) one of the two genes can acquire a new, positively selected function (neofunctionalization), (ii) both genes can share the function of the one common ancestor (subfunctionalization) or (iii) one of the two genes is lost, thus restoring the ancestral situation (degeneration). Which of the three possibilities is happening is difficult to say for X*mrk*. Certainly X*mrk* is a highly potent transforming oncogene. However, it can hardly be called a hazardous gene, because it displays its malignant function only in certain hybrids. Such hybrids are laboratory products and have never been found in the wild. In very rare instances, X*mrk*-induced pigmentary lesions, including melanoma, are found in non-hybrid fish. Some have even been collected in the natural habitats. In these fish somehow the control of *R* over X*mrk* has been impaired, for instance by somatic mutations (Schartl *et al.*, 1995). However, such melanoma are usually found in old, post-reproductive animals. This, together with the rarity of the phenomenon, indicates that X*mrk* does not confer a measurable negative impact on its bearers.

A physiological, positively selected function is also not known up to date. Moreover, many populations of Xiphophorus fish are polymorphic for X*mrk*. The gene frequency varies between the populations from less than 1% to more than 50%. In some species X*mrk* is totally absent. Thus, no general physiological function can be assigned to its gene product, which so drastically differs in biochemical features from the *egfr*-b-encoded RTK.

Is X*mrk* then on the path of degeneration, and do we find it still in some fish because the time since its duplication was too short to get wiped out totally? It has been described that X*mrk* can be easily lost. For instance large deletions that include X*mrk* or gene disruption by insertion of a transposable element have been observed in the laboratory (Schartl *et al.*, 1999). However, if the presence of X*mrk* in the different species of the genus is plotted on the DNA-sequence based phylogenetic tree (Meyer *et al.*, 1994), it becomes evident that the generation of X*mrk* was a monophyletic event that took place before radiation of the genus Xiphophorus, at least more than 5 million years ago (Weis and Schartl, 1998).

The question that remains is why X*mrk* has survived for such long times in the natural populations, encoding a fully functional – or even improved – RTK given that it has no obvious physiological function and that it is easily vulnerable. A solution might come from the observation that in all cases X*mrk* is found intimately linked to a locus, which encodes various pigment patterns of a certain type of melanophores, the so-called macromelanophores. This locus, *Mdl* (for macromelanophore-determining locus) is a polymorphic locus in Xiphophorus and encodes the genetic information for the development of this specific type of pigment cells as well as the pattern information (onset of pattern development, compartment and shape of the pattern). It has been shown that these *Mdl* patterns are important for kin recognition in the natural habitats, for instance under murky water conditions (Franck *et al.*, 2001). *Mdl* is an autonomous genetic entity that may be linked to X*mrk* or not, but certainly is not dependent on this linkage for its function (Weis and Schartl, 1998). A hypothesis for the evolutionary conservation of functional X*mrk* copies can be based on the fact that X*mrk*, however, may be dependent on the linkage to *Mdl*, in a way that it behaves like a hitchhiker. This theory is based on the observation that the frequency of a certain gene will be enhanced if one allele of a linked locus becomes selectively favoured (Maynard Smith and Haigh, 1974; Stephan *et al.*, 1992). This scenario can be extended to explain a situation in which a 'dispensable' or even mildly disadvantageous allele at one locus is maintained in a population as a result of its close linkage to a strongly positively selected allele at a second locus. In this context, it is interesting to note that the target for the transforming activity of X*mrk* are only the macromelanophores, specified by the different linked *Mdl* alleles, but no other type of pigment cells. The hitchhiking theory, however, cannot easily explain the negative purifiying selection that has obviously acted on X*mrk*.

Subfunctionalization of *mitf* duplicates in fish by degeneration of alternative regulatory and exonic sequences

Assessing the impact of gene and genome duplication on the complexity of the multiple cell biological and biochemical processes mediated by members of the EGF receptor family in fish requires the identification of genes encoding upstream (ligands) and downstream components of the different signal transduction cascades that have been also maintained as duplicates.

Recent evidence suggested that the Xmrk melanoma-inducing version of Egfr-b is able to prevent the transcriptional activation of melanocyte differentiation genes induced by one isoform of the microphthalmia-associated transcription factor MITF (Wellbrock *et al.*, 2002). MITF is a member of the basic helix-loop-helix leucine zipper (bHLH-Zip) protein family (for reviews, Tachibana, 2000; Goding, 2000). It plays a central role in the differentiation and/or survival of melanocytes, which is reflected by the correlation between a number of mutant pigmentation phenotypes and specific mutations in the *mitf* gene. Different mutations in the human gene can lead to Waardenburg syndrome type 2 or Tietz syndrome, both of which are associated with pigmentary changes. In mammals and birds several isoforms of MITF are known, which are encoded by a single gene. The different proteins are generated by the use of alternative promoters and 5′ exons (Yasumoto *et al.*, 1998; Udono *et al.*, 2000). The mammalian and avian MITF-m isoforms have a unique N-terminus of eleven amino acids not found in the other variants a, c and h. These last three isoforms have a common exon, B1b, directly upstream of exon 2, but differ in their N-terminus, encoded by the exons a, c and h, respectively. Differences are not limited to the primary structure of the protein, but are also found at the transcriptional level. Whereas *mitf*-m is expressed exclusively in melanocytes and melanoma cells, the other isoforms are expressed in a much wider spectrum of cells (Amae *et al.*, 1998, Udono *et al.*, 2000). MITF-m is the isoform involved in melanocyte differentiation.

Recently, two different isoforms of the transcription factor MITF, MITF-m and MITF-b, have been described in different teleost fish species (Fig. 4; Lister *et al.*, 2001; Altschmied *et al.*, 2002). Interestingly, each isoform is encoded by a separate gene, providing additional evidence for the high frequency of duplicate genes in fish genomes. The presence of both *mitf* duplicates in zebrafish, pufferfish and platyfish indicated that they were the result of an ancient duplication having arisen at least 100 million years ago. The duplication of *mitf* in the fish lineage was supported by some (but not all) phylogenetic methods, but confirmed by an evolutionary analysis of intron dynamics in the *mitf* genes during vertebrate evolution (Altschmied *et al.*, 2002). The average number of substitutions per site at third codon positions between the different fish *mitf* paralogues was 1.0 (compared to 0.6 between *mitf* orthologues from chicken and mammals) and therefore, here again, consistent with the value (1.02, s.d. = 0.24) reported by Taylor *et al.* (2001) for duplicates proposed to be the result of an ancient fish-specific genome duplication. Both *mitf* duplicates evolved under purifying selection (Altschmied *et al.*, 2002).

Structure and expression analysis of fish *mitf* paralogues provided evidence for a subfunctionalization where two alternative transcripts of a same gene with

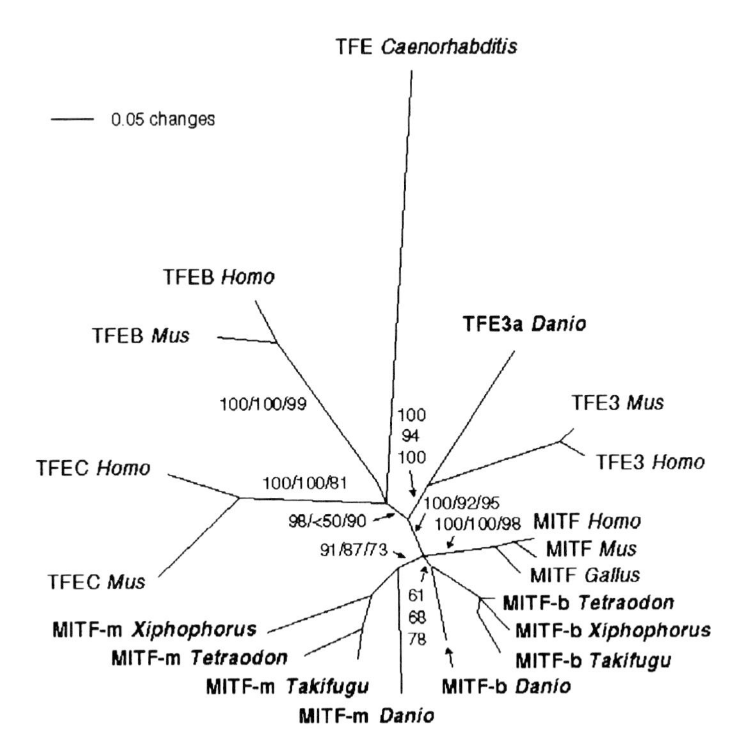

Figure 4. Phylogeny of MITF-related proteins.
Bootstrap values (%) using neighbour-joining (1,000 replicates, first value) and maximum parsimony analyses (100 replicates, second value), as well as the reliability values (%) for maximum likelihood analysis (quartet puzzling, 10,000 puzzling steps, thirst value) are given. The tree (neighbour-joining) is unrooted. Branches with less than 50% support have been collapsed. Protein regions encoded by alternative exons 1a/1b/1m have been removed for analysis. Only maximum parsimony supported the presence of a fish-specific group containing the MITF-b and MITF-m sequences (bootstrap value 86%). Accession numbers: MITFm *Danio rerio* AAD48371; MITFb *Danio rerio* AAK95588; MITFm Xiphophorus PSM cell line AF475090; MITFb *Xiphophorus maculatus* AF475091; MITF *Mus musculus* AAF81266; MITF *Homo sapiens* I38024, MITF *Gallus gallus* AAF67467, TFE3a *Danio rerio* AAK95589, TFE3 *Homo sapiens* P19532, TFE3 *Mus musculus* Q64092, TFEB *Homo sapiens* CAB54146, TFEB *Mus musculus* NP_035679, TFEC *Mus musculus* AF077742, TFEC *Homo sapiens* D43945, TFE *Caenorhabditis elegans* AAB37997. *Takifugu rubripes* MITF-b and MITF-m sequences have been deduced from data provided by the the Fugu Genome Consortium (FT:T000263 and FT:T008421, respectively).

different functions (in higher vertebrates) have been replaced by two distinct genes (in fish). In Xiphophorus, *mitf*-b is expressed apparently ubiquitously, since specific transcripts were detected by RT-PCR in all cell types analyzed. In contrast, *mitf*-m RNA was only detected in tissues containing melanocytes like eyes and skin and in melanomas, as well as in a Xiphophorus melanoma cell line, thus paralleling the expression pattern of *mitf*-m in higher vertebrates. Accordingly, mutations in the zebrafish *mitf*-m (also called *mitf*-a) perturb neural crest melanocytes but not eye development (Lister *et al.*, 1999) whereas in the *microphtalmia* mutant of the mouse the non-functional *mitf* gene affects the neural crest derived pigment cells and the eye development simultaneously. The expression of the *mitf-m* and *mitf-b* duplicates is partially overlapping in the zebrafish embryo, but only *mitf-m* is expressed in neural crest melanoblasts (Lister *et al.*, 2001).

Duplication of *mitf* and subsequent subfunctionalization of the duplicates was associated with the degeneration of at least one alternative exon (Fig. 5): the 1m-exon specific for the m-form of MITF in higher vertebrates was detected in the *mitf*-b gene of both pufferfishes but showed degenerated coding sequences and splicing sites (Altschmied *et al.*, 2002). As observed in other cases of subfunctionalization of duplicate fish genes, fish *mitf*-m and *mitf*-b also presented different expression patterns. Hence, subfunctionalization of *mitf* genes has been probably also associated with the degeneration of specific promoters and/or regulatory sequences (Fig. 5). The

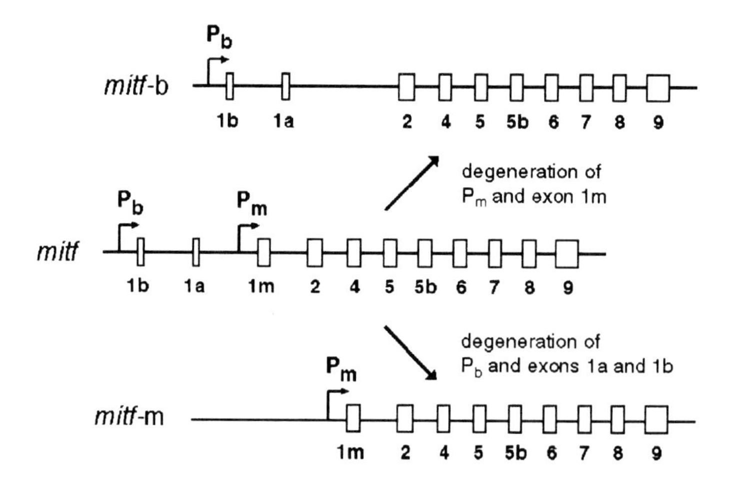

Figure 5. Simplified model for the subfunctionalization of fish *mitf* paralogues by differential degeneration of alternative introns and promoters.
After duplication of the ancestral *mitf* gene, promoter Pb and exons 1b and 1a necessary for the expression of MITF-b will degenerate in *mitf-m*, and vice-versa promoter Pm and exon 1m specific of the MITF-m isoform will be eliminated in *mitf-b*. Exons are numbered according to the human *mitf* gene, showing the loss of exon 3 and the gain of exon 5b in the fish genes (Altschmied *et al.*, 2002).

order of degenerative events remains to be determined: either mutations have first altered the coding sequence and/or the splicing site of an alternative exon in duplicate A (for example exon 1m in *mitf*-b) but not in duplicate B. By this way the selection for functionality on the promoter driving the expression of the transcript containing the alternative exon in duplicate A would have been abolished. Alternatively, alteration of promoter or specific regulatory sequences necessary for the expression of an alternative transcript in duplicate A but not in duplicate B would allow accumulation in duplicate A of non-conservative mutations in sequences being specific for this alternative transcript.

Conclusions

Analysis of genes encoding members of the EGF receptor family confirms the status of fish as an outstanding model for the study of the impact of duplication on gene function and organism complexity, and provides additional evidence for an ancient genome duplication in the ray-finned fish lineage. To the best of our knowledge, fishes possess the highest number of genes encoding such receptors ever reported (seven in teleosts, eight in some Xiphophorus species). New studies based on the different fish genome projects will also allow to assess if genes encoding ligands and downstream targets of these RTKs have been also maintained after duplication. A higher number of EGF receptor-related proteins might increase the different possible combinations of heterodimers and lead to new regulations and even new biochemical pathways. On the other hand, predominance of subfunctionalization might lead to a higher compartmentalization and specialization of the signal transduction pathways existing in higher vertebrates rather than to innovation. Clearly, a huge quantity of work is now required at the functional level to assess the impact of gene and genome duplication on the different pathways regulated by the members of the EGF receptor family in fish and other vertebrates.

Finally, these observations underline the complexity of fish models and the caution necessary for the transfer of knowledge from a fish model to higher vertebrates. There is now substantial evidence that some fish gene duplicates have evolved new functions divergent from those of the corresponding unique gene in higher vertebrates. Such caution is particularly true for signal transduction pathways involving multiple components, since a high number of these components are likely to be encoded by gene duplicates having experienced sub- and/or neofunctionalization during fish evolution. Nevertheless, especially in the case of subfunctionalization, specific mutants not viable in other animal models might be obtainable in fish. This should allow studying independently different pathways not separable in higher vertebrates.

Acknowledgements

This work is supported by grants to J.N.V. from the BioFuture program of the German Bundesministerium für Bildung und Forschung, and to M.S. from the Deutsche Forschungsgemeinschaft through the SFB 465 ('Entwicklung und Manipulation pluripotenter Zellen') and the Fonds der Chemischen Industrie. Takifugu data have been provided freely by the Fugu Genome Consortium for use in this publication only.

References

Abe, Y., Odaka, M., Inagaki, F., Lax, I., Schlessinger, J. and Kohda, D. (1998) Disulfide bond structure of human epidermal growth factor receptor. *J. Biol. Chem.*, **273**, 11150–11157.

Adam, D., Dimitrijevic, N. and Schartl, M. (1993) Tumor suppression in Xiphophorus by an accidentally acquired promoter. *Science*, **259**, 816–819.

Adamson, E.D. and Wiley, L.M. (1995) The EGFR gene family in embryonic cell activities. *Curr. Top. Dev. Biol.*, **35**, 71–120.

4Altschmied J., Delfgaauw, J., Wilde, B., Duschl, J., Bouneau, L., Volff, J.-N. and Schartl, M. (2002) Subfunctionalization of duplicate *mitf* genes associated with differential degeneration of alternative exons in fish. *Genetics*, **161**, 259–267.

5Altschmied, J., Ditzel, L. and Schartl, M. (1997) Hypomethylation of the *Xmrk* oncogene promoter in melanoma cells of Xiphophorus. *Biol. Chem.*, **378**, 145714–145766.

Amae, S., Fuse, N., Yasumoto, K., Sato, S., Yajima, I., Yamamoto, H., Udono, T., Durlu, Y.K., Tamai, M, Takahashi, K. and Shibahara, S. (1998) Identification of a novel isoform of microphthalmia-associated transcription factor that is enriched in retinal pigment epithelium. *Biochem. Biophys. Res. Commun.*, **247**, 710–715.

Amores, A., Force, A., Yan, Y.L., Joly, L., Amemiya, C., Fritz, A., Ho, R.K., Langeland, J., Prince, V., Wang, Y.L., Westerfield, M., Ekker, M. and Postlethwait, J.H. (1998) Zebrafish *hox* clusters and vertebrate genome evolution. *Science*, **282**, 1711–1714.

Aparicio, S. (2000) Vertebrate evolution: recent perspectives from fish. *Trends Genet.*, **16**, 54–56.

Baudler, M., Duschl, J., Winkler, C., Schartl, M. and Altschmied, J. (1997) Activation of transcription of the melanoma inducing *Xmrk* oncogene by a GC box element. *J. Biol. Chem.*, **272**, 131–137.

Elgar, G., Clark, M.S., Meek, S., Smith, S., Warner, S., Edwards, Y.J., Bouchireb, N., Cottage, A., Yeo, G.S., Umrania, Y., Williams, G. and Brenner, S. (1999) Generation and analysis of 25 Mb of genomic DNA from the pufferfish *Fugu rubripes* by sequence scanning. *Genome Res.*, **9**, 960–971.

Force, A., Lynch, M., Pickett, F.B., Amores, A., Yan, Y.L. and Postlethwait, J. (1999) Preservation of duplicate genes by complementary, degenerative mutations. *Genetics*, **151**, 1531–1545.

Franck, D., Dikomey, M. and Schartl, M. (2001) Natural selection and the maintenance of a macromelanophore colour pattern polymorphism in the green swordtail (*Xiphophorus helleri*). *Behaviour*, **138**, 467–486.

Froschauer, A., Körting, C., Bernhardt, W., Nanda, I., Schmid, M., Schartl, M. and Volff, J.-N. (2001) Genomic plasticity and melanoma formation in Xiphophorus. *Marine Biotechnol.*, **3**, S72–80.

Froschauer A., Körting, C., Katagiri, T., Aoki, T., Asakawa, S., Shimizu, N., Schartl, M. and Volff, J.-N. (2002) Construction and initial analysis of BAC contigs from the sex-determining region of the platyfish *Xiphophorus maculatus*. *Gene*, **295**, 247–254.

Goding, C.R. (2000) Mitf from neural crest to melanoma: signal transduction and transcription in the melanocyte lineage. *Genes Dev.*, **14**, 1712–1728.

Gómez, A., Wellbrock, C., Gutbrod, H., Dimitrijevic, N. and Schartl, M. (2001) Ligand-independent dimerization and activation of the oncogenic Xmrk receptor by two mutations in the extracellular domain. *J. Biol. Chem.*, **276**, 3333–3340.

Gutbrod, H., and Schartl, M. (1999) Intragenic sex-chromosomal crossovers of *Xmrk* oncogene alleles affect pigment pattern formation and the severity of melanoma in Xiphophorus. *Genetics*, **151**, 773–783.

Guy, P.M., Platko, J.V., Cantley, L.C., Cerione, R.A. and Carraway, K.L. 3rd (1994) Insect cell-expressed p180erbB3 possesses an impaired tyrosine kinase activity. *Proc. Natl. Acad. Sci. USA*, **91**, 8132–8136.

Hughes, A.L. (1999) Phylogenies of developmentally important proteins do not support the hypothesis of two rounds of genome duplication early in vertebrate history. *J. Mol. Evol.*, **48**, 565–576.

Kumar, S. and Hedges, S.B. (1998) A molecular timescale for vertebrate evolution. *Nature*, **392**, 917–920.

Lister, J.A., Close, J. and Raible, D.W. (2001) Duplicate *mitf* genes in zebrafish: complementary expression and conservation of melanogenic potential. *Dev. Biol.*, **237**, 333–344.

Lister, J.A., Robertson, C.P., Lepage, T., Johnson, S.L. and Raible, D.W. (1999) nacre encodes a zebrafish microphthalmia-related protein that regulates neural-crest-derived pigment cell fate. *Development*, **126**, 3757–3767.

Maynard Smith, J. and Haigh, J. (1974) The hitchhiking effect of a favourable gene. *Genet. Res.*, **23**, 23–35.

Meyer, A. and Schartl, M. (1999) Gene and genome duplications in vertebrates: the one-to-four (-to-eight in fish) rule and the evolution of novel gene functions. *Curr. Opin. Cell. Biol.*, **11**, 699–704.

Meyer, A., Morrissey, J.M. and Schartl M. (1994) Recurrent origin of a sexually selected trait in Xiphophorus fishes inferred from a molecular phylogeny. *Nature*, **368**, 539–542.

Moriki, T., Maruyama, H. and Maruyama, I.N. (2001) Activation of preformed EGF receptor dimers by ligand-induced rotation of the transmembrane domain. *J. Mol. Biol.*, **311**, 1011–1026.

Nanda, I., Volff, J.-N., Weis, S., Körting, C., Froschauer, A., Schmid, M. and Schartl, M. (2000) Amplification of a long terminal repeat-like element on the Y chromosome of the platyfish, *Xiphophorus maculatus*. *Chromosoma*, **109**, 173–180.

Ohno, S. (1970) *Evolution by Gene Duplication*. Springer Verlag, New York.

Ohno, S. (1999) The one-to-four rule and paralogues of sex-determining genes. *Cell. Mol. Life Sci.*, **55**, 824–830.

Postlethwait, J.H., Woods, I.G., Ngo-Hazelett, P., Yan, Y.L., Kelly, P.D., Chu, F., Huang, H., Hill-Force, A. and Talbot, W.S. (2000) Zebrafish comparative genomics and the origins of vertebrate chromosomes. *Genome Res.*, **10**, 1890–1902.

Robinson-Rechavi, M., Marchand, O., Escriva, H., Bardet, P.L., Zelus, D., Hughes, S., and Laudet, V. (2001) Euteleost fish genomes are characterized by expansion of gene families. *Genome Res.*, **11**, 781–788.

Roest Crollius, H., Jaillon, O., Bernot, A., Dasilva, C., Bouneau, L., Fischer, C., Fizames, C., Wincker, P., Brottier, P., Quetier, F., Saurin, W. and Weissenbach, J. (2000) Estimate of human gene number provided by genome-wide analysis using *Tetraodon nigroviridis* DNA sequence. *Nat. Genet.*, **25**, 235–238.

Schartl, M. (1995) Platyfish and swordtails: a genetic system for the analysis of molecular mechanisms in tumor formation. *Trends Genet.*, **11**, 185–189.

Schartl, M., Hornung, U., Gutbrod, H., Volff, J.-N. and Wittbrodt, J. (1999) Melanoma loss-of-function mutants in Xiphophorus caused by *Xmrk*-oncogene deletion and gene disruption by a transposable element. *Genetics*, **153**, 1385–1394.

Schlessinger, J. (2000) Cell signaling by receptor tyrosine kinases. *Cell*, **103**, 211–225.

Sierke, S.L., Cheng, K., Kim, H.H., and Koland, J.G. (1997) Biochemical characterization of the protein tyrosine kinase homology domain of the ErbB3 (HER3) receptor protein. *Biochem. J.*, **322**, 757–763.

Stephan, W., Wiehe, T.H.E. and Lenz, M.W. (1992) The effect of strongly selected substitutions on neutral polymorphism: analytical results based on diffusion theory. *Theor. Popul. Biol.*, **41**, 237–254.

Tachibana, M. (2000) MITF: a stream flowing for pigment cells. *Pigment Cell Res.*, **13**, 230–240.

Tajima, F. and Nei, M. (1984) Estimation of evolutionary distance between nucleotide sequences. *Mol. Biol. Evol.*, **1**, 269–285.

Taylor, J.S., Van de Peer, Y., Braasch, I. and Meyer, A. (2001) Comparative genomics provides evidence for an ancient genome duplication event in fish. *Philos. Trans. R. Soc. Lond. B Biol. Sci.*, **356**, 1661–1679.

Udono, T., Yasumoto, K., Takeda, K., Amae, S., Watanabe, K., Saito, H., Fuse, N., Tachibana, M., Takahashi, K., Tamai, M., Shibahara, S. (2000) Structural organization of the human microphthalmia-associated transcription factor gene containing four alternative promoters. *Biochim. Biophys. Acta*, **1491**, 205–219.

Ullrich, A. and Schlessinger, J. (1990) Signal transduction by receptors with tyrosine kinase activity. *Cell*, **61**, 203–212.

Van de Peer, Y., Taylor, J.S., Braasch, I. and Meyer, A. (2001) The ghost of selection past: rates of evolution and functional divergence of anciently duplicated genes. *J. Mol. Evol.*, **53**, 436–446.

Volff, J.-N. and Schartl, M. (2001) Variability of genetic sex determination in poeciliid fishes. *Genetica*, **111**, 101–110.

Weis, S. and Schartl, M. (1988) The macromelanophore locus and the melanoma oncogene *Xmrk* are separate genetic entities in the genome of Xiphophorus. *Genetics*, **149**, 1909–1920.

Wellbrock, C., Weisser, C., Geissinger, E., Troppmair, J. and Schartl, M. (2002) Activation of p59(Fyn) leads to melanocyte dedifferentiation by influencing MKP–1-regulated mitogen-activated protein kinase signaling. *J. Biol. Chem.*, **277**, 6443–6454.

Wittbrodt, J., Adam, D., Malitschek, B., Maueler, W., Raulf, F., Telling, A., Robertson, S.M. and Schartl, M. (1989) Novel putative receptor tyrosine kinase encoded by the melanoma-inducing *Tu* locus in Xiphophorus. *Nature*, **341**, 415–421.

Wittbrodt, J., Meyer, A. and Schartl, M. (1998) More genes in fish? *BioEssays*, **20,** 511–515.

Yamamoto, T., Nishida, T., Miyajima, N., Kawai, S., Ooi, T. and Toyoshima, K. (1983) The *erbB* gene of avian erythroblastosis virus is a member of the *src* gene family. *Cell*, **35**, 71–78.

Yasumoto, K., Amae, S., Udono, T., Fuse, N., Takeda, K. and Shibahara, S. (1998) A big gene linked to small eyes encodes multiple Mitf isoforms: many promoters make light work. *Pigment Cell Res.*, **11**, 329–336.

A. Meyer, Y. Van de Peer (eds.), Genome Evolution, 151-159.

The role of gene duplication in the evolution and function of the vertebrate Dlx/distal-less bigene clusters

Kenta Sumiyama, Steven Q. Irvine[†] & Frank H. Ruddle[*]
Department of Molecular, Cellular, and Developmental Biology, Yale University, PO Box 208103, New Haven, CT 06520, USA
[*]*Author for correspondence (E-mail: frank.ruddle@yale.edu)*
[†]*Present address: Department of Biological Sciences, University of Rhode Island, 100 Flagg Road, Kingston, RI 02881, USA*

Received 23.07.2002; Accepted in final form 29.08.2002

Key words: *Dlx/Distal-less* genes; Dlx3–7 cluster, enhancer sharing, evolution of *cis* elements, genomic regulation

Abstract

The Dlx gene family controls developmental patterning principally in the pharyngeal and cranial regions. We review the structure and function of these genes in the vertebrates and relate these properties to their evolution. We particularly focus on the Dlx3-7 bigene cluster which we postulate to be more derived phylogenetically and functionally than the other two bigene clusters, Dlx1-2 and Dlx5-6. We stress the transcriptional control of the Dlx3-7 bigene cluster, and postulate its control by Dlx1-2.

Argument

Darwin proposed that evolution is mediated by the generation of phenotypic diversity in life forms. The different phenotypes are subsequently selected by environmental conditions to yield species adapted to conditions prevalent in a particular period. The process is continuous: generating a succession of life forms through time. The rediscovery of Mendel's laws established the genetic basis of phenotypic diversity and implicated the gene as the functional unit. Thus variation in gene structure and function provides the basis of phenotypic variation.

Ohno introduced an important concept regarding the way genetic diversity might arise and mediate the evolutionary process. He postulated that gene duplication followed by gene diversification could be a significant source of genetic and concomitant phenotypic diversity (Ohno, 1970). The analysis of a broad spectrum of genetic systems shows that gene duplication generated by a number of mechanisms is ubiquitous and continuous and has resulted in the generation of families of related but diverse genes.

Ohno argued that immediately following duplication, the daughter genes would be essentially redundant each preserving the functional role of the parent. It can be presumed that any existing gene mediates at least one essential function. If so, following duplication, one gene is free to undergo diversification, while its unmodified counterpart supports the vital function. In instances where a gene supports multiple discrete essential functions, the process can be more complex with the assortment of different vital functions to both daughters and the concomitant diversification of both genes.

All genes are subject to duplication, but some may influence the evolutionary process more than others. Recently, homeotic genes have been identified that program pattern formation during development and play a significant role in determining body plan. These patterning genes are relatively few in number, constitute well-defined gene families, and are highly conserved throughout the animal kingdom. The homeotic genes are outstanding candidates for testing the validity of the Ohno gene duplication hypothesis.

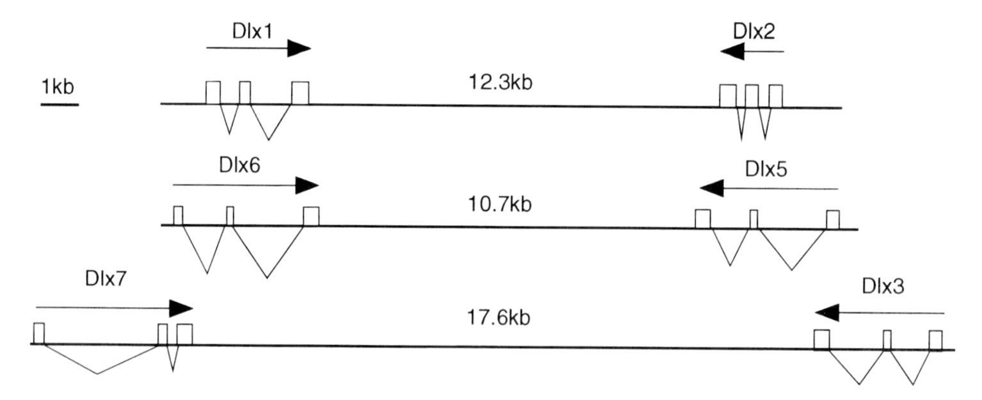

Figure 1. Approximate drawings for human Dlx clusters. Interegenic sizes (stop codon to stop codon) are shown. Empty boxes indicate coding regions. Reference genome sequences: AC023925 for Dlx1-2, AC004774 for Dlx5-6, and AC452638 for Dlx3-7 respectively.

Several well defined pattern formation gene systems have been reported. The Hox system which governs anterior-posterior axis and proximo-distal appendicular axis patterning is well known and serves as an excellent system for analysis. However, the Hox system is relatively complex consisting of 39 genes organized into four clusters in higher vertebrates (Ruddle *et al.*, 1999). Therefore, we have focused on the less complex Distal-less gene system consisting of only six genes divided into three bigene clusters. We believe that this gene system can be used profitably to provide insight into the Ohno gene duplication hypothesis.

The *Dlx/Distal-less* bigene cluster system

Evolutionary history

The distal-less system consists of three bigene clusters, termed Dlx1-2, Dlx5-6, and Dlx3-7 (Figure 1), that are closely linked within approximately one megabase to the Hox gene clusters, respectively, Hox D, A, and B on different chromosomes. Sequence comparisons between the mouse clusters shows a close relationship between Dlx1-2 and Dlx5-6 and a more distant relationship of both to Dlx3-7 (Figure 2). Orthologous gene clusters are also seen in the zebrafish. Phylogenetic analysis indicates that the Dlx gene family increased first by a tandem gene duplication, most probably in the primitive chordates, and then by several genome duplications in the vertebrates to give rise to the three bigene clusters seen today in man and mouse (Stock *et al.*, 1996).

Protostome invertebrates appear to have a single true Dlx gene, as exemplified by *Distal-less* in *Drosophila*, or *ceh-43* in *C. elegans* (Figure 3). Outside of the vertebrates, the only animal known to have a Dlx bigene cluster is the ascidian *Ciona intestinalis* (DiGregorio *et al.*, 1995). Ascidians are members of

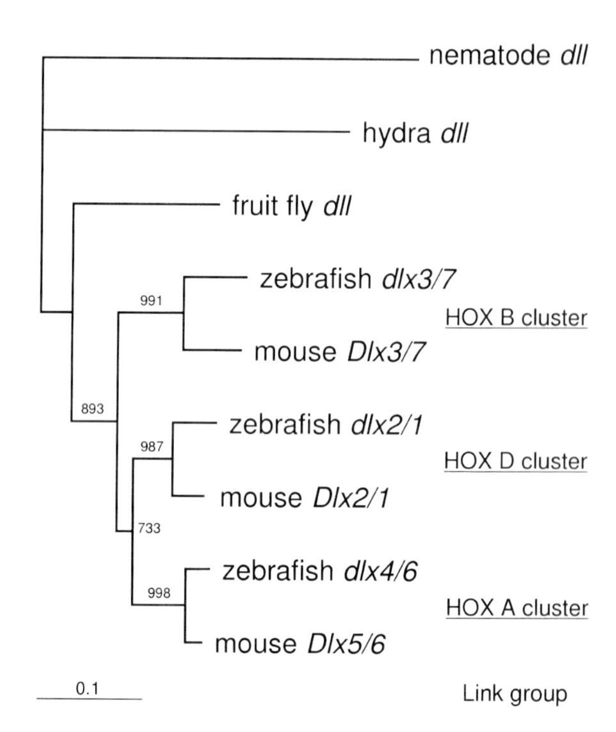

Figure 2. A neighbour-joining tree for Dlx clusters. Paired Dlx homeodomains are concatenated to represent each Dlx clusters for mouse and zebrafish. Single dlx gene in vertebrate ancestor is assumed. 1000 replication for bootstrap analysis have been done. Values more than 500 are shown in figure.

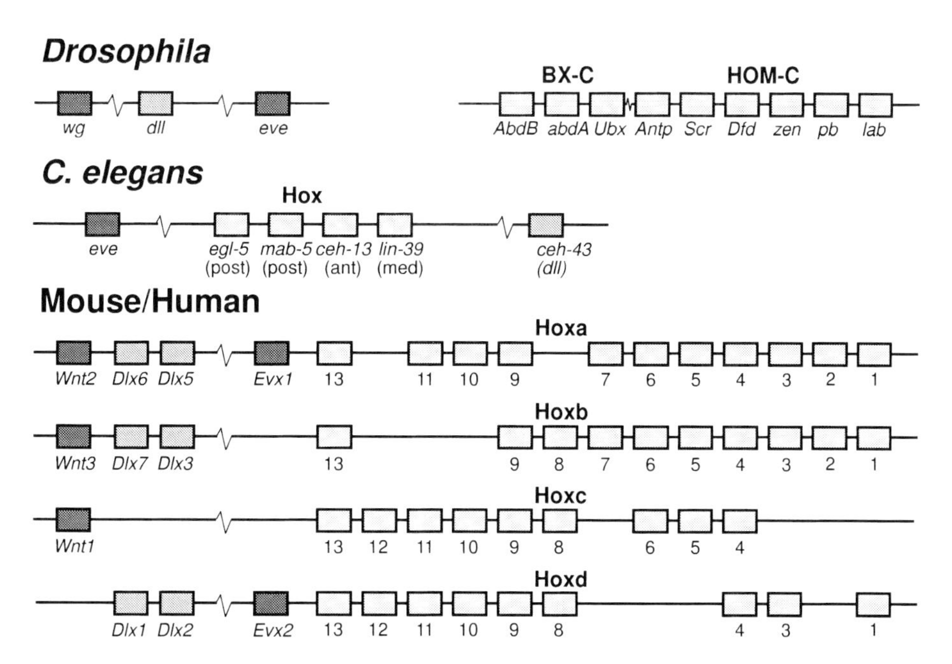

Figure 3. Hox and Dlx cluster organization in Drosophila, C.elegans and mammals (mouse and human). Boxes represent transcription units, and continuous horizontal lines connecting boxes represent linkage on the same chromosome. Hatch marks interrupting lines indicate a large gap between the genes represented. Note that not all intervening genes are shown. Observe also that the Dlx ortholog in C. elegans (ceh-43) is located on the opposite side of the Evx/Hox cluster relative to the pattern in mammals. In Drosophila the Dll and the Hox cluster are not linked.

the Urochordata, thought to be the most primitive of the chordate subphyla (Cameron *et al.*, 2000; Swalla *et al.*, 2000). Phylogenetic analysis of the protein sequences of the the clustered *Ci-Dll-A* and *Ci-Dll-B* from *C. intestinalis* (*CiDllA-B*) indicates that they are not closely related to the Dlx genes of vertebrates (Stock *et al.*, 1996). However, the similarity of their genomic organization to that of the vertebrate clusters, with a convergent transcriptional orientation suggests that *CiDllA-B* is homologous to the vertebrate Dlx clusters, which would put the origin of the tandem duplication producing the ancestral Dlx cluster prior to the common ancestor of all the extant chordates (Figure 4). (In this case one must assume that the placements of the ascidian genes on gene trees are not congruous with the true phylogeny of the Dlx clusters, likely due to unequal rates of evolution.)

An apparently unclustered Dlx gene has also been found in the ascidian *C. intestinalis* (*Ci-Dll-C*;, Caracciolo *et al.*, 2000), which may have arisen by a single duplication event, or by duplication of the cluster and subsequent loss of a gene. Curiously, only one Dlx gene, termed *AmphiDll* has been reported in the cephalochordate amphioxus, which is presumed to be more closely related to vertebrates than the ascidians. If another Dlx gene is not present in amphioxus, as indicated by genomic southern data (Holland *et al.*, 1996), and the ascidian cluster is truly homologous with those of the vertebrates, a Dlx gene must have been lost in the cephalochordate lineage.

In mouse and human, the three Dlx clusters are linked to Hox clusters A, B, and D. Thus the proto-vertebrate Dlx cluster was linked to the ancestral Hox cluster prior to the duplications leading to the present mammalian Hox clusters. Similar linkage patterns are apparent in the more numerous zebrafish clusters, although there may be more loss of Dlx genes in the 'supernumerary' clusters of teleosts (Figure 4; Amores *et al.*, 1998). Because of their inheritance from an ancestral cluster, the extant vertebrate Dlx genes form two clades, with *Dlx2*, *Dlx3*, and *Dlx5*, in one clade, and *Dlx1*, *Dlx6*, and *Dlx7*, in the other (Stock *et al.*, 1996). Tracing the history back to the most primitive extant vertebrate which has been examined leads to the agnathan lamprey *Petromyzon marinus*. Four Dlx genes have been found in *P. marinus*, three grouping with the *Dlx2/3/5* clade and one grouping with *Dlx1/6/7* in phylogenetic gene trees (Neidert *et al.*, 2001). There are probably three or four Hox clusters in this animal (Force *et al.*, 2002;

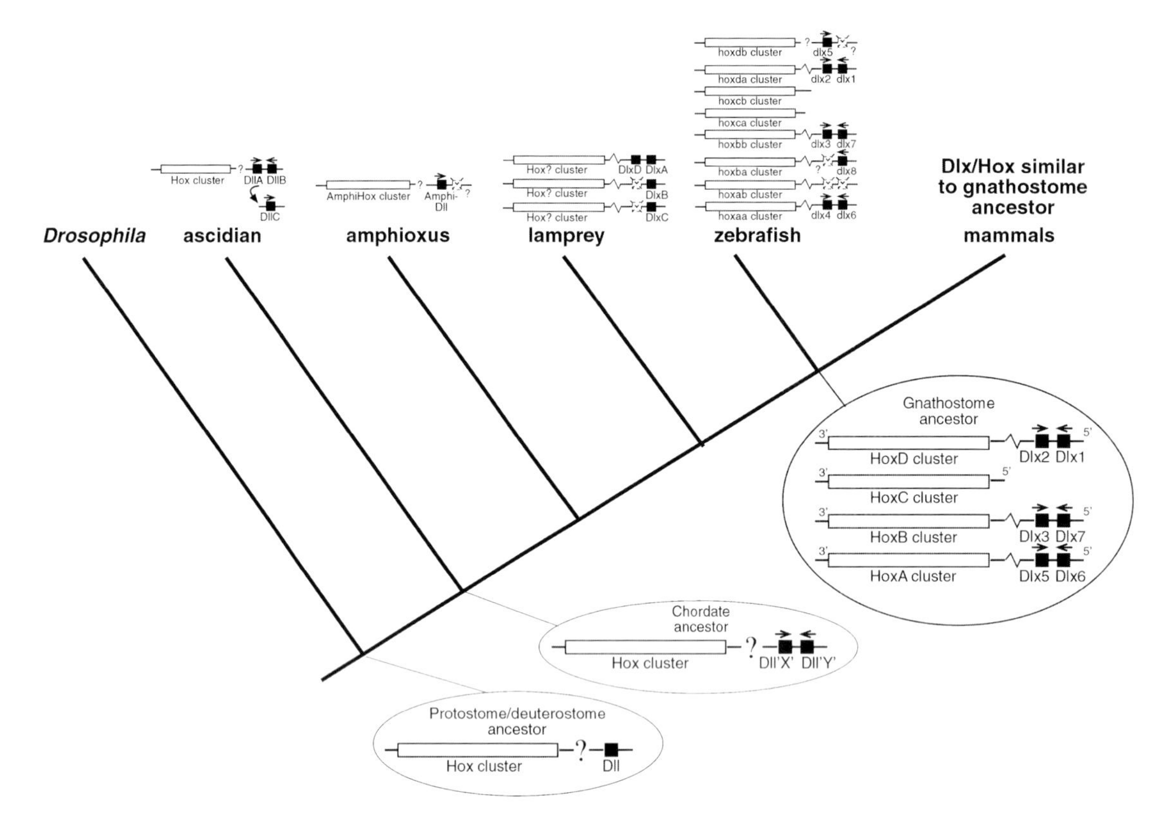

Figure 4. A hypothesis of Hox and Dlx cluster duplication history in animal evolution. Predicted Hox/Dlx cluster organization at ancestral nodes is shown in the ovals, while that for terminal taxa is shown above the name of the group. Hox clusters are shown as open rectangles, and Dlx genes are represented by black boxes, with their transcriptional orientation shown by arrows. Questionable linkage relationships are noted by question marks (all linkage shown for the lamprey Hox and Dlx clusters is conjectural). Inferred gene losses are shown by dashed boxes. Note that in this scenario the ancestral Hox/Dlx organization is similar to that of extant mammals.

Irvine *et al.*, 2002), so if the trees are correct, and the Dlx genes are linked to the Hox clusters, a likely model for the genomic arrangement in lampreys is shown in Figure 4. Here only one Dlx cluster remains, with two losses from two other presumed clusters. This scenario is consistent with the idea that the common ancestor of all the extant vertebrates had a complete, or nearly complete complement of Dlx clusters, as well as of Hox clusters (Force *et al.*, 2002; Irvine *et al.*, 2002). In this view, these ancestral clusters were inherited more or less intact by tetrapods, and were further duplicated in higher fishes.

Dlx patterns of control and expression

The genes within a cluster are transcribed convergently. In the case of the Dlx3-7 cluster this divides the cluster into a 5′ region private to Dlx3, a 3′ intermediate region of 14 kb common to both, and a 5′ region private to Dlx7. The Dlx1-2 and Dlx5-6 clusters have a similar organization. All genes have a three exon coding structure.

The Dlx genes are expressed in the cranial region from the hind brain region where they overlap with the anterior expressing Hox genes to the most anterior limit of the head including the regions of the mesencephalon and telencephalon. All three bigene clusters are expressed in the AER of developing limb buds and the first and second visceral arches (Ellies *et al.*, 1997; Qiu *et al.*, 1997). The Dlx3-7 genes show a divergent pattern of expression as compared to the expression patterns of Dlx1-2 and Dlx5-6 which are fairly similar to one another. For example, the Dlx3-7 genes are not expressed in the CNS while the other clusters are actively expressed in the brain. Moreover, Dlx3-7 is uniquely expressed in the placenta (Quinn *et al.*, 1998; Morasso *et al.*, 1999) and erythropoietic lineages (Takeshita *et al.*, 1993). These differences in expression between Dlx3-7 and the other bigene clusters are consistent with the phylogenetic relationships seen in Figure 2.

Certain patterns of expression can be associated with the structural organization of the distal-less genes. This is particularly so with respect to expression patterns in the visceral arches. For example, all three clusters show different patterns of expression in visceral arches one and two (Robinson and Mahon, 1994; Qiu *et al.*, 1997; Zhao *et al.*, 2000). However, the genes within clusters (i.e., genes 3 and 7 in the bigene cluster Dlx3-7) show highly similar patterns of expression. This concordant expression of genes within clusters suggests that they may be governed by shared cis-control elements in their shared intermediate 3′ domains (Ellies *et al.*, 1997). We have recently described conserved elements in the intermediate domain of Dlx3-7 by sequence comparisons between mouse and human Dlx3-7 clusters (Sumiyama *et al.*, 2002). Functional analysis of these motifs by transgenic mouse technology is consistent with their identity as cis-control elements. Moreover, our studies indicate that elements proximal to the Dlx3 coding region regulate expression in limbs primarily, while elements proximal to the Dlx7 gene are indispensable to the visceral arch expression of Dlx3.

Discussion and speculations

Duplication origins of the Dlx bigene clusters

The duplication history is poorly defined because of the paucity of Dlx mapping data in extant species. As noted above, the bigene cluster most probably arose in the early chordates, since a Dlx bigene cluster has been reported in the urochordate Ciona. Multiple Dlx bigene clusters probably arose subsequently by genome amplification in the vertebrate radiation. It is interesting that the Distal-less clusters are closely linked to the Hox clusters suggesting that the Hox and Dlx clusters duplicated simultaneously by common autopolyploidy events. In a previous report based on cluster sequence comparisons, we suggested that the Hox cluster topology relationships were A : D :: B : C and that the A : D clusters were the more primitive (Bailey *et al.*, 1997). Our analysis of the A and D clusters in the shark also supported the primitive nature of the A cluster (Kim *et al.*, 2000). The topology relationships of the Distal-less clusters, 1-2 : 5-6 :: 3-7 also show a consistency with the proposed Hox topology based on linkage relationships between Hox and Dlx clusters. The combined topology relationships can be written as Dlx1-2,Hoxd: Dlx5-6,Hoxa :: Dlx3-7,Hoxb : XXX,Hoxc, where XXX represents a dropout of a Dlx cluster. This topology suggests that Dlx1-2 and Dlx5-6 are more similar to one another and primitive, while Dlx3-7 is more divergent than the former and recent. Dlx expression patterns are consistent with this view, since Dlx1-2 and Dlx5-6 are extensively expressed in the brain while Dlx3-7 is not, and Dlx3-7 is uniquely expressed in more recently acquired mammalian organs and tissues such as placenta and formed elements of the blood (Quinn *et al.*, 1998; Morasso *et al.*, 1999; Takeshita *et al.*, 1993). These relationships are internally consistent but additional data is required to prove these relationships definitively.

Transcriptional regulation between Dlx clusters

The phylogenetic relationships between the Dlx bigene clusters takes on additional significance when we consider the possibility of trans regulation between the clusters. Previous studies have identified an enhancer motif in Dlx5-6 that is responsive to the Dlx2 transcription factor (Zerucha *et al.*, 2000). Dlx1-2 is expressed early in development and Dlx5-6 subsequent to its expression. We have recently reported on an element in the intermediate non-coding domain of Dlx3-7 that shows a nucleotide similarity to the Dlx5-6 enhancer motif. It will be of interest to determine if this motif is responsive to Dlx2 regulation. One might speculate on the possible connection between the phylogenetic and trans-regulatory relationships among the Dlx bigene clusters. One scenario might be that shown in Figure 5. Initially there exists a single preDlx1-2 that is auto-regulating. Following genome duplication there exists two clusters: a preDlx1-2 and a dupDlx1-2 where trans regulation exists initially between the two clusters. Following a second polyploid event, preDlx1-2 becomes Dlx1-2, and Dlx5-6, while dupDlx1-2 becomes Dlx3-7. The trans-control of Dlx2 is retained with respect to Dlx5-6 and Dlx3-7.

Sequence conservation between Dlx bigene cluster orthologs

As expected, the coding regions show a high level of sequence similarity between the Dlx orthologs in comparisons within the mammalia and within the ver-

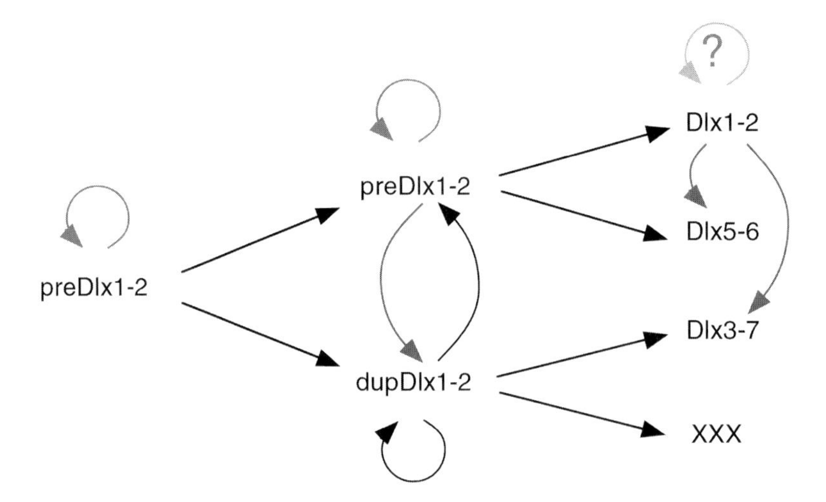

Figure 5. A model for duplication history of the Dlx clusters and their change in regulation. Arrows indicate putative regulational relationships.

tebrata. More surprisingly, a high level of similarity is also seen in non-coding regions. Human/mouse comparisons of the Dlx3/7 cluster shows a high sequence similarity in both 5′ and 3′ regions of each gene (Sumiyama *et al.*, 2002). Five major 3′ conserved elements were found in the shared intermediary region termed I37-1 through I37-5. The five conserved elements were tested functionally by transgene analysis. Results indicated that these motifs serve as control elements regulating the expression of the Dlx3-7 cluster genes.

The I37-1 element located proximal to the Dlx3 coding region showed high nucleotide sequence similarity between man and mouse (90%), but also between mouse and zebrafish (72%). This high level of similarity suggests an associated vital control function. Preliminary findings suggest a role in early limb development.

Element I37-2 shows second highest similarity and is located proximal to the Dlx7 gene. When this element is deleted visceral arch expression is lost. This element shows no similarity with the zebrafish even though Dlx expression in the visceral arches is much the same in mouse and zebrafish.

Element I37-5 is found in the center of the intergenic region. It has a partial sequence similarity with the adjacent element I37-1 (80% over 40 bp), possibly caused by a tandem duplication within the intermediary region. Considering the high level of conservation of I37-5 between human and mouse, we consider element I37-5 is most probably functional, suggesting that duplication is a significant factor not limited to coding domains, but also important in cis-element evolution.

Sequence conservation between Dlx bigene cluster paralogs

We have performed large-scale sequence comparisons between each of the paralogous Dlx gene clusters in both man and mouse. The coding regions show a high level of conservation. In contrast, the non-coding regions show little similarity. This is particularly interesting, since the Dlx genes express in overlapping fashion and might be expected to have common enhancer elements derived from their non-duplicated pre-bigene clusters. This result indicates that sequence evolution of cis-elements in non-coding regions is rapid in terms of large-scale sequence similarity, possibly due to the dynamic rearrangement of small protein binding elements within each functional cis-regulatory unit. It will be interesting to see whether these hypothecated small protein-binding motifs are shared between the paralogous Dlx clusters.

Enhancer sharing

The Dlx genes within the respective bigene clusters have highly similar overlapping patterns of expression. In regard to Dlx3-7, both Dlx3 and Dlx7 share highly similar expression patterns in the visceral-

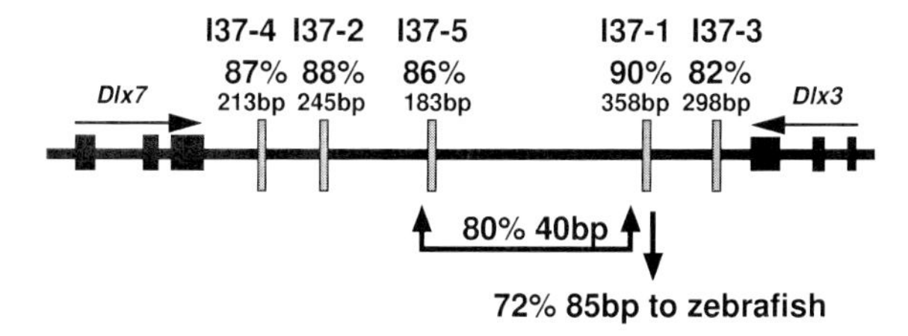

Figure 6. Five major conservation motifs in Dlx3-7 cluster elucidated by human and mouse comparison. I37-1 has further homology to Dlx3-7 intergenic region of zebrafish (Ekker, personal communication). I37-1 and I37-5 show similarity that suggests that those elements were generated by internal duplication.

alarches and in the limb. This raises the possibility that the paired Dlx genes may share enhancers that program common patterns of gene expression. Shared enhancers most probably exist in the common 3′ region shared by both genes (Ellies *et al.*, 1997). Our Dlx3-7 results have identified enhancers that regulate expression in both limb and visceral arches. Preliminary findings suggest that these enhancers may be shared between Dlx3 and Dlx7.

Shared enhancer elements could be either repressors or activators of gene expression. The repressor elements would be more difficult to identify by transgenic reporter assay, the usual method for cis-regulatory element identification, since their contribution to the shared expression pattern might not be readily apparent in comparing the effects of different pieces of flanking (regulatory) DNA on expression of reporters inserted into the two genes individually. One way around this situation is to examine the expression of different reporters inserted into the two genes on the same large genomic reporter construct. In this case, the contribution of a particular piece of cis-regulatory DNA can be assessed for the two genes simultaneously. Shared repressors would be apparent if the same expansion of the expression domains of the two reporters was seen simultaneously.

Rates of evolutionary change between Dlx3 and Dlx7

Although Dlx3 and Dlx7 share similar patterns of expression in visceral arches and limb buds, the rates of evolutionary change are significantly different. The Dlx7 gene showed approximately 10 times higher rates of evolutionary change at non-synonymous sites than that of Dlx3 in comparisons between man and mouse (Sumiyama *et al.*, 2002). This finding indicates that Dlx3 is under high levels of constraint, while Dlx7 is more freely evolving. Possibly, Dlx3 has retained original functions prior to cluster duplication, whereas Dlx7 is in the continuing process of acquiring new or modified functions. Since both fish and mammals have retained the Dlx7 ortholog, we assume that Dlx7 is indispensable irrespective of its capacity to further undergo modification. It will be of interest to define the common indispensable functions of Dlx7 postulated for both fish and mammals as well as determining its novel functions, especially in terms of the control of overlapping expression patterns common to Dlx7 and the more highly conserved Dlx3.

Blueprints for future research

Life expectancy of convergently duplicated genes

Enhancer sharing by paired genes may present an interesting special case in the life of duplicated genes. Normally, duplicated redundant genes will be deleted rapidly if they do not acquire an adaptive function. Convergently duplicated genes may represent an exception to this rule. In the initial duplicated state both genes may be dependent on shared enhancer elements in their common 3′ domains. Thus drastic modifications of these elements either by base pair change or by methylation may impair the function of both genes. Such changes would be selected against ensuring the coexistence of both genes over extended periods of time, enhancing the probability of adaptive diversification of both.

Tandem duplication of convergently duplicated genes is presumably a rare event, compared with 'in-line' tandem duplications, since it requires a double crossover of looped unequally paired homologs. (A double crossover is required to avoid an inversion of the whole remaining chromosome distal to the point of synapse.) However, convergently transcribed gene pairs are relatively common in the genomes of vertebrates. This observation suggests that these bigene clusters have particular constraints on gene loss following the duplication event, which cause both of the initially redundant genes to be retained long enough to accumulate mutations which confer unique functions. If the two genes had slightly different expression patterns after duplication, due to differences in their response to cis-regulatory elements, with even a small but necessary function in the non-overlapping spatial or temporal repertoire of expression, then both genes would have to be retained in spite of largely overlapping redundant functions (Sumiyama *et al.*, 2002). This would be a mechanism for 'buying time' to allow for more significant functional differences between the genes to arise, further reinforcing the necessity for the retention of both in the genome. It will be interesting to attempt to determine the probability of cis-gene duplication by parallel versus inverted events. If such measures were available, it would then be possible to compare the frequency of cis-parallel and cis-inverted gene pairs in the genome, providing an estimate of their relative life expectancy.

Evo-devo engines

In the Hox cluster genes, the general pattern is for the cis-paralogs within a cluster to have a greater difference in function and expression pattern than trans-paralogs. These functional differences are reflected in the greater sequence divergence between cis- than trans-paralogs. In the Dlx clusters, one sees a different pattern: here cis-paralogs have largely similar patterns of expression and function, while trans-paralogs show more differences – even though, like the Hox cluster genes the greater sequence divergence between cis- and trans-paralogs reflects their evolutionary history (Stock *et al.*, 1996). One explanation for this observation is that in the Dlx bigene clusters there might be a higher level of sharing of cis-regulatory elements between cis-paralogs than is present within the Hox gene clusters. This enhancer sharing would account for the similar patterns of expression between cis-paralogs.

Developmental control genes such as the clustered Hox and Dlx gene families exemplify unexpected patterns of gene duplication and diversification. Among orthologs there is conservation in both coding and non-coding domains. However, among trans-paralogs there is conservation in coding domains and high levels of diversification in the non-coding, regulatory domains. This is exemplified in the Dlx system where orthologs are highly similar in the coding and non-coding domains between human and mouse and to a lesser degree to the more distant Zebrafish. In contrast, trans-paralogs within mouse and human while conserved in coding domain are highly diversified in the non-coding domains. It is surprising that control elements identified in Dlx3-7 are not present in Dlx1-2 and Dlx5-6, even though the expression patterns of the three clusters in the visceral arches and limb buds are overlapped. These findings implicate the non-coding control elements as important agents of gene diversification, most probably playing an important role in evolution.

How may one explain non-coding control sequence revolution in the face of retention of similar expression patterns among paralogs? A key consideration is that in addition to similarities there are vital expression differences between the paralogs. These may be subtle, but still essential diversifications. One may theorize that to realize these differences comprehensive sequence changes must be made in the control domains, because of a high degree of cis-element interaction within these regions. The high level of conservation of gene cluster organization itself supports this assumption. One possibility is that small protein binding elements are retained, but simply rearranged. Identifying the specific control elements in each paralog and comparing each for the presence of common protein binding sites can test this explanation.

Acknowledgments

The study was supported by grants from the National Science Foundation (grant no. IBN-9630567) and National Institutes of Health (grant no. GM09966) to FHR.

References

Amores, A., Force, A., Yan, Y.-L., Joly, L., Amemiya, C., Fritz, A., Ho, R.K., Langeland, J., Prince, V., Wang, Y.-L., Westerfield, M., Ekker, M. and Postlethwait, J.H. (1998) Zebrafish *hox* clusters and vertebrate genome evolution. *Science*, **282**, 1711–1714.

Bailey, W.J., Kim, J., Wagner, G.P. and Ruddle, F.H. (1997) Phylogenetic reconstruction of vertebrate Hox cluster duplications. *Mol. Biol. Evol.* **14**, 843–853.

Cameron, C.B., Garey, J.R. and Swalla, B.J. (2000) Evolution of the chordate body plan: New insights from phylogenetic analyses of deuterostome phyla. *Proc. Natl. Acad. Sci. USA*, **97**, 4469–4474.

Caracciolo, A., Gregorio, A.D., Aniello, F., Lauro, R.D. and Branno, M. (2000) Identification and developmental expression of three *Distal-less* homeobox containing genes in the ascidian *Ciona intestinalis*. *Mech. Dev.* **99**, 173–176.

DiGregorio, A., Spagnuolo, A., Ristoratore, F., Pischetola, M., Aniello, F., Branno, M., Cariello, L. and DiLauro, R. (1995) Cloning of ascidian homeobox genes provides evidence for a primordial chordate cluster. *Gene*, **156**, 253–257.

Ellies, D.L., Stock, D.W., Hatch, G., Giroux, G., Weiss, K.M. and Ekker, M. (1997) Relationship between the genomic organization and the overlapping embryonic expression patterns of the zebrafish *dlx* genes. *Genomics*, **45**, 580–590.

Force, A., Amores, A. and Postlethwait, J.H. (2002) Hox cluster organization in the jawless vertebrate Petromyzon marinus. *J. Exp. Zool. (Mol. Dev. Evol.)*, **294**, 30–46.

Holland, N.D., Panganiban, G., Henyey, E.L. and Holland, L.Z. (1996) Sequence and developmental expression of *AmphiDll*, an amphioxus *Distal-less* gene transcribed in the ectoderm, epidermis and nervous system: insights into evolution of craniate forebrain and neural crest. *Development*, **122**, 2911–2920.

Irvine, S.Q., Carr, J.L., Bailey, W.J., Kawasaki, K., Shimizu, N., Amemiya, C.T. and Ruddle, F.H. (2002) Genomic analysis of Hox clusters in the sea lamprey Petromyzon marinus. *J. Exp. Zool. (Mol. Dev. Evol.)*, **294**, 47–62.

Kim, C., Amemiya, C., Bailey, W., Kawasaki, K., Mezey, J., Miller, W., Minoshima, S., Shimizu, N., Wagner, G. and Ruddle, F. (2000) Hox cluster genomics in the horn shark, Heterodontus francisci. *Proc. Natl. Acad. Sci. USA*, **97**, 1655–1660.

Morasso, M.I., Grinberg, A., Robinson, G., Sargent, T.D. and Mahon, K.A. (1999) Placental failure in mice lacking the homeobox gene *Dlx3*. *Proc. Natl. Acad. Sci. USA*, **96**, 162–167.

Neidert, A.H., Virupannavar, V., Hooker, G.W. and Langeland, J.A. (2001) Lamprey Dlx genes and early vertebrate evolution. *Proc. Natl. Acad. Sci. USA*, **98**, 1665–1670.

Ohno, S. (1970) *Evolution by Gene Duplication*. Springer Verlag, New York.

Qiu, M., Bulfone, A., Ghattas, I., Meneses, J.J., Christensen, L., Sharpe, P.T., Presley, R., Pedersen, R.A. and Rubenstein, J.L.R. (1997) Role of the Dlx homeobox genes in proximodistal patterning of the branchial arches: mutations of Dlx-1, Dlx-2, and Dlx-1 and -2 alter morphogenesis of proximal skeletal and soft tissue structures derived from the first and second arches. *Dev. Biol.* **185**, 165–184.

Quinn, L.M., Latham, S.E. and Kalionis, B. (1998) A distal-less class homeobox gene, DLX4, is a candidate for regulating epithelial-mesenchymal cell interactions in the human placenta. *Placenta*, **19**, 87–93.

Robinson, G.W. and Mahon, K.A. (1994) Differential and overlapping expression domains of *Dlx-2* and *Dlx-3* suggest distinct roles for *Distal-less* homeobox genes in craniofacial development. *Mech. Dev.* **48**, 199–215.

Ruddle, F.H., Carr, J.L., Kim, C.-B., Ledje, C., Shashikant, C.S. and Wagner, G. (1999) Evolution of chordate Hox gene clusters. *Ann. NY Acad. Sci.* **870**, 238–248.

Stock, D.W., Ellies, D.L., Zhao, Z., Ekker, M., Ruddle, F.H. and Weiss, K.M. (1996) The evolution of the vertebrate Dlx gene family. *Proc. Natl. Acad. Sci. USA*, **93**, 10858–10863.

Sumiyama, K., Irvine, S.Q., Stock, D.W., Weiss, K.M., Kawasaki, K., Shimizu, N., Shashikant, C.S., Miller, W. and Ruddle, F.H. (2002) Genomic structure and functional control of the Dlx3-7 bigene cluster. *Proc. Natl. Acad. Sci. USA*, **99**, 780–785.

Swalla, B.J., Cameron, C.B., Corley, L.S. and Garey, J.R. (2000) Urochordates are monophyletic within the deuterostomes. *Syst. Biol.* **49**, 52–64.

Takeshita, K., Bollekens, J.A., Hijiya, N., Ratajczak, M. and Ruddle, F.H. (1993) A homeobox gene of the Antennapedia class is required for human adult erythropoiesis. *Proc. Natl. Acad. Sci. USA*, **90**, 3535–3538.

Zerucha, T., Stuhmer, T., Hatch, G., Park, B.K., Long, Q., Yu, G., Gambarotta, A., Schultz, J.R., Rubenstein, J.L.R. and Ekker, M. (2000) A highly conserved enhancer in the Dlx5/Dlx6 intergenic region is the site of cross-regulatory interactions between Dlx genes in the embryonic forebrain. *J. Neurosci.* **20**, 709–721.

Zhao, Z., Stock, D., Buchanan, A. and Weiss, K. (2000) Expression of Dlx genes during the development of the murine dentition. *Dev. Genes Evol.* **210**, 270–275.

A. Meyer, Y. Van de Peer (eds.), Genome Evolution, 161-176.

Dopamine receptors for every species: Gene duplications and functional diversification in Craniates

Stéphane Le Crom, Marika Kapsimali, Pierre-Olivier Barôme & Philippe Vernier*
Développement, Evolution et Plasticité du Système Nerveux, UPR 2197, Institut de Neurobiologie A. Fessard, CNRS, Avenue de la Terrasse, F-91118 Gif-sur-Yvette, France;
**Author for correspondence: E-mail: vernier@iaf.cnrs-gif.fr*

Received 09.08.2002; Accepted in final form 29.08.2002

Key words: central nervous system, encephalization, G protein-coupled receptors, teleost, vertebrates

Abstract

The neuromodulatory effects of dopamine on the central nervous system of craniates are mediated by two classes of G protein-coupled receptors (D1 and D2), each comprising several subtypes. A systematic isolation and characterization of the D1 and D2-like receptors was carried out in most of the Craniate groups. It revealed that two events of gene duplications took place during vertebrate evolution, before or simultaneously to the emergence of Gnathostomes. It led to the conservation of two-to-four paralogous receptors (subtypes), depending on the species. Additional duplication of dopamine receptor gene occurred independently in the teleost fish lineage. Duplicated genes were maintained in most of the vertebrate groups, certainly by the acquisition of a few functional characters, specific of each subtypes, as well as by discrete changes in their expression territories in the brain. The evolutionary scenario elaborated from these data suggests that receptor gene duplications were the necessary conditions for the expansion of vertebrate forebrain to occur, allowing dopamine systems to exert their fundamental role as modulator of the adaptive capabilities acquired by vertebrate species.

Introduction

Dopamine is the prototype of a modulatory neurotransmitter. Dopamine is synthesized in specific neurons by hydroxylation and decarboxylation of the aromatic amino acid tyrosine, and is certainly used as a neurotransmitter in the whole Bilateria phylum. Although our current knowledge about dopamine is elusive or absent in many groups of species, especially in Protostomes, it can be taken as granted that the general role of dopamine is to control or to regulate specialized functions of the nervous system. A major effect of dopamine is certainly the modulation of sensory perception (visual and olfactory for example), since it has been demonstrated in both Protostomes (insects, molluscs) and in vertebrates. Most of the other known actions of dopamine have been observed in vertebrates, animals in which dopamine contributes to the regulation of hormone release in the pituitary gland, to the control of body temperature and food intake or the tuning of sensori-motor programs. Finally, in vertebrates also, dopamine systems are required to maintain or to manifest reward values of life experiences and, thus, motivation or aversion. The large and ancient interest manifested by scientists for dopamine came from the demonstration that dopamine neurotransmission is altered in several human pathologies, essentially Parkinson's disease, addiction to drugs of abuse, disorders of mood and affect (schizophrenia), prolactinomas of the anterior pituitary gland or hypertension.

The various effects of dopamine on organs and cells are mediated by metabotropic membrane receptors coupled to heterotrimeric G proteins (GPCR). Molecular approaches showed that the diversity of dopamine effects relied upon the binding of the transmitter on multiple receptors, which exhibit similar sequences and cellular activity. In the early nineties, sequences of five related genes encoding different dopamine receptor subtypes have been isolated in mammals, suggesting they were generated by gene duplication events. Gene duplications are theoretically one of the means by which organisms may increase cell diversity, organizational complexity and body size. Large-scale gene duplications are thought to be at the origin of major transition in the evolution of plants, fungi or animals. Some degree of genetic redundancy will obviously provide organisms with more physiological flexibility to adapt specific functions and behaviour to different and changing environments. However, for the experimental biologist, the physiological necessity for a multiplicity of proteins playing more or less the same roles in cells is never obvious to decipher. Indeed, given the inherent limits of experimental investigations, it is difficult to predict what the consequences of observations made at the molecular level of organization could be at the level of the organ or the whole organism. It appeared to be the case for the vertebrate dopamine receptors.

Viewing dopamine receptors as molecules in the context of cells, organs and organisms required also looking at their precise functions, localization and regulation in dopamine-responsive cells. This organization especially underscores the importance of the differentiation process and of the spatial arrangement of cells in the organization of an organism (Conrad, 1990; Kirschner and Gerhart, 1998). Therefore, understanding how the genetic multiplicity of G protein coupled receptors has been generated, and which is its functional counterpart, required to undertake a detailed comparative analysis of the various dopamine receptors present in Craniates. Clear differences may exist between structures and functions of homologous molecules found in different species. These differences have been selected for adaptive purpose all along evolution. Comparative studies of the structure, of the physiological role and the anatomical localization of dopamine receptors in Craniates has brought about observations and hypothesis, which may be of general interest for our understanding of the role of gene diversification in the evolution of the central nervous system.

The vertebrate dopamine receptors

In vertebrates, dopamine receptors belong to two distinct classes of G protein-coupled receptors. The first proposal of the existence of two classes of receptors for dopamine was based on the observation that, depending on the tissue, dopamine was able to exert different effect on adenylyl cyclase activity (Kebabian and Calne, 1979). The D_1 receptors were defined as the entities that allow dopamine to activate adenylyl cyclase and to promote cAMP accumulation in cells, whereas dopamine D_2 receptors were those that decreased cAMP amount in experimental preparations. Due to the pathophysiological interest of dopamine receptors in human, many pharmacological compounds were produced. They provided the tools to discriminate these two receptor classes and to establish the concept of multiple receptors for dopamine (reviewed in Seeman and Van Tol, 1994; Sokoloff and Schwartz, 1995; Neve and Neve, 1996).

As for the other monoamine receptors coupled to G proteins, molecular cloning and sequence analysis of dopamine receptors confirmed that D_1 and D_2 receptors constituted two structurally and functionally different classes of dopamine receptors. Based on sequence similarity and structural peculiarities, dopamine receptors belong to the group 'A' of the G protein-coupled receptor superfamily (Valdenaire and Vernier, 1997; Ji *et al.*, 1998; Bockaert and Pin, 1999). In mammals, two receptor subtypes have been assigned to the D_1 receptor class (D_{1A}/D_1 and D_{1B}/D_5) and the three others to the D_2 class (D_2, D_3, D_4). Nomenclature of dopamine receptor subtypes remains inadequate, because despite the recommendation of the IUPHAR Committee for Receptor Nomenclature (Vanhoutte *et al.*, 1996; Alexander and Peters, 2000), common usage has established a misleading naming of the receptors, such as D_1 and D_5 for the human D_{1A} and D_{1B} receptors. Similarly, in the D_2 receptor class, the logical nomenclature D_{2A}, D_{2B}, D_{2C}, instead of D_2, D_3, D_4, has never been adopted. In a recent past, the collation of the molecular data with the pharmacological tradition was problematic and often conflicting. Now, most of the contradictions have been resolved and the main characteristics of vertebrate dopamine receptors are summarized on Figure 1.

The D_1 receptor genes have no intron interrupting the coding sequence (although they have a small intron in the 5′ untranslated region) whereas D_2 receptor genes contain five to six introns in the coding region. Considering the overall protein structure and topology deduced from the receptor sequences, major differences exist between the two classes of dopamine receptors (reviewed in Civelli *et al.*, 1993; Neve and Neve, 1996; Missale *et al.*, 1998). The

bxsequences of the transmembrane segments involved in ligand binding and transduction exhibit sig-

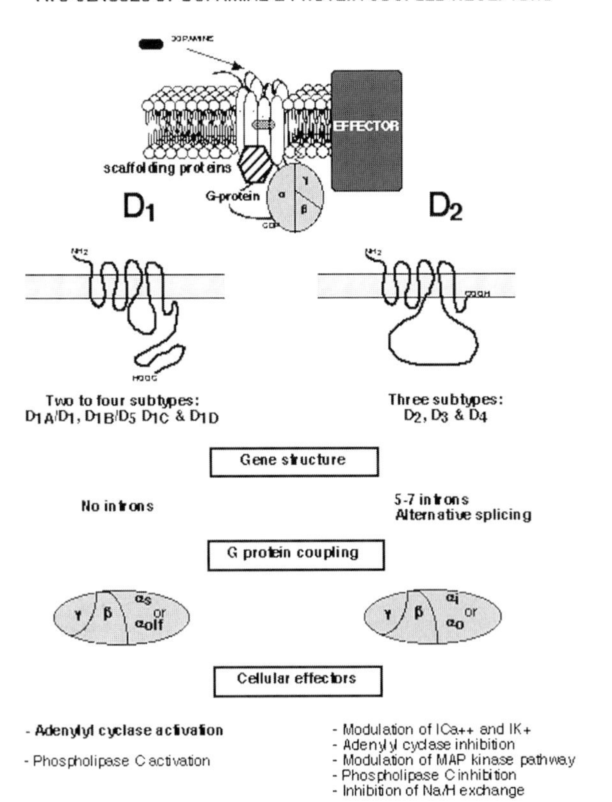

Figure 1. The main structural and functional features of the vertebrate dopamine receptors.
The two classes of dopamine receptors, named D_1 and D_2, belong to the G protein-coupled receptor superfamily. From a structural point of view, the D_1 and D_2 receptors are distinguished by different gene organization and by a few characteristics of the protein structure. D_1 receptors have a short third cytoplasmic loop and a long intracellular C-terminal stretch, whereas D_2 receptors have a long third cytoplasmic loop and a very short C-terminus anchored to the cytoplasmic side of the plasma membrane. Accordingly, the two classes of dopamine receptors are modulating differentially the activity of several intracellular signaling pathways via different classes of G proteins (mainly Gs/Golf proteins for the D_1 receptors and Go/Gi for the D_2 receptors).

nificant divergence between D_1 and D_2-related receptors. Only a few amino acid residues crucial for dopamine binding are shared by the members of the two dopamine receptor classes (Horn *et al.*, 2000). The D_2 receptors have a long third cytoplasmic loop and short cytoplasmic C-terminal end, whereas D_1 receptors exhibit a shorter third cytoplasmic loop but a very long C terminal tail, both segments being certainly involved in heterotrimeric G protein-coupling and receptor regulation (Figure 1). In mammals and birds, the D_2 receptor subtype exists in two isoforms generated by the alternative splicing of the pre-mRNA. This alternative splicing mechanism alters the length of the third cytoplasmic loop, raising the possibility for a specific activity for each of the isoforms.

In accordance to these structural differences, dopamine receptors of the D_1 and the D_2 classes interact with different classes of heterotrimeric G proteins. The D_1 receptors are primarily coupled to the Gs/Golf class of Gα proteins, whereas the D_2 receptors coupled essentially to Gi/Go proteins. However, this issue is not so simple, since for a given receptor subtype, coupling to G proteins may differ from one tissue to another, depending on G protein availability and interaction with other molecular partners (Lachowicz and Sibley, 1997; Sidhu and Niznik, 2000). The intracellular signaling pathways modulated by the two classes of receptors are also distinct. The major effect of $D_{1\text{-like}}$ receptors is to activate adenylyl cyclase. D_2-related receptors are generally able to inhibit adenylyl cyclase, but they also have many direct and major effects on voltage-gated ion channels and MAP-kinase pathways, to cite a few, well-documented examples (reviewed in Neve and Neve, 1996; Lachowicz and Sibley, 1997; Huff *et al.*, 1998; Missale *et al.*, 1998).

The very large body of data gathered over more than thirty years of investigation did not provide an integrated picture of the precise role of each of the dopamine receptor subtypes. Data obtained from *in vivo* and from *in vitro* studies were often ambiguous or conflicting. Moreover, when either the D_1 or the D_2 receptor subtypes were knocked out in mice, only the inactivation of the D_2 receptor subtype gave a conspicuous phenotype (a severe Parkinson's disease-like phenotype). The inactivation of the D_1 receptor subtype revealed unexpected role in motivational aspects of behaviour, as well as in learning and memory. No detectable or interpretable phenotype were obtained for the other dopamine receptor subtypes (Sibley, 1999). Thus, a comparative approach of the dopamine receptors was necessary to show when the dopamine receptors genes duplicated during vertebrates evolution, and for which adaptive purpose the multiplicity of dopamine receptors may have been conserved in Craniates (Fryxell, 1995; Vernier *et al.*, 1995).

Molecular phylogeny of dopamine receptors in Craniates

A molecular classification based on computation of identities and differences of the known dopamine receptor sequences provided interesting clues on the evolution and on some of the peculiarities of the dopamine receptor structure. The results of the molecular phylogenetical methods used (distance – Neighbor Joining-, Parsimony – PAUP or Maximum Likelihood) were essentially the same. The dopamine receptors belong to the large group of vertebrate monoamine receptors, which encompasses all the classes of catecholamine receptors and all the classes of the metabotropic serotonine receptors, plus the recently identified class of trace amine receptors (Borowski *et al.*, 2001). Not too surprisingly, molecular phylogenies of the monoamine receptors clearly indicate that D_1 and D_2 receptor classes are unrelated, or at least not more closely related to each other than they are from other classes of bioamine receptors (Figure 2). It is consistent with the very distinct structural characteristics and functional properties of the D_1 and D_2 dopamine. It implicates that they have acquired independently and convergently the ability to bind dopamine. This is also the case of the other classes of monoamine receptors (α_1, α_2, β, or TA, or $5HT_1$, $5HT_2$, $5HT_4$, $5HT_5$, $5HT_6$, $5HT_7$).

When our laboratory undertook to study the evolution of dopamine receptors, almost only mammalian sequences were available. It was thus necessary to isolate the whole complement of dopamine receptors in species belonging to the main groups of craniates that diverged at different times. We chose animals belonging to the main clades of craniates that diverged one from each other during 500 million years of evolution. It was much easier to isolate D_1 receptor sequences that D_2 receptor sequences since D_1 receptor genes have no intron. Therefore, genomic DNA could be used, instead of cDNAs prepared from whole organisms, giving more confidence to the possibility of isolating all the dopamine receptor-related sequences in the different species.

To begin with Agnathans, only one D_1-like sequence has been found in lampreys (*Lampetra fluviatilis* and *Petromyzon marinus*) and in hagfish (*Myxine glutinosa*). It is thus very likely that only one type of D_1 receptor exist in these species that are descended from the earliest diverging Craniates. Interestingly, these receptor sequences are not related to any of the known D_1 receptor sequences from jawed

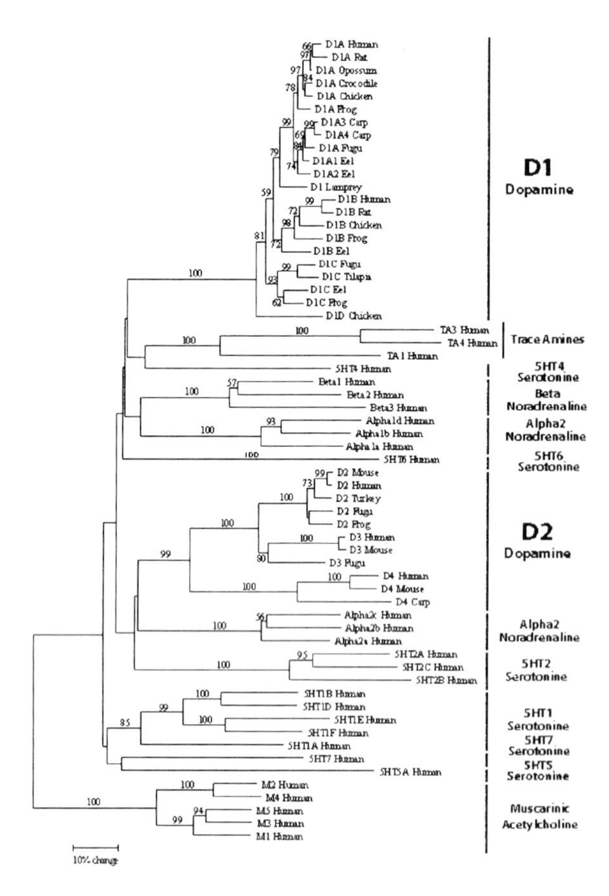

Figure 2. A phylogenetic tree of monoamine receptors in vertebrates with a special emphasis on dopamine receptors.
The tree has been generated by the Neighbor Joining method after a careful sequence alignment, build with the MUST package (Philippe, 1993). For the D_1 and D_2 dopamine receptor classes, receptor sequences have been chosen from species belonging to all the main groups of vertebrates. For the other classes of monoamine receptors, only the human sequences have been taken into account. The tree is rooted with the muscarinic acetylcholine receptors (M), which are the closest relatives of monoamine receptors in vertebrates. Bootstrap resampling method has been also carried out and only the significant values (percentage of one thousand resampling) are shown. In the D_1 receptor class, four categories of paralogous receptors are clearly identified, and correspond to the D_{1A}, D_{1B}, D_{1C} and D_{1D} subtypes of D_1 receptors. Each subtype comprised orthologous receptor sequences found in species that belong to the main groups of vertebrates, with the exception of the D_{1C} and D_{1D} receptor subtypes that do not comprise mammalian sequences (all the available sequences are not shown for the seek of clarity). Similarly, three subtypes are delineated in the D_2 receptor class, in Gnathostomes. Different apparent speeds of sequence divergence are clearly apparent on the tree. For example, the 'long branch effect' accounts for the deep branching of the D_4 receptor clade in the tree. Similar effect is responsible for the branching order in the D_1 receptor class. The consequence of this effect is that the tree does not reflect the evolutionary history of these receptors. The true phylogeny of dopamine D_1 receptors is described in Figure 3.

vertebrates, rendering plausible that they diverged from the ancestor of the D_1 receptor, before the duplications that gave rise to the different subtypes existing in later diverging vertebrates (see Figure 2).

Accordingly, in one cartilaginous fish (the electric ray *Torpedo marmorata*) three different D_1 receptor sequences have been found, whereas between three and five D_1 receptor sequences have been isolated in teleost fish, the earliest diverging jawed vertebrates we have studied (Cardinaud *et al.*, 1997, 1998). In *Xenopus*, a representative of anuran amphibians, the earliest diverging group of tetrapodes, three different D_1 receptor sequences have been isolated (Sugamori *et al.*, 1994). In all these species, it is easy to see from the phylogenetic tree that three of these sequences can be assigned to a precise subtype, namely, D_{1A}, D_{1B} and D_{1C} (Figure 2). In fish, additional D_1 receptor sequences have been found. Some of them are very closely related to D_1 subtypes such as the supplementary D_{1A} receptors found in eel, medaka, zebrafish or carp. In that case, which certainly corresponds to a gene duplication specific to fish, the two paralogous sequences have been logically named D_{1A1} and D_{1A2}. However, other sequences are distributed erratically in the phylogenetic tree and cannot be assigned to a given group of D_1 sequences. It is now very likely that teleost fish have undergone specific gene duplications, either in the very ancestral fish species or independently in several fish lineage (Wittbrodt *et al.* 1998; Aparicio, 2000; Robinson-Rechavi *et al.*, 2001; Taylor *et al.*, 2001). In that case, it is also probable that these sequences are mostly redundant and thus free to diverge fast. Evidence for that is the basal position of the sequences in the phylogenetic tree and the fact that they differ widely from one fish species to another, indicating also that some genes may have been 'lost' in several fish species (Taylor *et al.*, 2001). Incidentally, our classification based on molecular phylogenies helped to homogenize receptor nomenclature. For example, some of the first D_{1C} receptor sequences that were isolated in fish were misnamed D_{1B}, and the nomenclature of the sequences specifically duplicated in teleost fish were often illogical.

In amniotes, the situation is complicated either (Figures 2 and 3). A fourth D_1 receptor-related sequence has been isolated from chicken and designated as D_{1D} (Demchyshyn *et al.*, 1995). A more comprehensive search carried out in archosaurs representatives (birds and crocodiles) and lepidosaurs (lizzards and snakes), revealed that these species have

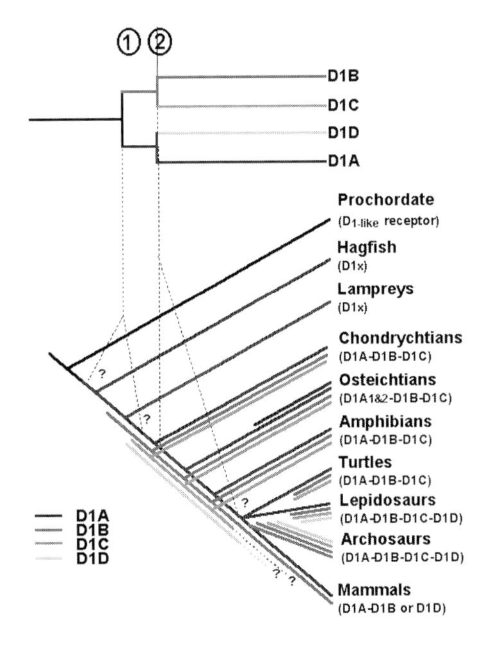

Figure 3. The history of duplication of the D_1 receptor genes in Craniates.
The putative emergence of the various D_1 receptor genes during evolution is shown relative to the established phylogeny of Craniates (lower panel). Note that the mammals, the Archosaurs (birds and crocodiles), the Lepidosaurs (snakes and lizzards) and the turtles diverged at about the same time. The upper panel presents the most probable phylogenetic relationships between the four known main subtypes of D_1 receptors. The four dopamine D_1 receptor subtypes have been generated after two steps of gene duplications. These two genetic events occurred before the emergence of jawed vertebrates since three subtypes (D_{1A}, D_{1B}, D_{1C}) with orthologous sequences present in most of the other vertebrate groups exist in cartilaginous fish, the earliest diverging Gnathostomes. Since only one D_1 receptor gene is present in hagfish and lampreys (the only living representative of jawless vertebrates or Agnathans), it is impossible to know with confidence when these two duplications took place during early Vertebrate evolution. In addition, and although it remains unlikely (based on the genomic data available), the possibility remains that the gene duplication that gave birth to the D_{1D} sequence occurred with the mergence of Amniotes (only this group of species possesses this receptor subtype). Specific gene duplications certainly occurred in the bony fish lineage, most of these animals being largely tetraploids. Several loss of D_1 receptor genes are evidenced here, the D_{1D} receptor gene in Anamniotes and in turtles, and the D_{1C} receptor gene in mammals, for example.

indeed four D_1 receptor-related genes, assigned to D_{1A}, D_{1B} D_{1C} and D_{1D} receptor subtypes. In turtles, D_{1A} and D_{1B} receptor-related sequences are present, but only a third sequence has been found, and it is distantly related to D_{1C} receptors. The most astonishing result came from mammals. Firstly, only D_{1A} and D_{1B} receptor-related sequences have been isolated ever in eutherian (placental) mammals, as well as in metatherian (marsupial) mammals. Secondly, in the two main species of prototherian mammals, platypus and echidnea, two D_1 receptor-related sequences have also been found, as in the other mammals, but they are assigned to the D_{1A} and D_{1D} subtypes. The most parsimonious interpretation of these data is that four different subtypes of D_1-related receptors existed in stem amniotes and that two of them disappeared in mammals, D_{1B} and D_{1C} in prototherian mammals and D_{1C} and D_{1D} in the other groups of mammals. Thus, mammals are an exception among vertebrates since the D_{1C} receptor subtype is present in all the groups of jawed vertebrates, the mammals being an exception, and the D_{1D} subtype present in all the amniotes, with the exception of placental and marsupial mammals.

A detailed analysis of the molecular phylogeny of the D_2 receptor class is not possible yet. However, based on currently available data, the picture of the $D_{2\text{-like}}$ receptor multiplicity looks very similar to that of the D_1 receptor class. Again, three subtypes are identified, namely the D2, D3 and D4 subtypes (Figure 2). These subtypes form three robust clades inside the D_2 receptor class, and they are accordingly found in most of the vertebrate groups where they have been investigated. For example, no ambiguity exists to assign fish, amphibian or bird receptors to the D2, D3 or D4 subtypes. Whether one or several $D_{2\text{-like}}$ receptors exist in Agnathans is not known at present, but it remains very likely that the evolutionary history that gave birth to the three D_2 receptor subtypes that exist in modern Gnathostomes is very similar to that of the D_1 receptor subtypes.

The molecular phylogeny of dopamine receptors revealed that the number of dopamine receptor subtypes in a given class varies from one group of vertebrates to another. However, taken all together, our data support the hypothesis of a double whole-genome duplication in the evolution of Craniates. It is however impossible to know when these duplications took place, except they mus have occurred before or simultaneously to the emergence of gnathostomes. A consequence of this statement is that bony fish and amphibians have lost one D_1 receptor sequence (D_{1D}), as well as turles, and the mammals have lost two of these genes. Another aspect of the evolution of the dopamine receptors is the clear differences of relative divergence (apparent 'evolutionary speed') exhibited by the sequences of the multiple subtypes of monoamine receptors. The 'long branch' effect is manifest for the D_4 receptor subtype in the D_2 class and for the D_{1C} and D_{1D} subtype in the D_1 class. This 'long-branch' effect conceals the 'true' evolutionary relationships of the D1 receptors in vertebrates. For example, the branch length for the mouse and human D_4 receptors is three to four time longer as compared to that of the D_2 or the D_3 receptor in the same species (see Figure 2). Interestingly, high degree of polymorphism exists for the human D_4 receptor gene (Van Tol *et al.*, 1992), in agreement with the fact that the D_4 receptor sequence has undergone a large drift, at least in mammalian species.

Functional characters of dopamine receptor subtypes in vertebrates

Dopamine receptors, as the other G protein-coupled receptors, basically play the role of an 'exchange factor' for the heterotrimeric G proteins, the activity of which is modulated by the ligand binding. Differences in both ligand binding affinity, basal activity, efficacy and potency for G protein activation, desensitization and internalization rates may be expected to be discriminating properties of the dopamine receptors. Indeed, acquisition of novel, derived functional characters is a mean by which duplicated genes will obviously gain non-redundant characters, leading to their conservation in the species where they are expressed. Whether such properties, specific of a given receptor subtype and conserved among species, existed within the dopamine receptor classes necessitated to be directly and specifically addressed. Such a comparison is currently only available for the D_1 receptor class, since no phylogenetical analysis of the D_2 receptor class has been undertaken yet (Niznik *et al.*, 1998).

Functional properties of the D_1 receptor class

To be able to compare directly the main functional properties of D_1 dopamine receptors, corresponding cDNA sequences were transfected in heterologous cell lines, allowing these molecules to be studied in the same cellular environment. Such a comparative approach as been carried out for the fish (European eel), amphibian (xenopus), bird (chicken) and mammalian (rat and human) D_1 receptor subtypes (Niznik *et al.*, 1998).

As far as ligand binding is concerned, the three D_1 receptor subtypes display differential affinity for the natural ligand, dopamine. The D_{1A}/D_1 receptor subtype has a five-to-ten fold lower affinity for dopamine than the D_{1B}/D_5 receptor subtype, whereas the fish and amphibian D_{1C} receptors also exhibit a lower dopamine affinity than the D_{1A}/D_1 subtype. The difference was very reproducibly found is every species tested. The affinity for artificial ligands, either agonists or antagonists, developed as drugs in human, reflects a specific kind of interaction with the binding pocket or even other part of the receptor structure. Comparison between the binding characteristics of the mammalian D_1 receptors with birds, amphibian of fish D_1 receptors revealed a large overall similarity (Figure 4).

Another important properties of G protein-coupled receptors is their so-called 'intrinsic' or 'basal' activity. This property is thought to reflect the allosteric character of these membrane proteins, which may oscillate spontaneously between different functional states. Thus, this intrinsic activity may be reflected in the basal level of GDP-GTP exchange, or in the basal level of 'second messengers' formation in cells where the analysed dopamine receptor is present as compared to cells that do not express the receptor. In the case of D_1 receptors, several authors have reported that the D_{1B}/D_5 receptor subtype displayed a significantly higher intrinsic activity than the D_{1A}/D_1 receptors. Depending on the species (European eel, xenopus, chicken, rat and human), and on the cell used, basal levels of cAMP are three-to-five times higher in D_{1B}/D_5 receptor-transfected cells that in cells expressing similar levels of D_{1A}/D_1 receptors (Tiberi and Caron, 1994; Cardinaud *et al.*, 1997; Sugamori *et al.*, 1998). Thus, it is very likely that a high intrinsic activity is an important specific property of D_{1B}/D_5 receptors, reflecting also the physiological peculiarity of this receptor subtype (Figure 4). The presence of this receptor in dopamine-target cells could provide them with a 'dopaminergic tone' that may be modified either by the binding to the receptor of different amount of dopamine or by the regulation of receptor activity through intracellular signaling mechanisms. The D_{1C} receptor sfrom xenopus and eel display a low basal activity, as the D_{1A}/D_1 receptors.

The last crucial functional property of G protein-coupled receptors is desensitization. Strictly speaking, desensitization qualifies the fact that, upon agonist exposure, the cell response elicited by a given receptor subtype progressively vanishes. The D_1 receptor subtypes have not been extensively compared for their desensitization properties. Neverthe-

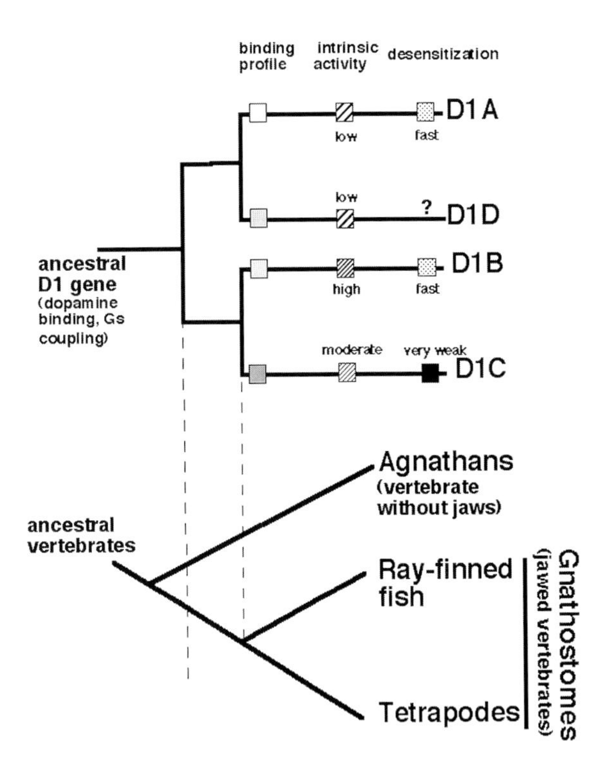

Figure 4. The duplication of D_1 receptor-encoding genes and the acquisition of subtype-specific functional properties.
The figure illustrates the acquisition of derived, specific characters by the different subtypes of dopamine D_1 receptors after the duplication of the ancestral genes. The two duplications that generate the current state of D_1 receptor multiplicity in vertebrates occurred before the emergence of Gnathostomes (jawed vertebrates). The hypothesis that is favored here is that the four subtypes of D_1 receptors known in modern Gnathostomes were generated by the same events of gene duplication (see Figure 3 and text). It implicates that the D_{1D} receptor subtype has been lost by the bony fishes, the amphibians and the mammals (except prototherian mammals). After the gene duplications that generated the four paralogous D_1 receptor subtypes, each of the receptor sequence acquired derived, subtype-specific properties. They are now characterized by a specific binding profile for ligand such as dopamine and artificial compounds developed as drugs for human diseases. The D_{1A} and D_{1D} receptors shared a relatively low affinity for dopamine, whereas the D_{1B} receptors and D_{1C} receptors have respectively a high and moderate affinity for the natural ligand. Basal, intrinsic activity of the D_{1B} receptor is the major discriminating property of this subtype, whereas the D_{1C} receptor is particularized by a very weak and transient process of agonist-induced desensitization. Based on sequence similarities, apparent sequence divergence and functional properties, and the fact that the D_{1A} receptor sequences are the most conserved in vertebrates, we propose that D_{1A} and D_{1D} receptors on the one hand, and D_{1B} and D_{1D} receptors, on the other hand shared a common ancestor.

less, for the D_{1A} receptors belonging to several vertebrate species, including European eel, xenopus, mouse, rat, monkey and human, the desensitization time course of the receptor is strikingly similar (Olson and Schimmer, 1992; Tiberi *et al.*, 1996; Lewis *et al.*, 1998; Jiang and Sibley, 1999). As a matter of fact, the D_{1A} receptor desensitizes very fast and up to 50% of the ability of the receptor to stimulate cAMP production is lost during the first five minutes of agonist exposure. The conservation of this fast desensitization rate by different species belonging to several vertebrate groups suggests it is an important discriminating property of the D_{1A} receptor subtype. The desensitization profile of the D_{1B} receptor is characterized by the fact that maximum cAMP levels raised by full agonists are always lower than those elicited by D_{1A} receptor activation (Jarvie *et al.*, 1993 and unpublished results). The desensitization time course of the D_{1B} receptor appears also attenuated relative to that of the D_{1A} receptor. It is probable that the D_{1B} receptor constitutive activity may favor a 'constitutive' desensitization of the receptor, as suggested for other catecholamine receptors (Pei *et al.*, 1994). It is probable that the high basal activity of the D_{1B} receptor and attenuated desensitization are specific D_{1B} receptor characters, conserved in most of the jawed vertebrates. The desensitization properties of the D_{1C} receptors have been studied with the eel and xenopus sequences (unpublished results). Interestingly, the D_{1C} receptors are not able to desensitize, except for a very short and weak attenuation of cAMP accumulation. Again, it can be proposed that the absence of desensitization is an important distinguishing property for the D_{1C} receptors.

Despite some uncertainties and lack of important data, the comparative approach to D_1 receptor functions brings new information about the functional properties that particularize the D_{1A}, D_{1B} and D_{1C} receptor subtypes that are expressed in most groups of jawed vertebrates. Since these functional properties are essentially conserved among phylogenetically distant vertebrate species, they have certainly been inherited from ancestral species in which these properties have been selected to fulfil physiological purpose (Figure 4). The D_{1A}/D_1 receptors exhibit the most highly conserved sequences in the D_1 receptor class and they are found in all groups of jawed vertebrate (gnathostomes) known so far. This subtype is characterized by a low basal activity, a relatively low affinity for dopamine and a fast and strong agonist-dependent desensitization. Such properties are fully compatible with a localization of the D_{1A} receptors close to the synaptic site of dopamine release (Hersch *et al.*, 1995). As also suggested by the phenotype of the targeted null mutation of the corresponding gene in mouse, the D_{1A}/D_1 receptor subtype is probably endowed by the most critical dopamine functions among the D_1 receptors. The D_{1B}/D_5 receptors are also present in all groups of Gnathostomes, but corresponding sequences are clearly less conserved than those of the D_{1A}/D_1 receptors. At the cellular level, D_{1B}/D_5 receptors are found on dendritic shafts, far from synapses (Bergson *et al.*, 1995; Hersch *et al.*, 1995). This localization may account for the high affinity and high intrinsic activity that characterized the vertebrate D_{1B}/D_5 receptors. Based on current data, D_{1C} receptors of fish and amphibians seem to be resistant to desensitization. Precise localization of D_{1C} receptors in neuronal networks is still unknown.

Functional properties of the D_2 receptor class

Even if a large amount of data have been obtained about the modulation of intracellular signalling pathways by $D_{2\text{-like}}$ receptors, the members of the D_2 receptor class have not been submitted to the same systematic analysis than the D_1 receptors. Moreover, functional and pharmacological data are available for mammalian species only. Thus, a thorough comparative analysis of the conserved and derived properties of the $D_{2\text{-like}}$ receptor subtypes is not really possible at this time. Some broad pictures of the conserved and derived characteristics of the D_2, D_3 and D_4 receptors may be presented however (Huff *et al.*, 1998; Missale *et al.*, 1998).

Firstly, each of the three D_2 receptor subtypes is characterized by a specific pharmacological profile with as general characteristic that some drugs are generic D_2 agonists or antagonists whereas more specific compounds have also been developed. Interestingly, dopamine has a significantly higher affinity (ten-to-thirty fold) for the D_3 and the D_4 than for the D_2 receptor subtypes. A lot of effort has been developed to obtain more specific drugs, and compounds such as BP 897 or nafadotride are now considered as reasonably specific D_3 receptor agonists and antagonists respectively (Sokoloff and Schwartz, 1995). In contrast, no usable discriminating compound is known yet for the D_4 receptor, although its higher affinity for clozapine has been widely popularised (Van Tol *et al.*, 1991).

Secondly, the inhibition of adenylyl cyclase activity has been historically the proposed discriminating and defining property of the D_2 receptor subtypes. It remains true that the modulatory inhibition of adenylyl cyclase activity is one of the most conserved characters of the D_2 receptor subtypes. Nevertheless, the most conspicuous activity of the D_2 receptor subtypes is the modulation of voltage-gated ionic channels. The D_2 receptor subtype has been consistently shown to inhibit the activity of several types of Ca^{2+} ionic channels. This effect is mediated by Goα protein in pituitary cells (Lledo *et al.*, 1992, 1994) and probably also in neurons from substantia nigra. Similarly, the D_3 and D_4 receptor subtypes are able to inhibit L and N-types of Ca^{2+} currents in transfected pituitary and neuronal cell lines (Seabrook *et al.*, 1994a, 1994b). Thus, the three D_2 receptor subtypes are able to efficiently modulate the activity of K^+ and Ca^{2+} channels in various types of excitatory cells, and this effect is probably the most consistent and most conserved effect of the D_2 receptors. Many other effects of the D_2 receptor subtypes have been demonstrated (mobilisation of intracellular calcium stores, PKC activation, release of arachidonic acid, cell toxicity or apoptosis, Na^+/H^+ exchange, coupling to glutamate-gated channels, inhibition of the Na^+/K^+ ATPase ... etc), but it remains difficult to know whether these effects are indeed a direct modulation of effectors by the α or $\beta\gamma$ subunits of G proteins or if it depends on indirect action or 'cross-talk' between intracellular signaling pathways (see Huff *et al.*, 1998; Missale *et al.*, 1998).

Thirdly, agonist-induced desensitization or other regulatory mechanisms may differ from one D_2 receptor subtype to another. Since the D_2 receptor subtype exist in two isoforms generated by alternative splicing, at least in mammals and birds, several authors have addressed the question of the regulation and functional consequences of this molecular variation. Sex steroid hormones modulate the splicing of the pre-mRNA in the rat, œstrogens favoring the production of the long isoform in the pituitary gland whereas androgens led to increase the amount of isoform in the olfactory bulb and the pituitary gland (Guivarc'h *et al.*, 1995, 1998). Such a regulation splicing may promote differential localization in the cells and coupling to signalling pathways. Finally, although the mechanisms of agonist-induced desensitization of the D_2 dopamine receptors have not been extensively analyzed, some degree of constitutive internalization may characterize the D_2 receptor subtype (Vickery and von Zastrow, 1999), and this is probably related to its significant intrinsic activity (Strange, 1999; Prou *et al.*, 2001).

To summarize, the lack of detailed comparative analysis of the functional properties of the D_2 receptor subtypes prevents to draw any firm conclusion about the selection of specific characters for each of these subtypes. It is however probable that the ancestral $D_{2\text{-like}}$ receptor exhibited canonical properties such as coupling to Go/Gi proteins, inhibition of cAMP production and the ability to modulate ion channels in neuronal and neuroendocrine cells. After the gene duplications that generated the three vertebrate D_2 receptor subtypes, each of the resulting paralogous genes derived some special ability related either to quantitative aspects of a pre-existing property (higher or lower dopamine affinity, higher or lower transduction efficacy) or to more qualitative aspects of the modulation of intracellular signaling pathways. For example, the neuron-specific effect of D_3 receptor on adenylyl cyclase, lead to anticipate that this receptor needs to interact with special components, which is present within neurons only. The constraint that a receptor will undergo to interact with a specific protein required for modulation of an effector in a specific cell type is certainly the kind of changes that are both easy to trigger from a genetical point of view, and 'necessary' from a physiological point of view.

Differential localization of dopamine receptor subtypes in the vertebrate nervous system

The role played by a modulatory neurotransmitter such as dopamine depend essentially on the localization of the receptors in the target areas of the nervous system as well as on their subcellular localization in the target cells. Beside the biochemical properties of receptors, the cells where receptors are expressed and the place in the cells where they are present, are major physiological feature that may constrain a newly duplicated receptor to be conserved. Indeed, as soon as two neurotransmitter receptors are differentially expressed in a cell type or tissue, the two receptors are no longer redundant, provided the tissue will remain submitted to the action of the transmitter (here, dopamine). In that case, they may be conserved even if they have identical effects in cells. Such phenomenon may occur after gene duplication either by acquisition of 'new' expression territories or by the split of the expression areas of the ancestral gene (Aparicio, 2000).

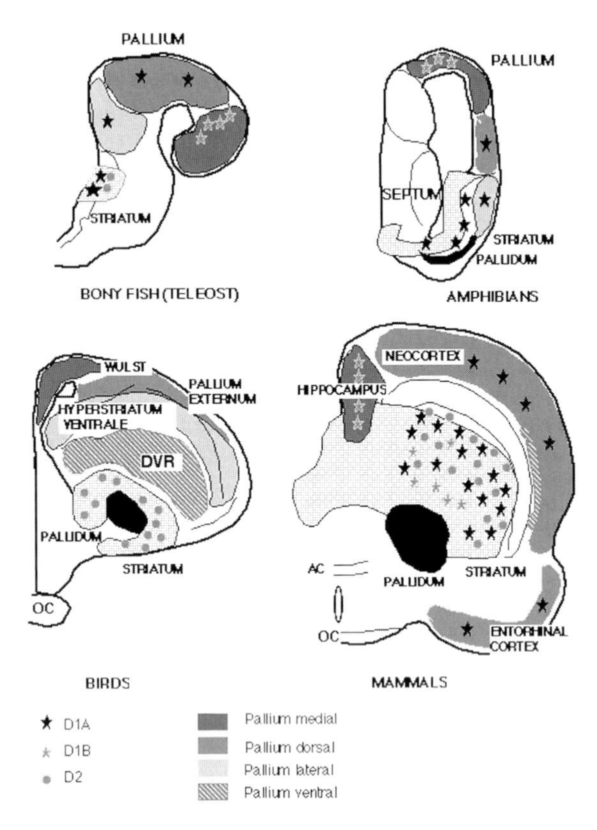

Figure 5. Schematic representation of the comparative distribution of the D_1 receptor subtypes in Vertebrates.
The overall distribution of these three dopamine receptor subtypes has been studied in several groups of vertebrates. Transversal section in the telencephalon of a bony fish (European eel), an amphibian (xenopus), a bird (chick) and a mammal (rat) are schematically represented. Homologous territories in the pallial-cortical region of the telencephalon exist in all the jawed vertebrates, as shown here by the grey shading of the different areas, but they display very different quantitative development. The relative distribution of D_{1A}, D_{1B} and D_2 dopamine receptor transcript has been obtained by *in situ* hybridization. Overall, the expression territories of the various dopamine receptor transcripts id overlapping, but specific location also exist, such as the hippocampus for the D_{1B} receptor subtype, a region that is probably homologous to the Dm-Dl junction in the fish pallium. Other homologous localization for each of these dopamine receptor subtypes have been evidenced in these four species, highlighting the conservative role that dopamine is playing in brain physiology.

Dopamine receptor distribution has been mainly studied by using *in situ* hybridization, which allows the detection of transcripts coding for the different receptor subtypes on tissue sections. Infortunately, very few studies provide a comparison of the localization of dopamine receptor subtypes, within a given species or between different species. In the nervous system, a potential problem in assessing the precise distribution of dopamine receptors is that receptors may be present on nerve terminals as well as on dendrites and cell bodies in the same area, where they exert very different roles. The recent availability of antibodies specific for dopamine receptor subtypes will permit to take into account the subcellular localization dopamine receptors.

The D_1 receptor subtypes are present in most of the known target areas of dopamine in the central nervous system of vertebrates. By compiling the data obtained by *in situ* hybridization in mammalian species (basically, rat, mouse and human; Tiberi *et al.*, 1991; Mansour *et al.*, 1992; Beischlag *et al.*, 1995; Meador-Woodruff *et al.*, 1996), in birds (Schnabel *et al.*, 1997; Sun and Reiner, 2000), in xenopus (unpublished data) and in fish (Kapsimali *et al.*, 2000), an overall picture of the conserved and derived features of the distribution of the dopamine D_1 receptors can be provided.

The tissue-specific expression of the D_1 receptor subtypes differs in three aspects: The relative abundance of the transcripts and proteins, their comparative distribution and the time course of the receptor expression throughout life. Interestingly, in all the studied species, namely human, rat mouse, chick, pigeon, xenopus, eel, medaka, zebrafish and carp the D_{1A} receptor subtype is the most abundantly expressed. On the opposite, the D_{1C} receptor in only weakly present in the brain (and not in any other organ) of fish (eel, medaka, and zebrafish) and amphibians (xenopus). The $_{D1B}$ is also only weakly abundant in mammals, birds amphibians and fish. This characteristic fits well with the higher conservation of D_{1A} sequences as compared to D_{1B} or D_{1C} sequences. The only exception is the D_{1D} receptor subtype, which seems to be rather abundant in the pigeon and turkey brain, although its sequence is also rather divergent (Schnell *et al.*, 1999). This may be accounted for by a specific role of this transcript in birds and reptiles.

The expression territories of each of the D_1 receptor subtypes have a rather widespread distribution in the brain with largely overlapping territories in the brains of mammalian and non-mammalian species (Figure 5). Both the D_{1A}/D_1 and D_{1B}/D_5 subtypes are found in telencephalic areas, mainly the dorsal telencephalon (dorsal pallium or neocortical areas) and in the medial pallial-hippocampal areas[1] of mammals. This is also true for the fish dorsal telencephalon, although the identification of homologous territories with mammals is difficult (Kapsimali *et al.*, 2000). Incidentally, the localization of dopamine receptors helped in identifying homologous brain regions in mammals, amphibians and fish. For example, in the case of the hippocampus, only the D_{1B}/D_5 receptor is present in mammals (in the Ammon horns). Localization of the D_{1B} receptor transcript in the medial pallium of amphibians and fish (the Dm-Dl junction in fish; Figure 5) led us to conclude that these brain areas where homologous in gnathostomes, which was confirmed by other anatomical and embryological data. Similarly, the expression of the D_{1A}/D_1 and D_{1B}/D_5 subtypes overlap in striatal and other subpallial areas territories, in several thalamic nuclei, in the preoptic nuclei and the dorsal and ventral periventricular areas of the hypothalamus, the optic tectum, and most of the cranial nerve-associated nuclei, both in rat and fish (Tiberi *et al.*, 1991; Mansour *et al.*, 1992; Meador-Woodruff *et al.*, 1996; Kapsimali *et al.*, 2000). It should be noticed that no D_1 receptor transcripts are present in the substantia nigra although the corresponding proteins are abundant, demonstrating the transport of the receptors from the striatum, along the axonal projections

In a few areas, either the D_{1A}/D_1 or D_{1B}/D_5 subtypes are expressed alone (Figure 5). For example, the the D_{1A} subtype only is present in the olfactory bulb, (this is true for all the jawed vertebrates where it has been examined), in the basolateral, lateral and central nuclei of amygdala, whereas the D_{1B} receptor transcripts are found in the anterior pretectal nuclei, the lateral mammillary nuclei and the parafascicular nucleus of the thalamus (Tiberi *et al.*, 1991; Meador-Woodruff *et al.*, 1992), and at homologous localizations in fish and amphibians Kapsimali *et al.*, 2000 and unpublished results).

As a general feature, when both the D_{1A} and D_{1B} receptors are present in a given anatomical area, a true co-localization of the transcripts is probable since, in many areas the cells labeled by the different RNA probes are strikingly similar. However, in the areas where most of the D_1 receptor subtypes are detected, the number of cells labeled for each of the subtypes often varied from one region to another. This is exemplified at best in fish where the experiments have been systematically conducted. In the ventromedial thalamic nucleus, D_{1A} and D_{1B} subtypes are abundant, but the D_{1A2} receptor mRNA is only weakly detected. In the ventral hypothalamus, D_{1A} transcript was more abundant than the D_{1A1}, D_{1B} and D_{1C} transcripts. As a matter of fact, D_{1C} receptor transcripts are found in a restricted area of the dorsal hypothalamus and in the cerebellum in cells that also contain the D_{1A} receptor mRNA. Thus, although the distribution of the four D_1 receptor transcripts is similar, the expression territories of the different D_1 receptor subtypes vary in a given area. The D_{1A} receptor subtypes are generally the most widely expressed, whereas the D_{1C} receptor subtype has a particularly restricted distribution. Based on our data, the role of the D_{1C} receptor may be related to adaptive neuroendocrine functions in fish, amphibians and most amniotes, functions that have been lost in mammals, together with the D_{1C} receptors.

Very similar observations may be made for the receptors of D_2 class. For example, the D_2 receptor subtype is by far the most highly expressed in the brain, overall, as compared to the D_3 and D_4 receptors. Moreover, the distribution of the D_4 receptor transcript strikingly overlaps with that of the D_2 receptor, in the dorsal striatum and cortical areas for example. This distribution is probably conserved in many vertebrate species, since the distribution pattern of the D_2 receptor subtype is very similar in birds and mammals. In contrast, the D_3 receptor is, on average, expressed at a lower level than the D_2 receptor subtype, but it is found in areas where no or very D_2 receptors are present, such as the islands of Calleja and the nucleus accumbens. Again, the distribution of the three paralogous D_2 receptor subtypes displays a picture of large overlap, differential abundance and a few specific localizations.

These observations have important implications to understand how the various subtypes of the D_1 or D_2 receptors have evolved after the duplication of the ancestral genes in vertebrates. The similarities of the expression territories of the D_1 receptor genes implicates that the main *cis*-regulatory elements controlling tissue-specific transcription of genes have been conserved after the gene duplication process. However, a few -but significant- differences are observed among regional distributions of the D_1 subtypes,

[1]The dorsal region of the telencephalon or pallium is divided in four main compartments, dorsal, medial, lateral and ventral pallium. These pallial areas form a true cortex (a layered organization) almost only in mammals. The medial pallium is probably homologous to the mammalian hippocampus, the lateral pallium to entorhinal cortex, the dorsal pallium to the neocortex and the ventral pallium to the claustral complex.

probably sustaining enough functional relevance in order to be conserved.

An evolutionary scenario of the evolution of dopamine systems in Craniates

As a neuromodulatory transmitter, dopamine has certainly played a crucial role in the adaptation of animal behaviours all along the evolution of Craniates. Obviously, the multiplication of dopamine receptors by gene duplications and the conservation of several receptor subtypes were instrumental for these adaptations. Based on the data and observations that have been gathered during this last few years a plausible scenario for the emergence and evolution of receptors in the vertebrate nervous system can be proposed. By building such a scenario, it is possible to take into account all the available elements of knowledge, to integrate them in a few operational and functional hypothesis, which can be tested and refined as soon as new findings will be made (Gans, 1989).

Although the precise phylogenetical relationships between the craniate D_1 and D_2 classes of dopamine receptors and putative homologues in Prochordates and Protostomians remain to be firmly established, there is little doubt that the origin of the D_1 and of the D_2 dopamine receptors is more ancient that Craniates. Homologues of vertebrate D_1 receptors and D_2 receptors exist in Protostomians such as insects – *Drosophila* –, nematods – *C. elegans* – and molluscs – *Lymnea* – (unpublished data). It is probable that the two classes of dopamine receptors acquired specific characteristics to fulfil the requirement of the demanding new functions, which accompanied the emergence of Craniates. Indeed, this major evolutionary event was accompanied with the development of new sensory organs and of the derivatives of the neural crest such as the sympathetic peripheral nervous system or the partitioned heart (Gans and Northcutt, 1983; Northcutt and Gans, 1983). After these rather dramatic changes, the characteristics of the dopamine receptors of ancestral vertebrates were stabilized and remained essentially unchanged over times in the whole vertebrate phylum. These characteristics include the structural determinants of ligand binding, of the coupling to a given set of G protein, of specific localization in cells and some general features of receptor desensitization and regulation.

In the Craniate phylum, the occurrence of two major steps of gene duplications that generated the present state of gene multiplicity in the majority of the gene families is still debated. These duplications probably corresponded to whole genome tetraploidisation (Holland, 1999), although alternate mechanisms may have also occurred. Data obtained in the comparative study of dopamine receptors supports the occurrence of these gene duplications. They certainly took place between the emergence of the first Craniates and that of the first jawed vertebrates, without more accurate information. These two events of gene duplication were certainly at the origin of the three or four receptor subtypes, which have been found in each class of dopamine receptor, in modern Gnathostomes except mammals. These two major steps of gene duplications have also been evidenced for many developmental genes expressed in the vertebrate nervous system. We think they correspond to the dramatic genetic changes that underlie the remodelling of organogenesis and accompanied the emergence of vertebrates from ancestral chordates (Holland *et al.*, 1994). To account for the number of receptor genes that are observed in the different group of these jawed vertebrates, one has also to admit episodes of gene loss. Most cartilaginous and bony fish have lost one gene (the D_{1D} in the D_1 receptor class and a putative 'D_6' receptor in the D_2 class), as well as amphibians and turtles. In contrast, most of the amniotes have kept the four genes in the D_1 class but not in the D_2 class, as all the other jawed vertebrate groups. The mammals have lost two receptor genes in the D_1 class, either the D_{1C} and D_{1D} or the D_{1C} and D_{1B}, depending on the mammalian phylum. Thus, our current hypothesis states that the new duplicated receptors were recruited subsequently to the changes driven by the huge remodelling of the nervous system at the origin of vertebrates. They were used to achieve the increased demand in information processing and in neuromodulation that underlie processes such as motivation, emotion, motor programming and working memory.

Based on the analysis of the anatomical distribution and of the functional characteristics of the dopamine receptors we have made, it is obvious that both, two main conclusions can be drawn. The first one is that a few molecular properties easily distinguish between the different subtypes (Figure 4). The D_{1A}/D_1 receptor subtype exhibit a relatively low affinity for dopamine, a low intrinsic activity and a fast rate of agonists induced desensitization. It is by far the most conserved subtype of the D_1 class, and the hagfish and lamprey sequences (although clearly not of the D_{1A} subtype) are closer to it than from the other D_1 receptor subtypes. It is thus plausible that D_{1A} receptor subtype still resembles to the ancestral vertebrate D_1 receptor. In contrast, the D_{1B}/D_5 receptor has a higher affinity for dopamine and a high intrinsic activity that influence the intensity of its desensitization. Its sequence is less conserved in all the vertebrate species analysed so far. Interestingly, it has not been found in prototherian mammals (echidnea and platypus), in which it seems to be substituted by the D_{1D} subtype. As far as the D_{1C} receptor subtype is concerned, it resembles the D_{1A} receptor, but it is particularized by a weak and transient desensitization upon agonist stimulation. It is the less conserved of the D_1 receptor subtypes and that it has disappeared in mammals. The D_{1D} receptor subtype may be attributed the same kind of statements, although its functional properties have been even less investigated than those of the D_{1C} receptors. Inside the D_2 receptor class, the D_2 receptor subtype is the most conserved and it is present in all the Gnathostomes investigated up to now. Its general role may well be to control cell excitability as well as release of neurotransmitter and hormones by activating K^+ channels and inhibiting Ca^{2+} current. The D_3 receptor subtype is certainly more 'neuron specific' that the D_2 receptor. It exhibits a higher affinity for dopamine than the D_2 receptor, consistent with an 'extrasynaptic' localization and fucntion. The D_4 receptor is much less conserved and may play an accessory role only in the control of pallial-cortical activity, at least in mammals (Sibley, 1999).

The second important conclusion is that no major changes in expression territories have been evidenced between the dopamine receptor subtypes in the few – but phylogenetically diverse – vertebrate species that have been studied (Figure 5). The D_{1A}/D_1 receptor is always the most highly expressed. The subcellular localisation of D_{1A}/D_1 receptors at the basis of the dendritic spines, in close proximity to the dopamine release sites in the mammalian basal ganglia, certainly account for its 'low' affinity and fast desensitization rate. The D_{1A} receptor is thus mainly a 'synaptic' receptor. Both in mammals and fish, the tissue distribution of D_{1B}/D_5 receptors overlap with that of the D_{1A}/D_1 receptors, but with significant specific features. The D_{1B}/D_5 receptor subtype is the only one present in a few areas such as hippocampus (Tiberi *et al.*, 1991; Kapsimali *et al.*, 2000). At the cellular level, D_{1B}/D_5 receptors are found on dendritic shafts, far from synapses, accounting for their high affinity and high intrinsic activity. These properties fit with a 'non-synaptic' neurohormonal type of receptor. In contrast, based on current data, D_{1C} receptors of fish and amphibians seem to be resistant to desensitization. Precise localization of D_{1C} receptors in neuronal networks is still unknown, but the D_{1C} receptor mRNA has a highly restricted distribution in the eel brain, being present mainly in the dorsal and medial hypothalamic nuclei and ventral habenula (Kapsimali *et al.*, 2000). We suggest that the role of the D_{1C} receptor is related to adaptive neuroendocrine functions in fish, amphibians and most amniotes, functions that have been lost in mammals, together with the D_{1C} receptors. Similar suggestions can be made for the D_2 receptor subtypes with the D_2 subtype being also closer to the dopamine release sites than the D_3 receptor, this latter playing a central role at the end of the development of the nervous system and during adaptation of the limbic areas to environmental pressure.

These observations may suggest that changes in expression territories were not a major reason to keep the duplicated genes away from inactivation. Instead, the molecular properties of each of the receptor subtypes could be the main driving force for the conservation of the duplicated dopamine receptor subtypes. Each of the D_1 receptor subtypes has indeed distinct functional properties that have obvious adaptive consequences for the modulation of neuronal activity in regulatory networks. Nevertheless, even modest expression changes may be sufficient to drive gene maintenance. This contention fits well with the hypothesis pushed forward by Force and colleagues who analyzed the massive gene duplications conserved in teleost fish (especially in cyprinids, such as zebrafish, carps and goldfish; Force *et al.*, 1999). Accordingly, compensation for a loss of either of the two gene copies will be never complete, even if their functional characteristics and expression territories are not significantly different. It is precisely the case of the dopamine D_1 and D_2 receptor subtypes. One parsimonious statement would be that both functional properties and differential expression were sufficiently different among the dopamine receptor subtype to be conserved, with significant differences, however, among the vertebrates.

Finally, the most unexpected inference that came out this overview of dopamine systems in vertebrates is that no clear correlation exists between the conservation or the loss of a given subtype of dopamine receptor and one of the major anatomical or physiological characteristics of the vertebrate groups. For example, although ancestral amphibians were the first vertebrates to leave an aquatic environment to come onto land, the fish-amphibian transition was not accompanied by any changes in the number or the properties of the dopamine receptors. Similarly, in hominids, the significant increase in the proportion of the neocortex (and more especially the prefrontal cortex) did not justified any modification of the anatomical and functional characteristics of the dopamine receptors. Instead, mammals including human have only two D_1 receptor subtypes instead of three or four in the other vertebrate groups. In sharp contrast, the neuronal circuits that synthesize dopamine are very well conserved in all the Craniate groups (Smeets and Reiner, 1994). Similarly, all the molecular components of the catecholamine synthesizing pathways have been remarkably conserved. Indeed, no duplications of the catecholamine synthesizing enzymes (TH, D/TβH and PNMT) have been recorded in Craniates (and even in the whole Bilateria phylum, as far as we know.

The idea here would be that the general role and organization of dopamine systems have not been significantly changed during evolution in the whole vertebrate phylum. The role of dopamine is modulatory in essence, and it affects many functions from sensory perception to vegetative functions, emotions and memory. However, if the overall organization of the central nervous system of Craniates has been conserved throughout evolution, significant changes occurred in the relative size and connections of several brain areas in the different groups of Craniates. It is particularly the case of the pallial-cortical regions that appear extremely specialized in birds and mammals as compared to amphibians and fish, for example. The relationships of the pallial areas with the subpallial basal ganglia or the thalamic nucleus have been also accordingly remodelled. To accommodate such huge changes in the way sensory perception and motor action are elaborated, regulatory pathways such as the dopamine systems needed to be adapted. The role played by dopamine is basically non-specific, but it applies to functions that, in many instances, are extremely specific of a given species.

Thus, if conservation of duplicated receptor genes would have been adaptive, it is more by allowing some general 'flexibility' or 'evolvability' to Craniate species, for changing the functional structure of their nervous system. Since the overall organization of the brain has not been modified in Gnathostomes, but only the quantitative relationships between areas controlled by dopamine, there was no need for dramatic changes in the expression territories and functional properties of the dopamine receptors all along vertebrate evolution. However, the conservation of duplicated genes with slightly different properties and expression territories provided a higher degree of tolerance and adaptability to the dopamine regulation of target areas.

References

Alexander, S.P.H. and Peters, J.A. (2000) Receptor and ion channel nomenclature. *Trends Pharmacol. Sci.*, Supplement.

Aparicio, S. (2000) Vertebrate evolution: recent perspectives from fish. *Trends Genet.* **16**, 54–56.

Beischlag, T.V., Marchese, A., Meador-Woodruff, J.H., Damask, S.P., O'Dowd, B.F., Tyndale, R.F., van Tol, H.H., Seeman, P. and Niznik, H.B. (1995) The human dopamine D5 receptor gene: cloning and characterization of the 5'-flanking and promoter region. *Biochemistry*, **34**, 5960–5970.

Bergson, C., Mrzljak, L., Smiley, J.F., Pappy, M., Levenson, R. and Goldman-Rakic, P.S. (1995) Regional, cellular, and subcellular variations in the distribution of D1 and D5 dopamine receptors in primate brain. *J. Neurosci.* **15**, 7821–7836.

Bockaert, J. and Pin, J.P. (1999) Molecular tinkering of G protein-coupled receptors: an evolutionary success. *EMBO J.* **18**, 1723–1729.

Borowski, B., Adham, N., Jones, K.A., Raddatz, R., Artymyshyn, R., Ogozalek, K.L. and Durkin, M.M. (2001) Trace amines: Identification of a family of mammalian G protein-coupled receptors. *Proc. Natl. Acad. Sci. USA*, **98**, 8966–8971.

Cardinaud, B., Sugamori, K.S., Coudouel, S., Vincent, J.D., Niznik, H.B. and Vernier, P. (1997) Early emergence of three dopamine D1 receptor subtypes in vertebrates. Molecular phylogenetic, pharmacological, and functional criteria defining D1A, D1B, and D1C receptors in European eel Anguilla anguilla. *J. Biol. Chem.* **272**, 2778–2787.

Cardinaud, B., Gibert, J.M., Liu, F., Sugamori, K.S., Vincent, J.D., Niznik, H.B. and Vernier, P. (1998) Evolution and origin of the diversity of dopamine receptors in vertebrates. *Adv. Pharmacol.* **42**, 936–940.

Civelli, O., Bunzow, J.R. and GRandy, D.K. (1993) Molecular diversity of the dopamine receptors. *Annu. Rev. Pharmacol. Toxicol.* **32**, 281–307.

Conrad, M. (1990) The geometry of evolution. *Biosystems*, **24**, 61–81.

Demchyshyn, L.L., Sugamori, K.S., Lee, F.J.S., Hamadanizadeh, S.A. and Niznik, H.B. (1995) The dopamine D1D receptor. *J. Biol. Chem.* **270**, 4005–4012.

Force, A., Lynch, M., Pickett, F.B., Amores, A., Yan, Y.L. and Postlethwait, J. (1999) Preservation of duplicate genes by complementary, degenerative mutations. *Genetics*, **151**, 1531–1545.

Fryxell, K.J. (1995) The evolutionary divergence of neurotransmitter receptors and second- messenger pathways. *J. Mol. Evol.* **41**, 85–97.

Gans, C. (1989) Stages in the origin of vertebrates: analysis by means of scenarios. *Biol. Rev. Camb. Philos. Soc.* **64**, 221–268.

Gans, C. and Northcutt, R.G. (1983) Neural crest and the origin of vertebrates: A new head. *Science*, **220**, 268–274.

Gerhart, J. and Kirschner, M. (1997) *Cells, embryos and evolution*, Blackwell Science, Malden MA.

Guivarc'h, D., Vernier, P. and Vincent, J.D. (1995) Sex steroid hormones change the differential distribution of the isoforms of the D2 dopamine receptor messenger RNA in the rat brain. *Neuroscience*, **69**, 159–166.

Guivarc'h, D., Vernier, P. and Vincent, J.D. (1998) Alternative splicing of the D2 dopamine receptor messenger ribonucleic acid is modulated by activated sex steroid receptors in the MMQ prolactin cell line. *Endocrinology* **139**, 4213–4221.

Hersch, S.M., Ciliax, B.J., Gutekunst, C.A., Rees, H.D., Heilman, C.J., Yung, K.K., Bolam, J.P., Ince, E., Yi, H. and Levey, A. I. (1995) Electron microscopic analysis of D1 and D2 dopamine receptor proteins in the dorsal striatum and their synaptic relationships with motor corticostriatal afferents. *J. Neurosci.* **15**, 5222–5237.

Holland, P.W. (1999) Gene duplication: past, present and future. *Semin. Cell Dev. Biol.* **10**, 541–547.

Holland, P.W., Garcia-Fernandez, J., Williams, N.A. and Sidow, A. (1994) Gene duplications and the origins of vertebrate development. *Development*, **Suppl. 1994**, 125–133.

Horn, F., van der Wenden, E.M., Oliveira, L., Ijzerman, A.P. and Vriend, G. (2000) Receptors coupling to G proteins: is there a signal behind the sequence? *Proteins*, **41**, 448–459.

Huff, R.M., Chio, C.L., Lajiness, M.E. and Goodman, L.V. (1998) Signal transduction pathways modulated by D2-like dopamine receptors. *Adv. Pharmacol.* **42**, 454–457.

Jarvie, K.R., Tiberi, M., Silvia, C., Gingrich, J.A. and Caron, M.G. (1993) Molecular cloning, stable expression and desentization of the human dopamine D1B/D5 receptor. *J. Recept. Res.* **13**, 573–590.

Ji, T.H., Grossmann, M. and Ji, I. (1998) G protein-coupled receptors. I. Diversity of receptor-ligand interactions. *J. Biol. Chem.* **273**, 17299–17302.

Jiang, D. and Sibley, D.R. (1999) Regulation of D(1) dopamine receptors with mutations of protein kinase phosphorylation sites: attenuation of the rate of agonist-induced desensitization. *Mol. Pharmacol.* **56**, 675–683.

Kapsimali, M., Vidal, B., Gonzalez, A., Dufour, S. and Vernier, P. (2000) Distribution of the mRNA encoding the four dopamine D(1) receptor subtypes in the brain of the European eel (*Anguilla anguilla*): comparative approach to the function of D(1) receptors in vertebrates. *J. Comp. Neurol.* **419**, 320–343.

Kebabian, J.W. and Calne, D.B. (1979) Multiple receptors for dopamine. *Nature*, **277**, 93–96.

Kirschner, M. and Gerhart, J. (1998) Evolvability. *Proc. Natl. Acad. Sci. USA*, **95**, 8420–8427.

Lachowicz, J.E. and Sibley, D.R. (1997) Molecular characteristics of mammalian dopamine receptors. *Pharmacol. Toxicol.* **81**, 105–113.

Lewis, M.M., Watts, V. J., Lawler, C.P., Nichols, D.E. and Mailman, R.B. (1998) Homologous desensitization of the D1A dopamine receptor: efficacy in causing desensitization dissociates from both receptor occupancy and functional potency. *J. Pharmacol. Exp. Ther.* **286**, 345–353.

Lledo, P.M., Homburger, V., Bockaert, J. and Vincent, J.D. (1992) Differential G protein-mediated coupling of D2 dopamine receptors to K^+ and Ca^{2+} currents in rat anterior pituitary cells. *Neuron*, **8**, 455–463.

Lledo, P.M., Vernier, P., Kukstas, L.A. and Vincent, J.D. (1994) Coupling of dopamine receptors to ionic channels in excitable tissues. In *Dopamine receptors and Transporters* (Ed. Nizni, H.B.), Marcel Dekker, New York, NY, pp. ...-...

Mansour, A., Meador-Woodruff, J.H., Zhou, Q., Civelli, O., Akil, H. and Watson, S.J. (1992) A comparison of D1 receptor binding and mRNA in rat brain using receptor autoradiographic and in situ hybridization techniques. *Neuroscience*, **46**, 959–971.

Meador-Woodruff, J.H., Mansour, A., Grandy, D.K., Damask, S.P., Civelli, O. and Watson, S.J., Jr. (1992) Distribution of D5 dopamine receptor mRNA in rat brain. *Neurosci. Lett.* **145**, 209–212.

Meador-Woodruff, J.H., Damask, S.P., Wang, J., Haroutunian, V., Davis, K.L. and Watson, S.J. (1996) Dopamine receptor mRNA expression in human striatum and neocortex. *Neuropsychopharmacology*, **15**, 17–29.

Missale, C., Nash, S.R., Robinson, S.W., Jaber, M. and Caron, M. G. (1998) Dopamine receptors: from structure to function. *Physiol. Rev.* **78**, 189–225.

Neve, K.A. and Neve, R.L. (1996) *The Dopamine Receptors*, Humana Press, Totowa, NJ.

Niznik, H.B., Liu, F., Sugamori, K.S., Cardinaud, B. and Vernier, P. (1998) Expansion of the dopamine D1 receptor gene family: defining molecular, pharmacological, and functional criteria for D1A, D1B, D1C, and D1D receptors. *Adv. Pharmacol.* **42**, 404–408.

Northcutt, R.G. and Gans, C. (1983) The genesis of neural crest and epidermal placodes: a reinterpretation of vertebrate origins. *Q. Rev. Biol.* **58**, 1–28.

Olson, M.F. and Schimmer, B.P. (1992) Heterologous desensitization of the human dopamine D1 receptor in Y1 adrenal cells and in a desensitization-resistant Y1 mutant. *Mol. Endocrinol.* **6**, 1095–1102.

Pei, G., Samama, P., Lohse, M., Wang, M., Codina, J. and Lefkowitz, R.J. (1994) A constitutively active mutant ß2-adrenergic receptor is constitutively desensitized and phosphorylated. *Proc. Natl. Acad. Sci. USA*, **91**, 2699–2702.

Philippe, H. (1993) MUST, a computer package of Management Utilities for Sequences and Trees. *Nucleic Acids Res.* **21**, 5264–5272.

4rou, D., Gu, W. J., Le Crom, S., Vincent, J.D., Salamero, J. and Vernier, P. (2001) Intracellular retention of the two isoforms of the D(2) dopamine receptor promotes endoplasmic reticulum disruption. *J. Cell Sci.* **114**, 3517–3527.

Robinson-Rechavi, M., Marchand, O., Escriva, H. and Laudet, V. (2001) An ancestral whole-genome duplication may not have been responsible for the abundance of duplicated fish genes. *Curr. Biol.* **11**, R458-R459.

Schnabel, R., Metzger, M., Jiang, S., Hemmings, H.C., Jr., Greengard, P. and Braun, K. (1997) Localization of dopamine D1 receptors and dopaminoceptive neurons in the chick forebrain. *J. Comp. Neurol.* **388**, 146–168.

Schnell, S., You, S., Foster, D.N. and El Halawani, M.E. (1999) Molecular cloning and tissue distribution of an avian D2 dopamine receptor mRNA from the domestic turkey (*Meleagris gallopavo*). *J. Comp. Neurol.* **407**, 543–554.

Seabrook, G.R., Kemp, J.A., Freedman, S.B., Patel, S., Sinclai, r. H.A. and McAllister, G. (1994a) Functional expression of human D3 receptor in differentiated neuroblastoma X glioma NG 108–15 cells.. *Br. J. Pharmacol.* **111**, 391–393.

Seabrook, G.R., Knowles, M., Brown, N., Myers, J., Sinclair, H., Patel, S., Freedman, S.B. and McAllister, G. (1994b) Pharmacology of high-threshold calcium currents in GH4C1 pituitary cells and their regulation by activation of human D2 and D4 dopamine receptors. *Br. J. Pharmacol.* **112**, 728–734.

Seeman, P. and Van Tol, H.H. (1994) Dopamine receptor pharmacology. *Trends Pharmacol. Sci.* **15**, 264–270.

Sibley, D.R. (1999) New insights into dopaminergic receptor function using antisense and genetically altered animals. *Annu. Rev. Pharmacol. Toxicol.* **39**, 313–341.

Sidhu, A. and Niznik, H.B. (2000) Coupling of dopamine receptor subtypes to multiple and diverse G proteins. *Int. J. Dev. Neurosci.* **18**, 669–677.

Smeets, W. and Reiner, A. (1994) *Phylogeny and Development of Catecholamine Systems in the CNS of Vertebrates*, Cambridge University Press, Cambridge.

Sokoloff, P. and Schwartz, J.C. (1995) Novel dopamine receptors half a decade later [see comments]. *Trends Pharmacol. Sci.* **16**, 270–275.

Strange, P.G. (1993) New insights into dopamine receptors in the central nervous system. *Neurochem. Int.* **22**, 223–236.

Strange, P.G. (1999) Agonism and inverse agonism at dopamine D2-like receptors. *Clin. Exp. Pharmacol. Physiol. Suppl.* **26**, S3-S9.

Sugamori, K.S., Demchyshyn, L.L., Chung, M. and Niznik, H.B. (1994) D_{1A}, D_{1B}, and D_{1C} dopamine receptors from *Xenopus laevis*. *Proc. Natl. Acad. Sci. USA*, **91**, 10536–10540.

Sugamori, K.S., Scheideler, R., Vernier, P. and Niznik, H.B. (1998) Dopamine D1B receptor chimeras reveal modulation of partial agonist activity by carboxy terminal tail sequences. *.J Neurochem.* **71**, 2593–2599.

Sun, Z. and Reiner, A. (2000) Localization of dopamine D1A and D1B receptor mRNAs in the forebrain and midbrain of the domestic chick. *J. Chem. Neuroanat.* **19**, 211–224.

Taylor, J.S., Van de Peer, Y., Braasch, I. and Meyer, A. (2001) Comparative genomics provides evidence for an ancient genome duplication event in fish. *Philos. Trans. R. Soc. Lond. B Biol. Sci.* **356**, 1661–1679.

Tiberi, M. and Caron, M.G. (1994) High agonist-independent activity is a distinguishing feature of the dopamine D1B receptor subtype. *J. Biol. Chem.* **269**, 27925–27931.

Tiberi, M., Jarvie, K.R., Silvia, C., Falardeau, P., Gingrich, J.A., Godinot, N., Bertrand, L., Yang-Feng, T.L., Fremeau, R.T., Jr. and Caron, M.G. (1991) Cloning, molecular characterization, and chromosomal assignment of a gene encoding a second D1 dopamine receptor subtype: differential expression pattern in rat brain compared with the D1A receptor. *Proc. Natl. Acad. Sci. USA*, **88**, 7491–7495.

Tiberi, M., Nash, S.R., Bertrand, L., Lefkovitz, R.J. and Caron, M.G. (1996) Differential regulation of dopamine D1A receptor responsiveness by various G protein-couped receptor kinases. *J. Biol. Chem.* **271**, 3771–3778.

Valdenaire, O. and Vernier, P. (1997) G protein coupled receptors as modules of interating proteins: A family meeting. *Prog. Drug Res.* **49**, 458–482.

Van Tol, H.H., Bunzow, J.R. and H.C., G. (1991) Cloning of the gene for a human dopamine D4 receptor with high affinity for the antipsychotic clozapine. *Nature*, **350**, 610–614.

Van Tol, H.H.M., Wu, C.M., Guan, H.C., Ohara, K., Bunzow, J.R., Civelli, O., Kennedy, J., Seeman, P., Niznik, H.B. and Jovanovic, V. (1992) Multiple dopamine D4 receptor variants in the human population. *Nature*, **358**, 149–152.

Vanhoutte, P.M., Humphrey, P.P. and Spedding, M. (1996) International Union of Pharmacology recommendations for nomenclature of new receptor subtypes. *Pharmacol. Rev.* **48**, 1–2.

Vernier, P., Cardinaud, B., Philippe, H. and Vincent, J.D. (1997) Molecular phylogenies: How helpful are they for receptor classification?. *Ann. NY Acad. Sci.* **817**, 241–243.

Vernier, P., Cardinaud, B., Valdenaire, O., Philippe, H. and Vincent, J.-D. (1995) An evolutionary view of drug-receptor interactions: The example of the bioamine receptor family.. *Trends Pharmacol. Sci.* **16**, 375–381.

Vickery, R.G. and von Zastrow, M. (1999) Distinct dynamin-dependent and -independent mechanisms target structurally homologous dopamine receptors to different endocytic membranes. *J. Cell Biol.* **144**, 31–43.

Wittbrodt, J., Meyer, A. and Schartl, M.(1998) More genes in fish? *BioEssays*, **20**, 511–515.

A. Meyer, Y. Van de Peer (eds.), Genome Evolution, 177-184.

Nuclear receptors are markers of animal genome evolution

Héctor Escrivá García, Vincent Laudet & Marc Robinson-Rechavi*
Laboratoire de Biologie Moléculaire et Cellulaire, Ecole Normale Supérieure de Lyon, 46 allée d'Italie, F–69364 Lyon cedex 07, France
Author for correspondence: E-mail: marc.robinson@ens-lyon.fr

Received 25.07.2002; Accepted in final form 29.08.2002

Key words: animal evolution, domain, duplication, hormone, phylogeny, receptor, transcription factor, vertebrate evolution

Abstract

Nuclear hormone receptors form one evolutionary related super-family of proteins, which mediate the interaction between hormones (or other ligands) and gene expression in animals. Early phylogenetic analyses showed two main periods of gene duplication which gave rise to present-day diversity in most animals: one at the origin of the family, and another specifically in vertebrates. Moreover this second period is composed itself by, probably, two rounds of duplication, as proposed by Susumu Ohno at the origin of vertebrates. There are indeed often two, three or four vertebrate orthologs of each invertebrate nuclear receptor, in accordance with this theory. The complete genome of *Drosophila melanogaster* contains 21 nuclear receptors, compared to 49 in the human genome. In addition, many nuclear receptors have more paralogs in the zebrafish than in mammals, and a genome duplication has been proposed at the origin of ray-finned fishes. Nuclear receptors are a very good model to investigate the dating and functional role of these duplications, since they are dispersed in the genome, allow robust phylogenetic reconstruction, and are functionnaly well characterized, with different adaptations for different paralogs. We illustrate this with examples from differents nuclear receptors and different groups of species.

Introduction

Gene and genome duplications are thought to have been motor forces in shapping the evolution of organisms (Ohno, 1970). There are different ways to study these duplications, illustrated by various papers in this volume. The in-depth study of gene families has the major advantage of linking duplications to functional adaptations. Moreover, the use of genes from various species, combined with phylogenetic analysis, allows rigourous dating of duplication events relative to speciations. The main focus of such studies in the animal kingdom has been the *hox* and related gene clusters (Málaga-Trillo and Meyer, 2001; Holland, 2002). The advantages of these genes are well known. They cluster in a conserved manner, facilitating homology statements. Expression is colinear, concerns major features of animal development, and conserved between animal phyla such as insects and chordates. On the other hand, the colinearity constraint means that for each cluster, only one independent locus is sampled in the genome. And the short conserved domain of hox proteins makes them poor phylogenetic markers, hampering tests of duplication history.

Nuclear hormone receptors are a major class of transcriptional regulators in metazoans, in which they regulate functions as diverse as reproduction, differentiation, development, metabolism, metamorphosis, or homeostasis (Laudet and Gronemeyer, 2002). They function as ligand activated transcription factors, thus providing a direct link between signaling molecules that control these processes and transcriptional responses. Nuclear receptors form a superfamily of phylogenetically related proteins, which share a common structural organization (Figure 1): a variable N-terminal region (A/B domain), a central well conserved DNA binding domain (DBD, C domain), a non conserved hinge (D domain) and a carboxyl-terminus, moderately conserved ligand binding domain (LBD, E domain) (Laudet and Gronemeyer, 2002). The

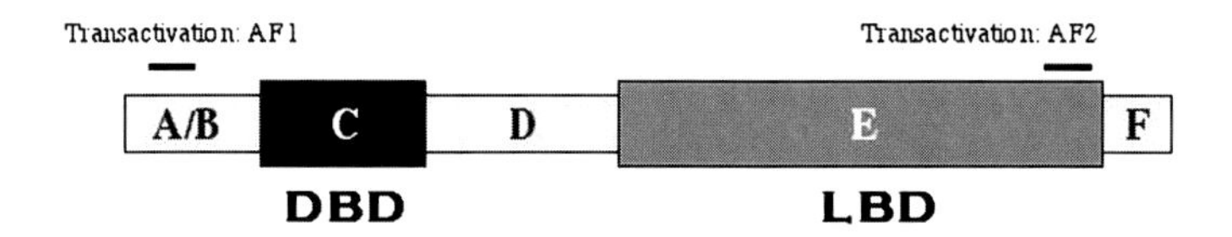

Figure 1. Organization of a typical nuclear receptor
A/B, C, D, E, F: regions as defined in estrogen receptors; F may be absent in some receptors. DBD: DNA binding domain. LBD: ligand binding domain. AF–1 and AF–2: transcription activation functions.

superfamily includes receptors for hydrophobic molecules such as steroid hormones (estrogens, glucocorticoids, progesterone, mineralocorticoids, androgens, vitamin D, ecdysone, oxysterols, bile acids, etc.), retinoic acids (all-*trans* and 9-*cis* isoforms), thyroid hormones, fatty acids, leukotrienes and prostaglandins (Escriva *et al.*, 2000). A large number of nuclear receptors have also been identified by homology with the conserved DBD and LBD, but have no identified natural ligand, and are referred to as 'orphan receptors'. Nuclear receptors control many key functions, notably in development and metabolism (Laudet and Gronemeyer, 2002). Since nuclear receptors are promising targets for numerous human diseases, such as cancer or diabetes, a large arsenal of functional data has been generated on the super-family. In this chapter, we will review the advantages of the superfamily of nuclear receptors as markers of genome evolution.

Nuclear receptors are good phylogenetic markers

Our original work ten years ago (Laudet *et al.*, 1992) showed nuclear receptors to be strong phylogenetic markers at a variety of evolutionary distances, from diversification of metazoans to vertebrate-specific paralogs. This is due to the good conservation of the DNA- and ligand-binding domains, notably of key residues which facilitate alignment, and the large number of amino-acids sites which can be compared: typically more than 200 amino-acids after removal of all positions with a gap. This phylogenetic quality has been confirmed in a robust manner with diverse methods, at various evolutionary distances and with different numbers of sequences (Amero *et al.*, 1992; Escriva *et al.*, 1997; Laudet, 1997; Robinson-Rechavi *et al.*, 2001a). Nuclear receptors have thus been used as a case study for various methods based on phylogeny, such as comparative evolution of protein domains (Thornton and Desalle, 2000) or automatic classification of protein families (Wicker *et al.*, 2001). It has also allowed defining a robust nomenclature of nuclear receptors, with families and sub-families based on the phylogeny (Nuclear Receptors Nomenclature Committee, 1999). These six families include two very large groups, NR1 with notably thyroid hormone receptors, retinoic acid receptors and the ecdysone receptor, and NR2 with notably RXR and COUP-TF, as well as a more homogeneous family of mostly steroid receptors (NR3), and smaller families comprising NGFIB (NR4), FTZ-F1 (NR5), or GCNF1 (NR6).

This quality of nuclear receptors as phylogenetic markers provides us with a strong mean to determine relations of paralogy and orthology, and the relative timing of gene duplication and speciation events, independently of molecular clock considerations. Rooting is especially important to be able to order events, and another advantage of nuclear receptors in phylogenetic studies is that the phylogeny of one family can always be robustly rooted by outgroup sequences from other families of nuclear receptors. For example, thyroid hormone receptor (TR) sequences can be used to root the phylogeny of retinoic acid receptors (RAR). Thanks to high statistical support for the phylogeny, it is then possible to determine unambigously that both duplications that gave rise to RARα, RARβ and RARγ happened after the divergence of amphioxus and vertebrates, but before the divergence of actinopterygian fish and sarcopterygian tetrapodes.

Nuclear receptors are dispersed in the genome

Genomes are large, complex structures. This does not mean, luckily, that nothing can be said without complete genomes from all species we are interested in! But if apparently small samples can be informative, we must be very attentive to risks of bias. A bias occurs when sampling is done in such a way that increasing the sample size increases support for a false result. The first human chromosomes entirely

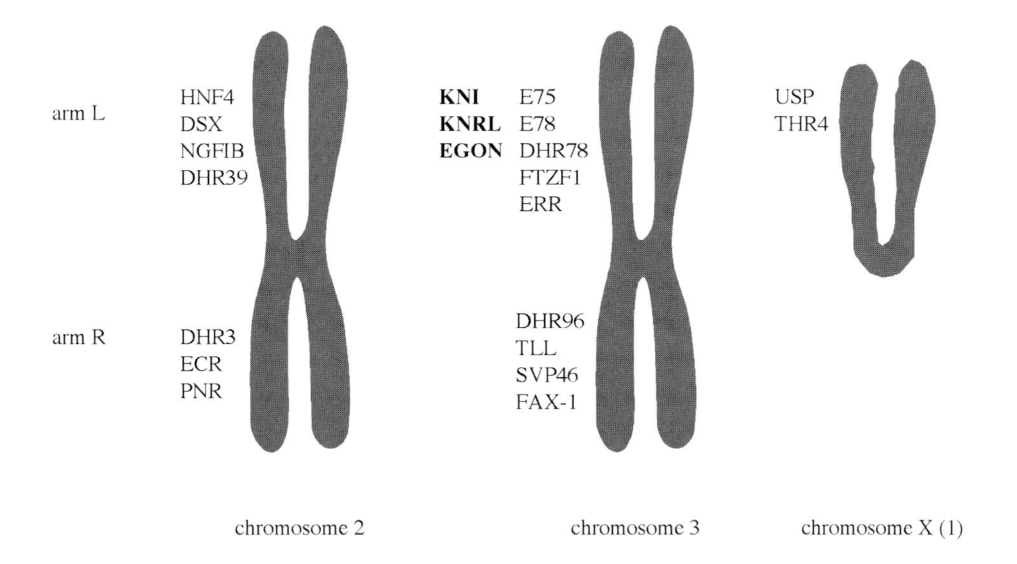

Figure 2. Distribution of nuclear receptors on *Drosophila* chromosomes.
Nuclear receptors are represented next to the chromosome arm where they map, in no special order. In bold, the three nuclear receptors without a LBD (group NR0B of (Nuclear Receptors Nomenclature Committee, 1999)). There are no nuclear receptors among the 85 predicted genes of chromosome 4, nor among the two predicted genes of chromosome Y, so these chromosomes are not represented.

sequenced, chromosome 22 (Dunham *et al.*, 1999) and 21 (Hattori *et al.*, 2000), happen to be the two smallest, for obvious technical reasons. Although they brought a wealth of information, they cannot be assumed to represent fairly the overall properties of the genome, for example in density of genes or expression (Caron *et al.*, 2001).

While the conservation of colinearity in *hox* clusters is one of the most exciting facts of Evo / Devo, it sets such a constraint that they constitute one locus in most invertebrates, three to seven in vertebrates (Holland, 2002). This means that sequencing more *hox* genes in a species usually does not increase much the sampling of the genome, only the information on a given locus. On the contrary, no functional relation between gene position and function or phylogeny of nuclear receptors is known, with the notable exception of the three receptors without a LBD in Drosophila, which are grouped on the same chromosome arm (Figure 2). All other nuclear receptor genes are randomly dispersed in the complete genomes of *Drosophila* and human (Robinson-Rechavi *et al.*, 2001b; Laudet and Gronemeyer, 2002). Overall there is no evidence of selection which would bias the evolution of nuclear receptor genes, compared to that of the whole genome. The 21 nuclear receptors of *Drosophila melanogaster*, the smallest complete set known in one animal, thus probably represent a fair sampling of its genome (Figure 2), and there is no reason to believe it is otherwise in other animals.

Nuclear receptors are good indicators of major events of genome evolution

The proof of the pudding is in the eating. How do nuclear receptors fit in the picture in cases where we have a good idea of genome evolution? Is their history consistent with that of the general gene population? Probably the most studied genome duplication hypothesis in animals is the '2R' theory (Holland, 2002; Hughes and Friedman, 2002; Lundin *et al.*, 2002). Although the details of what happened at the origin of vertebrates remain controversial, it is well established that many genes, maybe most, have two to four paralogs in vertebrates, all orthologous to a unique invertebrate gene. Whether these paralogs arose through genome or gene duplications, it is certainly a behaviour typical of the vertebrate 'gene population' in its early history. In full agreement with this observation, a majority of nuclear receptors have two to four paralogs in land vertebrates, with a unique ortholog in the fruit fly (Laudet *et al.*, 1992; Maglich *et al.*, 2001; Laudet and Gronemeyer, 2002b) (Figure 3). Consistently with the majority of genes studied so far (Holland, 2002; Escriva *et al.*, 2002), these

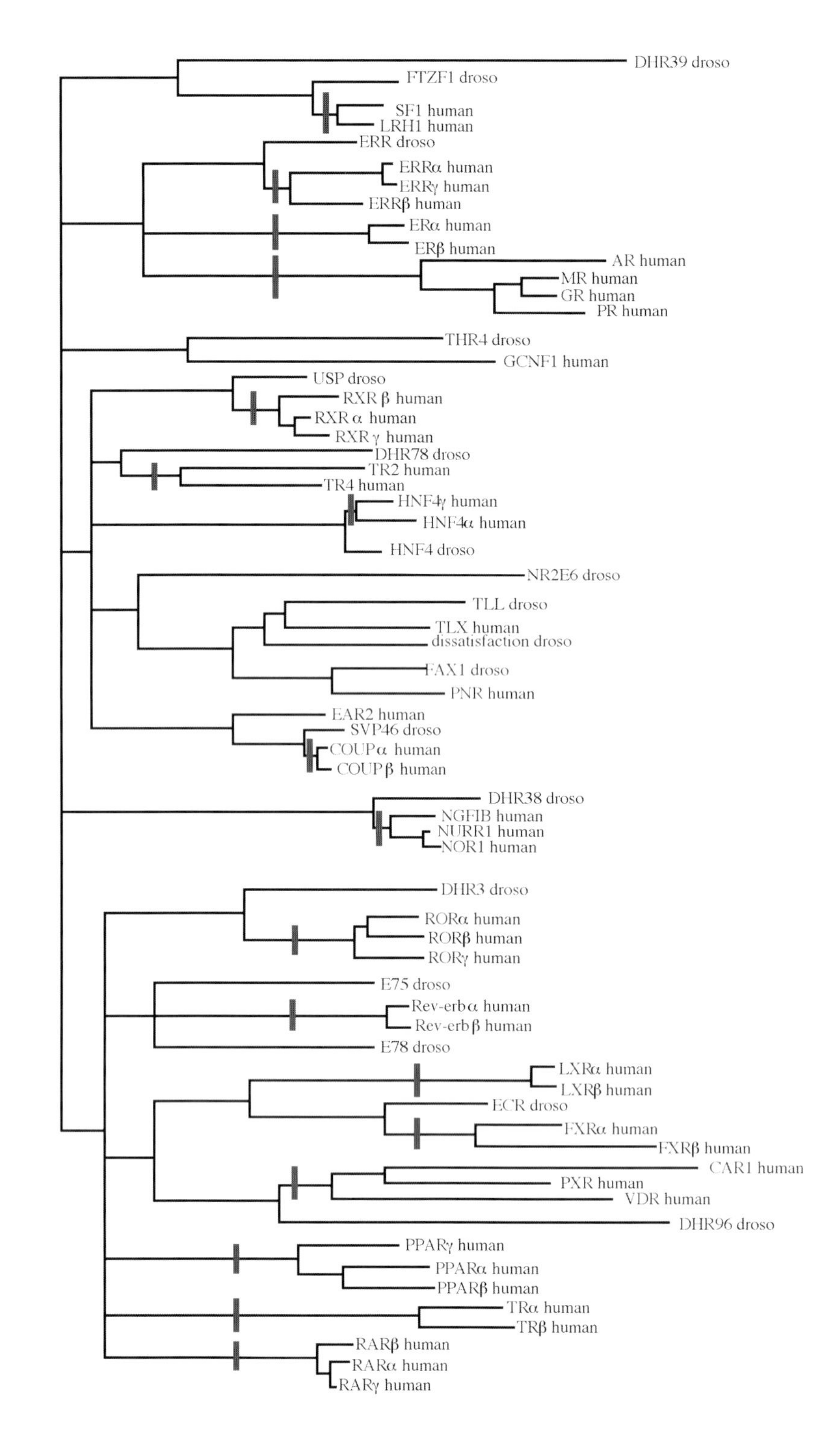

Figure 3. Phylogeny of human and fly nuclear receptors
Unrooted tree based on a Neighbor-Joining phylogeny with Poisson corrected amino-acid distances, with unsupported nodes collapsed in polytomies, as judged by bootstrap support and likelihood tests. Vertical grey bars indicate vertebrate-specific duplications.

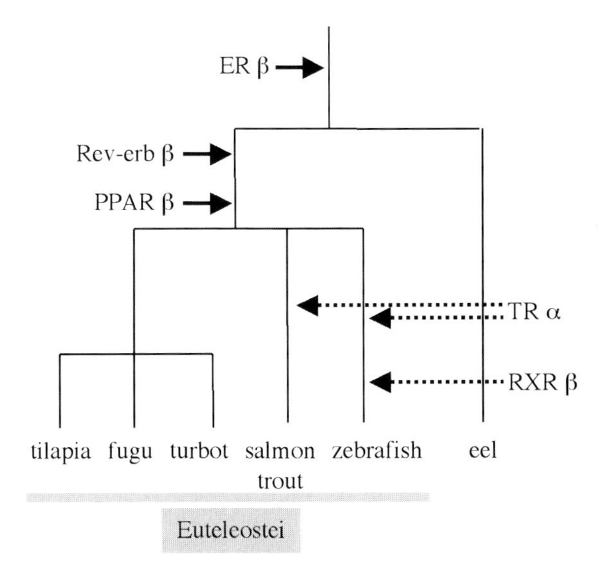

Figure 4. Fish-specific duplications of nuclear receptors
The tree represents a simplified phylogeny of the fishes in which nuclear receptors were characterized by RT-PCR. Branch lengths are arbitrary, and do not reflect evolutionary distance. Arrows with full lines indicate origin of duplications, as determined by phylogenetic analysis. Arrows with broken lines indicate possible origin of duplications, with weak support in the analysis. All duplications indicated are specific to actinopterygian fish.

duplications occured after the divergence between vertebrates and amphioxus, some before and some after the divergence between lamprey, hagfish and gnathostomes. And as for all genes studied so far, there is no vertebrate-specific series of ancient duplications leading to more than four paralogs of nuclear receptors. For early vertebrate evolution, nuclear receptors thus pass the test of representativity of the population of genes.

There is much interest in the possibility that actinopterygian fishes may have arisen following a complete genome duplication (Amores *et al.*, 1998; Postlethwait *et al.*, 2000; Robinson-Rechavi *et al.*, 2001c). What is certain is that many, maybe most, genes have two paralogs in fishes which are orthologous to a unique gene in mammals (Robinson-Rechavi *et al.*, 2001a; Taylor *et al.*, 2001a). Sampling of nuclear receptors in fishes by RT-PCR has shown that they once again completely reflect the distribution of the ensemble of genes (Robinson-Rechavi *et al.*, 2001a): for most, but not all, fish-specific gene duplicates can be found, although not all gene duplications appear simultaneous (Figure 4). Five nuclear receptors were found duplicated in fishes: PPARβ, Rev-erbβ, ERβ, RXRβ and TRα. On the other hand, only one ortholog, whose phylogeny matches that of the species, is found for TRβ and ERα, despite extensive RT-PCR searches, in 8 and 10 species respectively. The distribution of these nuclear receptor duplications in fish phylogeny (Figure 4) raises the question of a unique tetraploidization event at the origin of additional fish genes, as proposed on the basis of *hox* clusters (Amores *et al.*, 1998) and comparative mapping (Postlethwait *et al.*, 2000). Further phylogenetic examination of all genes characterized in at least three fish orders shows a diverse timing of gene duplication events, occuring often independently in different euteleost orders (Robinson-Rechavi *et al.*, 2001c). Although this observation is still controversial (Robinson-Rechavi *et al.*, 2001d; Taylor *et al.*, 2001b), it appears difficult to concile with a unique event of tetraploidization, and rather suggests many independent events of gene or chromosome duplication (Robinson-Rechavi *et al.*, 2001c).

On the opposite, the fruit fly *D. melanogaster* appears to have a rather minimal animal genome set, with the lowest known number of genes in a metazoan (Adams *et al.*, 2000). And indeed, no nuclear receptor is specifically duplicated in the lineage leading to the fly (Maglich *et al.*, 2001) (Figure 3). Similarly, there is no evidence for ancient genome duplication in the lineage of land vertebrates (Sarcopterygii), and no nuclear receptor is specifically duplicated in that lineage either (Robinson-Rechavi *et al.*, 2001b).

So the pudding was eaten, and it is tasty! When there are large scale events of gene/genome duplication, nuclear receptors are duplicated, in a massive and timely manner. When there are none, they are usually not. Nuclear receptors indeed appear to be faithful markers of genome evolution.

Nuclear receptor duplication is functionally relevant

A good marker of genome evolution should ideally also be able to tell us more about the function of gene and genome duplications in evolution, and especially in the appearance of major innovations. This is one of the most facinating features of *hox* genes, leading to their undisputed place in the center of animal evolution and development ("evo/devo"). Newcomers in the field, nuclear receptors need further functional characterization in evolutionary important species, such as amphioxus for chordate evolution, before their role can be fully understood. We will try to give leads here as to the part nuclear receptors may play in completing the picture of how gene and genome duplications have allowed functional innovation in the history of animals.

Differences in gene expression may be a major force in the specialization of duplicate genes (Force *et al.*, 1999). Amphibian thyroid hormone receptors (TR) provide an interesting example of how a differenciation in expression times can lead to important differences in function. There are two thyroid hormone receptors in land vertebrates, TRα and TRβ, products of an ancient vertebrate duplication, consistent with the 2R hypothesis. Although frog metamorphosis is trigered by thyroid hormone, TRα is expressed very early in frog metamorphosis, at an intermediate level, and its expression does not play a dramatic part in this event. On the opposite, TRβ level starts very low, and has a sudden peak that coincides with the climax of metamorphosis, when the major morphological changes take place: TRβ, and not TRα, is the vector of the hormonal message for this major change from an obligate aquatic to a land-dwelling animal (Fairclough and Tata, 1997). There is a dramatic difference in the roles of these two paralogs. Moreover, metamorphosis in flatfish is also triggered by thyroid hormone but it is apparently TRα and not TRβ, that is up-regulated during the climax (Yamano and Miwa, 1998). These examples show how the adaptation of each duplicated gene to an important biological function can differ from one evolutionary lineage to the other.

Another case of specialization of paralogs which arose through vertebrate-specific duplication is that of the three PPARs (Desvergne and Wahli, 1999): PPARα has a large but not ubiquitous spectrum of expression in mammals, and binds leukotrienes; PPARβ is expressed ubiquitously, at low levels, and binds fatty acids; and PPARγ is specifically expressed at high levels in fat tissues, and binds prostaglandines. Phylogenetic analysis and isolation of cDNA from lamprey and amphioxus confirm that the three vertebrate PPARs duplicated at the origin of vertebrates. The presence of a lamprey PPAR gene, phylogenetically localized before the vertebrate PPARα-PPARβ duplication but after the vertebrate PPARγ divergence, suggests (i) that one duplication took place before and one after the divergence of lamprey and gnathostomes and (ii) that a PPARγ ortholog gene should exist in lampreys. Functional characterization of the ligand binding capacity of the unique amphioxus PPAR and the two lamprey PPARs is hoped to bring forth a complete picture of how '2R' duplications have lead to functional specialization.

The presence of a conserved response element of 19 bp called IR7 in the 5' non coding region of the nuclear orphan receptor TR2 of *Xenopus*, rainbow trout, zebrafish and mouse suggests an important role for this element in vertebrate TR2 genes, but not in its paralog TR4 (Le Jossic and Michel, 1998). We studied the amphioxus ortholog gene to TR2 and TR4 (AmphiTR2/4) and we confirmed the presence of the IR7 element in its 5' non coding region. We also characterized the IR7 as a response element for the AmphiRAR (retinoic acid receptor) and for AmphiTR2/4 itself and we showed that both AmphiRAR and AmphiTR2/4 can transactivate gene expression throught the IR7 element. These findings strongly support the hypothesis of the conserved evolutionary role of the IR7 element in the control of gene expression. The most parsimonious explanation for this conservation is a key regulatory ancestral role that after the vertebrate specific gene duplication was lost for TR4 genes, allowing them to develop new functions (Escriva *et al.*, 2002a). This provides a nice illustration of the duplication-degeneration-complementation (DDC) theory of gene evolution (Force *et al.*, 1999), according to which duplicate genes are preserved by differential loss of regulatory functions.

There are many other such examples, which highlight the necessity of studying nuclear receptors in more diverse chordates, such as amphioxus and lamprey, but we should not forget that the basic diversity of nuclear receptors we observe nowadays in animals also results from gene duplications, much more ancient (Mendonça *et al.*, 1999; Escriva *et al.*, 2000). Thus there is probably a lot to be understood about the diversification of metazoans by comparative studies of nuclear receptors, as from other key gene families (Suga *et al.*, 1999).

Finally, it is extremely encouraging for studies integrating evolution, genomes and development that different model gene families can be linked functionally in this context. For example, HOX genes (which are present as four duplicated clusters in vertebrates and only one cluster in the cephalochordate, amphioxus), specify cell identity along the anterior-posterior axis of the embryo. In vertebrates, retinoic acid is implicated in the patterning of the anterior-posterior axis and the induction of HOX genes. In the closest relative to vertebrates, the amphioxus, retinoic acid also controls the induction of HOX genes. In vertebrates and amphioxus, the retinoic acid acts throught its binding to the retinoic acid receptors (RAR) (Shimeld, 1996). As for other nuclear receptors, there are three retinoic acid receptors in vertebrates (RARα, RARβ, and RARγ) but only one in amphioxus (AmphiRAR). Thus it may well be that duplicate nuclear receptors have co-evolved with the (duplicate) loci they regulate, as vertebrate development became increasingly diverse.

Conclusion

Nuclear receptors have several major characteristics that make them remarkable players in the study of gene and genome duplications: they are good phylogenetic markers, allowing robust timing of duplication *vs.* speciation events; they are dispersed in animal genomes, thus representing an unbiased sample of sequences; in the well studied duplication events at the origin of vertebrates and in the teleost fish lineage, their duplication history is totally consistent with that of the whole gene population, again demonstrating that they represent a good sampling of the genome; duplicate nuclear receptors are known to have functional relevance, notably to development and various physiological processes, including functional interplay with duplicated *hox* genes. Recent studies using nuclear receptors have allowed to challenge the hypothesis of a unique ancient origin of duplicated genes in teleost fish (Robinson-Rechavi *et al.*, 2001c), and to characterize the timing of duplications at the origin of vertebrates (Escriva *et al.*, 2002a).

In the future, characterization of nuclear receptors from species, such as amphioxus, lamprey and zebrafish, will allow us to relate genome duplications, gene duplications, morphological innovations, and physiological processes (Escriva *et al.*, 2000; Thornton, 2001). This will notably necessitate the characterization of the expression patterns, ligand-specificity and target genes of duplicated and non duplicated orthologues from these species. Moreover, new bioinformatic tools need to be developped to allow the comparison of structure, function and evolution on the scale of the super-family. Extensive study of a few well chosen gene families appears as the most promising path to link evolutonary and functional genomics in the future. Nuclear receptors are one of these families.

Acknowledgements

We thank Association de Recherche sur le Cancer, Centre National pour la Recherche Scientifique, Ecole Normale Supérieure de Lyon, Ministère de l'Education Nationale, and Région Rhône-Alpes for financial support.

References

Adams, M.D., Celniker, S.E., Holt, R.A., Evans, C.A., Gocayne, J.D., Amanatides, P.G., Scherer, S.E., Li, P.W., Hoskins, R.A., Galle, R.F., George, R.A., Lewis, S.E., Richards, S., Ashburner, M., Henderson, S.N., Sutton, G.G., Wortman, J.R., Yandell, M.D., Zhang, Q., Chen, L.X., Brandon, R.C., Rogers, Y.H., Blazej, R.G., Champe, M., Pfeiffer, B.D., Wan, K.H., Doyle, C., Baxter, E.G., Helt, G., Nelson, C.R., Gabor, G.L., Abril, J.F., Agbayani, A., An, H.J., Andrews-Pfannkoch, C., Baldwin, D., Ballew, R.M., Basu, A., Baxendale, J., Bayraktaroglu, L., Beasley, E.M., Beeson, K.Y., Benos, P.V., Berman, B.P., Bhandari, D., Bolshakov, S., Borkova, D., Botchan, M.R., Bouck, J., *et al.* (2000). The genome sequence of drosophila melanogaster. *Science* 287, 2185–2195.

Amero, S.A., Kretsinger, R.H., Moncrief, N.D., Yamamoto, K.R. and Pearson, W.R. (1992). The origin of nuclear receptor proteins: A single precursor distinct from other transcription factors. *Mol. Endocrinol.* 6, 3–7.

Amores, A., Force, A., Yan, Y.L., Joly, L., Amemiya, C., Fritz, A., Ho, R.K., Langeland, J., Prince, V., Wang, Y.L., Westerfield, M., Ekker, M. and Postlethwait, J.H. (1998). Zebrafish *hox* clusters and vertebrate genome evolution. *Science* 282, 1711–1714.

Caron, H., van Schaik, B., van der Mee, M., Baas, F., Riggins, G., van Sluis, P., Hermus, M.-C., van Asperen, R., Boon, K., Voûte, P.A., Heisterkamp, S., van Kampen, A. and Versteeg, R. (2001). The human transcriptome map: Clustering of highly expressed genes in chromosomal domains. *Science* 291, 1289–1292.

Desvergne, B. and Wahli, W. (1999) Peroxisome proliferator-activated receptors: nuclear control of metabolism. *Endocrine Rev.* 20, 649–688.

Dunham, I., Shimizu, N., Roe, B.A., Chissoe, S., Hunt, A.R., Collins, J.E., Bruskiewich, R., Beare, D.M., Clamp, M., Smink, L.J., Ainscough, R., Almeida, J.P., Babbage, A., Bagguley, C., Bailey, J., Barlow, K., Bates, K.N., Beasley, O., Bird, C.P., Blakey, S., Bridgeman, A.M., Buck, D., Burgess, J., Burrill, W.D., O'Brien, K.P. and et al. (1999). The DNA sequence of human chromosome 22. *Nature* 402, 489–495.

Escriva, H., Delaunay, F. and Laudet, V. (2000). Ligand binding and nuclear receptor evolution. *BioEssays* 22, 717–727.

Escriva, H., Holland, N.D., Gronemeyer, H., Laudet, V. and Holland, L.Z. (2002a). The retinoic acid signaling pathway regulates anterior/posterior patterning in the nerve cord and pharynx of amphioxus, a chordate lacking neural crest. *Development*, in press.

Escriva, H., Manzon, L., Youson, J., Laudet, V. (2002b) Analysis of lamprey and hagfish genes reveals a complex history of gene duplications during early vertebrate evolution. *Mol. Biol. Evol.*, **19**, 1440–1450.

Escriva, H., Safi, R., Hänni, C., Langlois, M.-C., Saumitou-Laprade, P., Stehelin, D., Capron, A., Pierce, R. and Laudet, V. (1997). Ligand binding was aquired during evolution of nuclear receptors. *Proc. Natl. Acad. Sci. USA* 94, 6803–6808.

Fairclough, L. and Tata, J.R. (1997) An immunocytochemical analysis of the expression of thyroid hormone receptor alpha and beta proteins during natural and thyroid hormone-induced metamorphosis in Xenopus. *Dev. Growth Differ.* 39, 273–283.

Force, A., Lynch, M., Pickett, F.B., Amores, A., Yan, Y.-l. and Postlethwait, J. (1999). Preservation of duplicate genes by complementary, degenerative mutations. *Genetics* 151, 1531–1545.

Hattori, M., Fujiyama, A., Taylor, T.D., Watanabe, H., Yada, T., Park, H.S., Toyoda, A., Ishii, K., Totoki, Y., Choi, D.K., Soeda, E., Ohki, M., Takagi, T., Sakaki, Y., Taudien, S., Blechschmidt, K., Polley, A., Menzel, U., Delabar, J., Kumpf, K., Lehmann, R., Patterson, D., Reichwald, K., Rump, A., Schillhabel, M., Schudy, A., Zimmermann, W., Rosenthal, A., Kudoh, J., Schibuya, K., Kawasaki, K., Asakawa, S., Shintani, A., Sasaki, T., Nagamine, K., Mitsuyama, S., Antonarakis, S.E., Minoshima, S., Shimizu, N., Nordsiek, G., Hornischer, K., Brant, P., Scharfe, M., Schon, O., Desario, A., Reichelt, J., Kauer, G., Blocker, H., Ramser, J., Beck, A., Klages, S., Hennig, S., Riesselmann, L., Dagand, E., Haaf, T., Wehrmeyer, S., Borzym, K., Gardiner, K., Nizetic, D., Francis, F., Lehrach, H., Reinhardt, R. and Yaspo, M.L. (2000). The DNA sequence of human chromosome 21. *Nature* 405, 311–319.

Holland, P. (2002) More genes in vertebrates? In *Genome Evolution*, (Eds. Meyer, A. Meyer and van de Peer, Y.), Kluwer Academic Publishers, Dordrecht, in press.

Hughes, A.L. and Friedman, R. (2002) 2R or not 2R: testing hypotheses of genome duplication in early vertebrates. In *Genome Evolution* (Eds. Meyer, A. and van de Peer, Y.), Kluwer Academic Publishers, Dordrecht, in press.

Laudet, V. (1997). Evolution of the nuclear receptor superfamily: Early diversification from an ancestral orphan receptor. *J. Mol. Endocrinol.* 19, 207–226.

Laudet, V. and Gronemeyer, H. (2002) *The Nuclear Receptors Factsbook*, Academic Press, London.

Laudet, V., Hänni, C., Coll, J., Catzeflis, C. and Stéhelin, D. (1992). Evolution of the nuclear receptor gene family. *EMBO J.* 11, 1003–1013.

Le Jossic, C. and Michel, D. (1998). Striking evolutionary conservation of a cis-element related to nuclear receptor target sites and present in tr2 orphan receptor genes. *Biochem. Biophys. Res. Commun.* 245, 64–69.

Lundin, L.-G., Larhammar, D. *et al.* (2002) Numerous groups of chromosomal regional paralogies strongly indicate two genome doublings at the root of vertebrates. In *Genome Evolution* (Eds. Meyer, A. and van de Peer, Y.), Kluwer Academic Publishers, Dordrecht, in press.

Maglich, J.M., Sluder, A.E., Guan, X., Shi, Y., McKee, D.D., Carrick, K., Kamdar, K., Willson, T.M. and Moore, J.T. (2001). Comparison of complete nuclear receptor sets from the human, *Caenorhabditis elegans* and *Drosophila* genomes. *GenomeBiology.com* 2, research0029.0021-0027.

Málaga-Trillo, E. and Meyer, A. (1) Genome duplications and accelerated evolution of Hox genes and cluster architecture in teleost fishes. *Amer. Zool.*, 41: 676–686.

Mendonça, R.L., Escriva, H., Vanacker, J.-M., Bouton, D., Delannoy, S., Pierce, R., Laudet, V. (1999) Nuclear hormone receptors and evolution. *Amer. Zool.* 39, 704–713.

Nuclear Receptors Nomenclature Committee (1999). A unified nomenclature system for the nuclear receptor superfamily. *Cell* 97, 161–163.

Ohno, S. (1970) *Evolution by Gene Duplication*. Springer-Verlag, New York, NY.

Postlethwait, J.H., Woods, I.G., Ngo-Hazelett, P., Yan, Y.L., Kelly, P.D., Chu, F., Huang, H., Hill-Force, A. and Talbot, W.S. (2000). Zebrafish comparative genomics and the origins of vertebrate chromosomes. *Genome Res.* 10, 1890–1902.

Robinson-Rechavi, M., Marchand, O., Escriva, H., Bardet, P.-L., Zelus, D., Hughes, S. and Laudet, V. (2001a). Euteleost fish genomes are characterized by expansion of gene families. *Genome Res.* 11, 781–788.

Robinson-Rechavi, M., Carpentier, A.-S., Duffraisse, M. and Laudet, V. (2001b). How many nuclear hormone receptors in the human genome? *Trends Genet.* 17, 554–556.

Robinson-Rechavi, M., Marchand, O., Escriva, H. and Laudet, V. (2001c). An ancestral whole-genome duplication may not have been responsible for the abundance of duplicated fish genes. *Curr. Biol.* 11, R458-R459.

Robinson-Rechavi, M., Marchand, O., Escriva, H. and Laudet, V. (2001d). Re: Revisiting recent challenges to the ancient fish-specific genome duplication hypothesis. *Curr. Biol.* 11, R1007–R1008.

Shimeld, S.M. (1996) Retinoic acid, HOX genes and the anterior posterior axis in chordates. *BioEssays* 18, 613–616.

Suga, H., Koyanagi, M., Hoshiyama, D., Ono, K., Iwabe, N., Kuma, K. and Miyata, T. (1999). Extensive gene duplication in the early evolution of animals before the parazoan-eumetazoan split demonstrated by g proteins and protein tyrosine kinases from sponge and hydra. *J. Mol. Evol.* 48, 646–653.

Taylor, J.S., Van de Peer, Y., Braasch, I. and Meyer, A. (2001a). Comparative genomics provides evidence for an ancient genome duplication event in fish. *Phil. Trans. R. Soc. Lond. Ser. B* 356, 1661–1679.

Taylor, J.S., van de Peer, Y. and Meyer, A. (2001b). Revisiting recent challenges to the ancient fish-specific genome duplication hypothesis. *Curr. Biol.* 11, R1005–R1007.

Thornton, J.W. (2001). Evolution of vertebrate steroid receptors from an ancestral estrogen receptor by ligand exploitation and serial genome expansions. *Proc. Natl. Acad. Sci. USA* 98, 5671–5676.

Thornton, J.W. and Desalle, R. (2000). A new method to localize and test the significance of incongruence: Detecting domain shuffling in the nuclear receptor superfamily. *Syst. Biol.* 49, 183–201.

Wicker, N., Perrin, G.R., Thierry, J.C. and Poch, O. (2001). Secator: A program for inferring protein subfamilies from phylogenetic trees. *Mol Biol Evol* 18, 1435–1441.

Yamano, K. and Miwa, S. (1998) Differential gene expression of thyroid hormone receptor α and β in fish development. *Gen. Comp. Endocrinol.* 109, 75–85.

A. Meyer, Y. Van de Peer (eds.), Genome Evolution, 185-194.

The fates of zebrafish Hox gene duplicates

Chris Jozefowicz[1], James McClintock[2] & Victoria Prince[1,2,3,*]
The Committees on [1]Evolutionary Biology and [2]Developmental Biology, [3]The Department of Organismal Biology and Anatomy, The University of Chicago, Chicago, IL 60637, USA;
[*]Author for correspondence: E-mail: vprince@midway.uchicago.edu

Received 29.01.2002; Accepted in final form 29.08.2002

Key words: duplication, hindbrain, Hox clusters, Hox genes; teleost, zebrafish

Abstract

In his 1970 book, Susumu Ohno stressed the importance of gene duplication in the evolution of the vertebrate genome and body plan. He elaborated the idea that duplication events provide novel genetic material on which evolution may act. Data are accumulating to show that extensive duplication events, perhaps incorporating the duplication of entire genomes, occurred in the lineage leading to teleost fishes. These duplications may have been pivotal in the explosive radiation of this highly successful vertebrate group. Thus, the teleosts provide us with an ideal opportunity to investigate the fates and functions of duplicated genes. A convenient system for these studies is the zebrafish, *Danio rerio*, which has become a popular genetic and embryological model.

Introduction

The evolutionarily conserved Hox genes play a key role in organizing the body plan of all animals. As a result of a wide-scale, possibly genome-wide, duplication event early in the ancestry of the teleosts, the zebrafish has at least 48 Hox genes arranged over seven clusters, as opposed to the 39 Hox genes in four clusters characteristic of mouse and human. Although many Hox duplicates have been lost in the lineage leading to zebrafish, there are ten cases where both duplicates have been retained. What mechanisms underlie retention of these duplicates and what are the implications for teleost body patterning? The wealth of data on Hox gene function in mouse and other tetrapod models provides an ideal comparison point for zebrafish studies. Here we discuss recent data that provide insight into the mechanisms underlying retention of some Hox duplicates in zebrafish.

Hox cluster duplications in the chordate lineage

The Hox genes were first characterized in *Drosophila melanogaster*, where eight linked *Antennapedia* class homeobox genes make up the Homeotic complex. These genes encode transcription factors that are characterized by their role in conferral of segmental identity along the anteroposterior (A-P) axis of the body (reviewed by McGinnis and Krumlauf, 1992). Subsequently, in every bilaterian in which they have been examined, Hox genes have been shown to play critical roles in determining A-P identity, and to furthermore exist as clustered genes, although gene and cluster number vary.

Comparative analysis of Hox cluster organization has revealed that variations in Hox gene number between species reflect an evolutionary history characterized by two types of duplication events: tandem duplication and whole cluster duplication. Current models suggest that single cluster organizations, like that of *D. melanogaster*, arose via the tandem duplication of ancestral Hox genes (Kappen *et al.*, 1989; Kmita-Cunisee *et al.*, 1998). A single cluster organization appears to be common to all protostomes, and a single cluster with seven genes was in place in the ancestor of all bilaterians (de Rosa *et al.*, 1999). A single Hox cluster is also assumed to be characteristic of primitive deuterostomes, with the cephalochordate amphioxus having the longest reported cluster comprising 14 genes (Ferrier *et al.*, 2000).

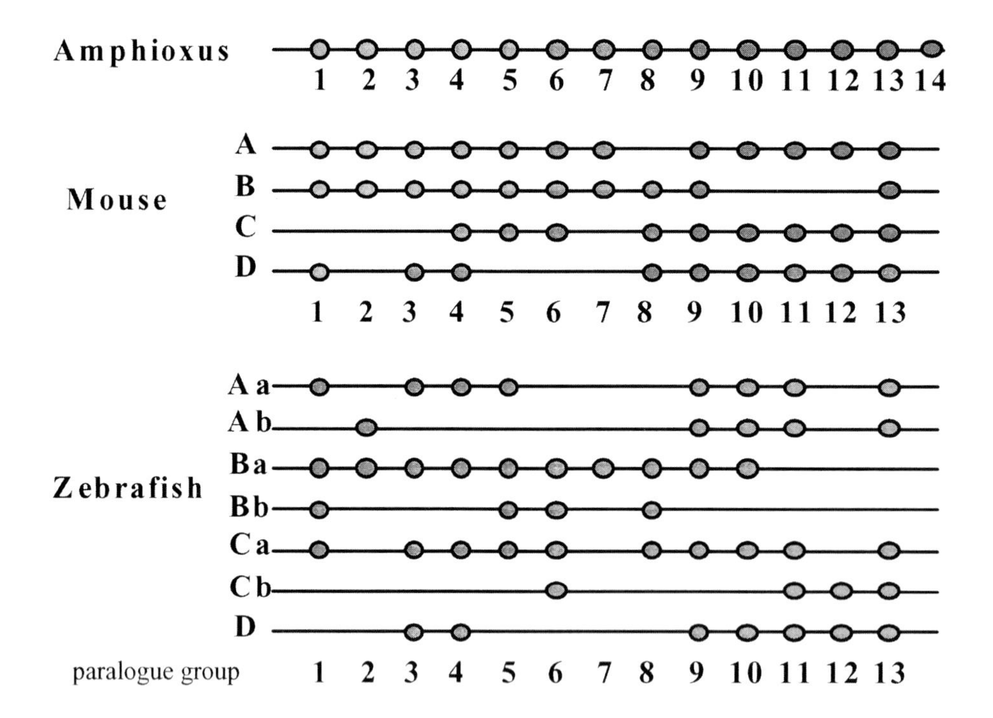

Figure 1. Schematic of Hox cluster organization in amphioxus, mouse and zebrafish. Note that there are a total of seven Hox clusters in the zebrafish (Aa, Ab, etc).

The origin of vertebrates is associated with major expansions in gene number, possibly as a result of two rounds of whole genome duplication (the '2R' hypothesis), which would result in duplication of entire Hox clusters (reviewed by Holland *et al.*, 1994; Sidow 1996). Consistent with this, tetrapod vertebrates have four clusters of Hox genes. Mouse and human have had their Hox cluster organization fully described; they share a 39 gene organization over four clusters, A-D (reviewed by McGinnis and Krumlauf, 1992). The genes fall into 13 paralogue groups, with most paralogue groups having less than a full complement of four genes as a result of secondary gene losses (Fig. 1). A large number of Hox genes have also been isolated from the tetrapod developmental model systems frog (*Xenopus laevis*) and chick (*Gallus gallus*), and in each case there is no evidence to suggest differences from the 39 gene mammalian organization (for instance, Godsave *et al.*, 1994).

The precise timing of the two postulated rounds of genome duplication at the origin of vertebrates remains uncertain. Amphioxus approximates a pre-duplication vertebrate ancestor, with the first 13 Hox genes in its single cluster representing ancestors of the 13 mammalian paralogue groups. Primitive, jawless vertebrates (the cyclostomes: hagfish and lampreys) represent potential models for vertebrate ancestors based on phylogenetic position (Fig. 2), and might be expected to fall into an intermediate state between a single, ancestral Hox cluster organization and a derived, four cluster organization. Recent analyses of the sea lamprey (*Petromyzon marinus*) Hox clusters (Force et al., 2002; Irvine et al., 2002) suggest a minimum of three Hox clusters, with a fourth cluster considered likely. However, the phylogenetic analyses of Force and colleagues (2002) suggest that only one duplication event occurred prior to the divergence of the agnathans and gnathostomes, with a second duplication event occurring within the lineage leading to lampreys.

The four Hox cluster organization of the mammals was initially assumed to be a general characteristic of Gnathostomata (jawed vertebrates), and early PCR-screen work on teleosts failed to uncover any major deviations from this expectation (Misof and Wagner, 1996; Misof *et al.*, 1996). However, more complete analysis of the zebrafish Hox genes, including complete linkage analysis, revealed at least 48 Hox genes arrayed over seven clusters in this ostariophysan teleost (Amores *et al.*, 1998). Based on both sequence

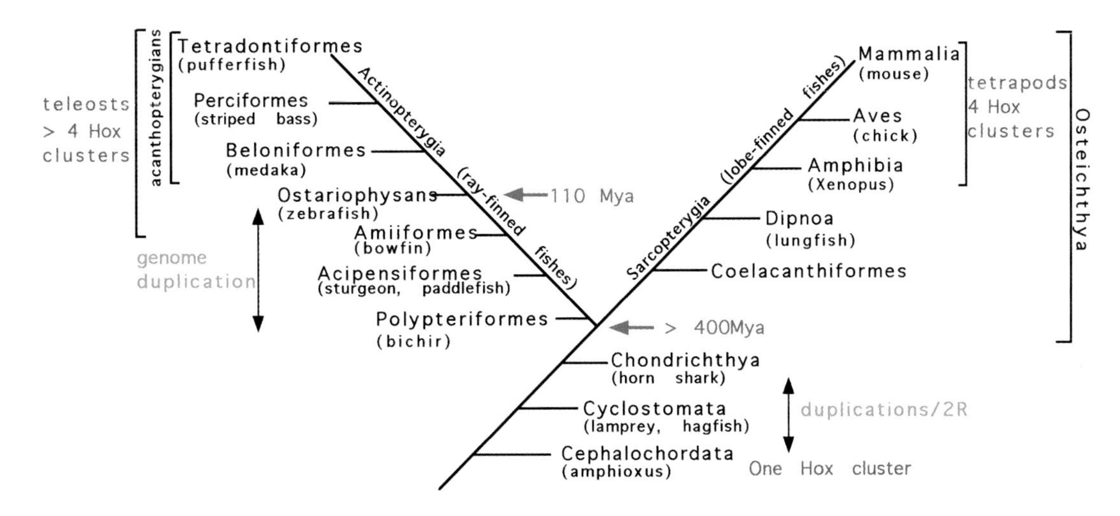

Figure 2. Vertebrate phylogeny showing Hox cluster number and putative duplication events.

and comparative linkage analysis to adjacent non-Hox genes, the seven clusters (Aa, Ab, Ba, Bb, Ca, Cb and D) have been assigned as duplicates of the four mammalian Hox clusters (Fig. 1); a duplicate D cluster was either lost during evolution or missed in the initial analysis. Taken together with other synteny relationship analyses between zebrafish and mammals, these data point towards a whole genome duplication event in the lineage leading to zebrafish (Amores *et al.*, 1998; Postlethwait *et al.*, 1998). A recent linkage map for the acanthopterygian teleost medaka (*Oryzias latipes*), similarly reveals seven clusters of Hox genes (Naruse *et al.*, 2000). These data have also allowed reinterpretation of the description of a four cluster organization for a pufferfish, the tetraodontiform fish *Fugu rubripes*: its four clusters appear to comprise two A clusters, one B and one C, with the likelihood of other clusters yet to be found (Aparicio, 2000). Similarly, although only four Hox clusters have been recognized in the perciform fish *Morone saxatilis*, these represent one A, two B and one C cluster (Ed Stellwag, pers. comm.).

Taken together, the presence of more than four Hox clusters in divergent teleost groups (the acanthopterygians and the ostariophysans) suggests that an entire Hox complement duplication event, relative to the four-cluster state seen in mammals, is common to the teleosts, and must have occurred in their common ancestor or earlier in more primitive actinopterygians (ray-finned fishes). This understanding is favored as it is more parsimonious than the possibility of multiple Hox cluster duplication events in these different teleost lineages. This indicates that the four cluster state seen in mammals may be a primitive feature of osteichthyans (bony gnathostomes).

In accordance with the comparative approach, the question of the ancestral osteichthyan Hox cluster organization has been approached by an examination of Hox clusters in Chondrichthyes (cartilaginous gnathostomes), a vertebrate group that is a sister group to Osteichthyes (Fig. 2). Investigations of the horn shark (*Heterodontus francisci*) have, to date, revealed the presence of only two Hox clusters, M and N, but the sequences and organization of the Hox genes within these clusters are entirely consistent with M and N representing the A and D clusters, respectively, as described in mammals (Kim *et al.*, 2001). It is possible that chondrichthyans and sarcopterygians (lobe-finned osteichthyans, including the tetrapods) will prove to share a similar four cluster Hox organization.

These findings leave open the question of when in the lineage leading to teleosts the genome duplication event occurred. The major vertebrate groups, Sarcopterygia and Actinopterygia diverged more than 400 Mya (Fig. 2; Carroll, 1998), thus the duplication must have occurred subsequent to this time. The teleosts have been radiating for roughly 110 My; this represents the most recent date at which the duplication could have occurred. A more accurate pinpointing of the time of the duplication will require analysis of basal actinopterygian fishes such as Polypteriformes (bichir), Acipensiformes (sturgeon, paddlefish), or Amiiformes (bowfin).

Mechanisms of retention of duplicated genes

The classical model of gene duplication holds that because duplicated genes are initially identical they can be considered functionally redundant, and stresses the role of the acquisition of novel function in the retention of gene duplicates. The assumption of redundancy may well be inaccurate in many cases - chimeric genes have been described that are the result of incomplete duplication of individual genes in the genomes of plants and *Drosophila* (e.g. Symonds and Gibson, 1992; Long *et al.*, 1996; Amador and Juan, 1999). However, this assumption may be safer within the vertebrates, where, as described above, large-scale (possibly genome-wide) duplications created loci that were initially identical. The model suggests that following duplication, one gene copy is under selection leaving the other free to drift, accumulate deleterious mutations, and finally lose function. As deleterious mutations are far more likely than beneficial ones (Lynch and Conery, 2000) the fixation of duplicated genes as a result of acquisition of some key novel function (neo-functionalization) is thought to be an extremely rare event. A much more common fate is the reduction of one of the duplicates to a pseudogene or its outright loss, a non-functionalization event. In this model, loss of duplicated genes is a common and relatively rapid evolutionary event. (In accordance with this model, the majority of duplicated Hox genes have been lost from the zebrafish clusters.) However, as vertebrate genomes appear to be rife with ancient gene duplicates (Nadeau and Sankoff, 1997), this classical model seems insufficient to explain the data. How then can we explain the retention of large numbers of duplicated genes without requiring the widespread evolution of novel functions?

Some insight has come from recent work by Force and colleagues (Force *et al.*, 1999; Lynch and Force, 2000). Based on the complexity of eukaryotic gene loci, the authors suggest that many genes may have multiple and separable functions. In particular, they suggest that the modular nature of eukaryotic gene enhancers may lead to a partitioning of gene functions following duplication, such that separate expression domains (spatial or temporal) are lost for each duplicate. Enhancers could also change with respect to the levels of gene expression, so that duplicates produce some lower amount of protein than did the ancestral, pre-duplicate gene. Such changes may lead to the duplicates retaining complementary functions (subfunctionalization) – both duplicates will then be required to recapitulate the original gene function (referred to as the duplication-degeneration-complementation, or "DDC" model). These complementary mutations ensure that both gene copies go to fixation in the genome. An important extension of this model is that once gene functions are parceled out between duplicates, each gene may be free to evolve along a novel trajectory.

The zebrafish Hox genes as a paradigm for the study of duplicate gene evolution

It has been established that Hox gene duplicates exist in the zebrafish, although the precise date of the duplication event remains obscure. The zebrafish provides a tractable model system in which we can examine the functional significance of Hox gene duplications. Using comparative expression and functional studies, we can begin to investigate what events have allowed retention of select pairs of gene duplicates. In an ideal scenario we would compare zebrafish Hox genes to those of a species that approximates the ancestral, pre-duplication condition. Unfortunately, the primitive fishes that are most likely to provide such a comparison group have not yet been studied.

However, Hox genes are highly conserved in their sequence, cluster organization, and regulation, which permits (even requires) comparisons to be made over wide evolutionary distances. Thus, informative comparisons can be made between zebrafish and such phylogenetically distant osteichthyans as mice, assuming that the four cluster organization, seemingly wide-spread in sarcopterygians, reflects the ancestral osteichthyan condition. Indeed, Hox genes are so conserved at the level of protein function that they can be functionally substituted for one another between distantly related species (for instance, Lutz *et al.*, 1996). Furthermore, the comparison of zebrafish to mouse takes advantage of the wealth of data concerning mouse Hox gene expression and function.

This general approach was used to investigate the zebrafish *hoxb5* duplicates (Bruce *et al.*, 2001). In this study, the expression patterns of zebrafish *hoxb5a* and *hoxb5b* were compared to that of the single mouse *Hoxb5* gene, and found to recapitulate its overall expression. The zebrafish *hoxb5* duplicates have different, but overlapping, expression patterns, yet appear to share identical biochemical functions as assessed by a gain-of-function approach. Thus, it appears that in this case, zebrafish *hoxb5a* and *hoxb5b* represent a partitioning of an expression domain with respect to the murine *Hoxb5* gene. Assuming that the murine *Hoxb5* gene reflects the ancestral osteichthyan state, the Hox duplication in the teleost lineage has lead to a sub-functionalization for these zebrafish *hoxb5* duplicates in accordance with the DDC model. Further tests of the model would include demonstrating that these two zebrafish genes are able to functionally substitute for one another, although it should be remembered that even when the DDC model is invoked to explain the fixation of gene duplications, this does not rule out subsequent neo-functionalization events which might obscure functional equivalence. It would also be of interest in this case to explore the regulatory sequences of the mouse *Hoxb5* and zebrafish *hoxb5a* and *hoxb5b* in order to identify changes in the zebrafish sequences that underlie the presumed partitioning of the ancestral expression domain.

Recent studies in our lab have pursued similar questions with respect to the four zebrafish Hox genes comprising paralogue group (PG) 1, which include a pair of duplicates with respect to the four-cluster state, *hoxb1a* and *hoxb1b*. However, in this case we have not found a straightforward example of sub-functionalization between the duplicates (*i.e.* a situation in which *hoxb1a* and *hoxb1b* together account for the expression and function established for the single mouse *Hoxb1* gene). Instead, the evolution of these zebrafish duplicates reflects a more complex situation of 'function shuffling' among the members of the paralogue group.

Function shuffling in paralogue group 1 genes

The PG1 genes are a particularly good system in which to investigate potential sub-functionalization because two of the three mouse genes have had both gene function and regulation studied in great detail. These experiments have shown that mouse *Hoxa1* and *Hoxb1* are necessary for proper development of the hindbrain. In zebrafish, as in mouse and chick, hindbrain morphology is conceptually simple, with overt segmentation dividing the hindbrain into seven compartments termed rhombomeres (r1-r7 from A to P; Fig. 3A; reviewed by Moens and Prince, 2002). This basic organization is conserved across the vertebrates.

There are a wealth of molecular and neuroanatomical markers that allow the identity of individual rhombomeres to be unambiguously recognized (reviewed by Moens and Prince, 2002). For example, in all vertebrates the cranial branchiomotor (BM) nerves have a rhombomere-specific disposition. Thus, in both mouse and zebrafish the Vth cranial nerve has cell bodies that lie in r2 and r3 and project axons through an exit point in r2 out to the adjacent first pharyngeal arch. In contrast, the VIIth nerve has cell bodies that originate in r4 and r5 but then undergo a characteristic posterior migration back to r6 and r7, while projecting axons anteriorly to exit through r4 and innervate the adjacent second pharyngeal arch (Fig. 3A). In zebrafish, the reticulospinal (RS) interneurons also have an obvious rhombomere-specific organization, with the large contralaterally projecting Mauthner neurons providing a particularly obvious marker of r4 identity (Fig. 3A).

Murine *Hoxa1* and *Hoxb1* are first expressed during gastrulation stages within the hindbrain primordium at the level of presumptive rhombomere 4. *Hoxa1* expression rapidly recedes towards the posterior as development progresses, while *Hoxb1* retains a high level of expression in r4 as the result of positive auto-regulation through a Hox/Pbx binding site (Murphy and Hill, 1991; Pöpperl *et al.*, 1995). (Mouse *Hoxd1* is not expressed during central nervous system development (Frohman and Martin, 1992), and we will not discuss its function further here). Null mutant analysis has shown that *Hoxa1* plays a somewhat unusual role for a Hox gene, in hindbrain segmentation, whereas *Hoxb1* plays the more expected "Hox-like" role of conferring segmental identity to r4.

Loss of *Hoxa1* function causes a radical reduction in the A-P extent of r4 and r5, with a concomitant reduction in the size of the adjacent otic vesicle (Chisaka *et al.*, 1992; Mark *et al.*, 1993). In addition, the motor neurons of the r4-derived VIIth and r5/6-derived VIth nerves fail to form. The Hoxa1 protein has also been shown to activate *Hoxb1* transcription through the Hox/Pbx binding site (diRocco *et al.*, 1997), and it has recently been shown that *Hoxa1* is required to set the appropriate anterior limit of *Hoxb1* expression (Barrow *et al.*, 2000).

In the absence of functional *Hoxb1*, the r4-derived VIIth BM nerve fails to form, although an unidentified population of cell bodies, which do not undergo the posterior migration characteristic of the VIIth nerve, does form transiently at the r4 level (Goddard

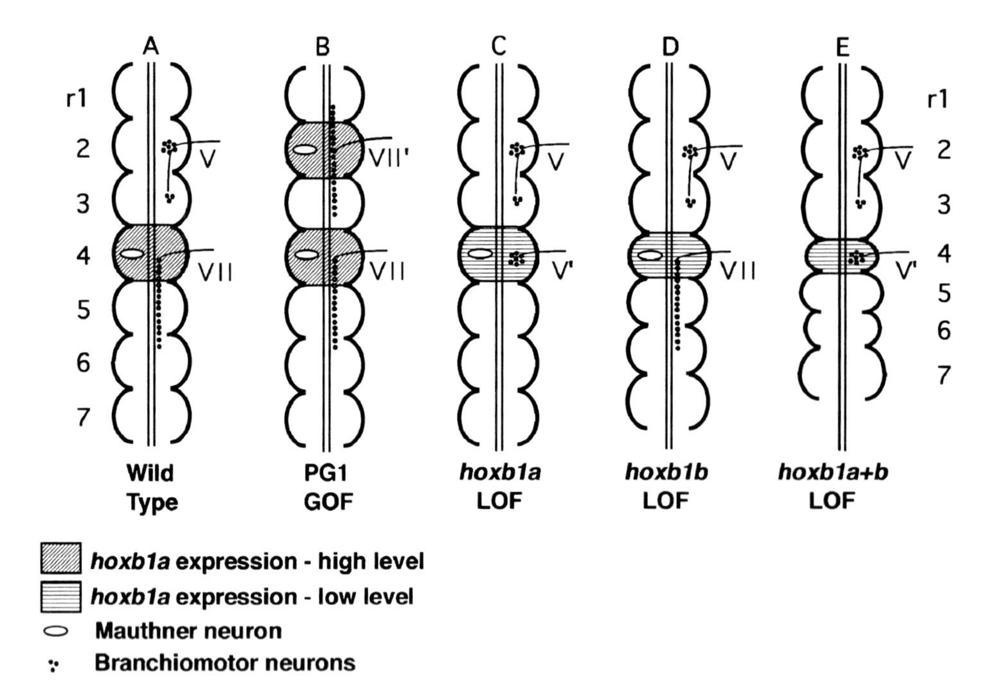

Figure 3. Summary of the results of functional assays with zebrafish Hox PG1 genes. (A) r4 of a wild-type embryo is characterized by Mauthner neurons, *hoxb1a* expression and the VIIth nerve axonal exit point. (B) In gain-of-function (GOF) experiments with PG1 genes r2 takes on properties of r4, a posteriorizing homeotic transformation. Thus, ectopic r4-characteristic Mauthner neurons form at the r2 level, VIIth nerve-like neurons form at the r2 level (VII') and *hoxb1a* is ectopically expressed at the r2 level. (C) In *hoxb1a* loss-of-function (LOF) experiments the branchiomotor neuron disposition is altered such that r4 neurons take on properties of r2 neurons (V'), an anteriorizing homeotic transformation. However, r4 Mauthner neurons are not altered in response to *hoxb1a* LOF. (D) The *hoxb1b* gene is important for normal segmentation of r4 through r6. (E) Reductions in rhombomere size are exacerbated in the double knock-downs, and only when *hoxb1a* and *hoxb1b* are both knocked-down are the Mauthner neurons lost.

et al., 1996; Gaufo *et al.*, 2000). In addition, breakdown of the usual auto-regulatory feedback loop leads to a gradual decline in *Hoxb1* transcription (Studer *et al.*, 1996). Mice mutant for both *Hoxa1* and *Hoxb1* show additional defects, such that r4 and r5 are further reduced or absent (Gavalas *et al.*, 1998; Studer *et al.*, 1998).

In the zebrafish, the two duplicates of *Hoxb1* do *not* together recapitulate the expression pattern of mouse *Hoxb1*. Instead, the zebrafish *hoxb1a* gene has a *Hoxb1*-like expression, but the zebrafish *hoxb1b* gene has expression like that of mouse *Hoxa1*. Thus, *hoxb1a* shows persistent expression within r4, exactly as described for *Hoxb1* expression in mouse, chick and *Xenopus* (reviewed by Prince 1998), whereas *hoxb1b* shows gastrula stage expression in presumptive r4, which then recedes towards the posterior (Alexandre *et al.*, 1996; McClintock *et al.*, 2001). The similarities of the *hoxb1b* gene expression pattern to that of mouse *Hoxa1* led the zebrafish gene to be initially assigned as a *Hoxa1* orthologue (Alexandre *et al.*, 1996), before linkage analysis revealed its true identity (Amores *et al.*, 1998).

Interestingly, although the zebrafish does have an orthologue of mouse *Hoxa1*, the *hoxa1a* gene, this gene is not expressed during hindbrain development. Rather, zebrafish *hoxa1a* is expressed at late neurulation stages in a small bilaterally located group of neurons in the ventral midbrain (McClintock *et al.*, 2001; Shih *et al.*, 2001). As midbrain expression has not generally been described for Hox genes, this domain seems at first observation to reflect a neo-functionalization event. However, our comparative analyses have demonstrated that midbrain expression is more likely a primitive characteristic of the vertebrate PG1 genes. Thus, we find expression of *Hoxa1* orthologues in a similar group of cells not only in another teleost, medaka, but also in the sarcopterygian chick (C.J., J.M., and V.P., manuscript in prep.). Furthermore, we have confirmed a previous description of midbrain expression for *Xenopus Hoxa1* (Kolm and Sive, 1995). As *Xenopus* and chick combine both hindbrain and midbrain expression domains of *Hoxa1*, we hypothesize that zebrafish *hoxb1b* has taken on the hindbrain patterning role of tetrapod *Hoxa1*, freeing *hoxa1a* to lose its hindbrain expression domain while retaining an ancestral midbrain patterning role.

To test the hypothesis that zebrafish *hoxb1a* and *hoxb1b* are the functional equivalents of mouse *Hoxa1* and *Hoxb1*, respectively, we have used both gain and loss-of-function approaches. Our gain-of-function experiments, using global mis-expression by mRNA injection, have shown that ectopic expression of any zebrafish PG1 gene, or even of the single mphioxus PG1 gene, can produce a classic posteriorizing homeotic transformation where r2 takes on properties of r4 (Fig. 3B; McClintock *et al.*, 2001). Alterations to r2 include expression of the r4 marker *hoxb1a* itself, and alterations to neuronal disposition: ectopic r4-characteristic Mauthner neurons form at the r2 level, and r2-derived branchiomotor neurons are relocated close to the floor-plate in an anteroposteriorly distributed line, as normally found for the r4-derived VIIth nerve neurons. Some differences were detectable in the functional capacities of the *hoxb1a* and *hoxb1b* genes using this approach: only *hoxb1a* mRNA causes more extensive posteriorizing transformations that affect tissues anterior to r2, hinting at some functional divergence between the duplicates. As *hoxb1a* transcription is activated in r2 by any of the mis-expressed PG1 genes, it is possible that the effects on rhombomere identity are mediated indirectly by *hoxb1a*. A more direct test of the functions of the individual genes was achieved using a loss-of-function approach (McClintock *et al.*, 2002).

Our loss-of-function studies have made use of stabilized antisense RNA oligonucleotides termed morpholinos. The morpholino is designed to complement the sequence around the start codon of an mRNA transcript, where it anneals to prevent translation of the message. This approach has been shown to act specifically and efficiently in a variety of systems to 'knock-down' gene function (for instance, Nasevicius *et al.*, 2000). We find that knockdown of *hoxb1a* leads to a robust anteriorizing transformation in the hindbrain, such that the r4-derived VIIth nerve BM neurons resemble those of the r2-derived Vth nerve. The transformed cells do not undergo the posterior migration characteristic of the VIIth nerve, although they retain their normal projection to the adjacent second pharyngeal arch (Fig. 3B; McClintock *et al.*, 2002). Consistent with our hypothesis that *hoxb1a* is the functional equivalent of mouse *Hoxb1*, this phenotype closely resembles that of the mouse *Hoxb1* knockout (Goddard *et al.*, 1996; Studer *et al.*, 1996; Gaufo *et al.*, 2000). Furthermore, the BM neuron phenotype is the precise reciprocal of the gain-of-function phenotype described above, although no changes were found in the r4-derived Mauthner neurons.

We have also found that knock-down of *hoxb1b* causes a phenotype very similar to the mouse *Hoxa1* knockout: r4, r5 and r6 are all reduced in A-P extent, as is the otic vesicle (Fig. 3D). Simultaneous knock-down of both zebrafish *hoxb1* duplicates leads to additional defects not observed in either single knock-down. The reduction in size of rhombomeres 4–6 is greatly exacerbated, and only when both genes are disrupted do the r4-derived Mauthner neurons fail to form (Fig. 3E; McClintock *et al.*, 2002). These results reveals a degree of functional redundancy between the duplicates, such that each can compensate for loss of the other. However, this redundancy is incomplete, as RNA rescue experiments reveal that only *hoxb1a* is able to properly pattern the VIIth nerve BM neurons. Thus, co-injection of *hoxb1a* mRNA with hoxb1a morpholino efficiently rescues the knock-down defect, but *hoxb1b* mRNA fails to rescue the phenotype. However, mouse *Hoxb1* RNA can also rescue the hoxb1a morpholino phenotype, consistent with our hypothesis that zebrafish *hoxb1a* and mouse *Hoxb1* have equivalent functions.

We can now interpret our gain-of-function experiments as revealing an indirect effect of *hoxb1b* and other PG1 genes on BM neuron patterning, mediated through activation of *hoxb1a* transcription, most likely by binding to a hox/pbx site equivalent to the one described for mouse *Hoxb1*. The inability of *hoxb1b* to compensate for loss of *hoxb1a* shows that, despite partial functional redundancy, the *hoxb1* duplicates have also evolved independent functions, and cannot be considered identical at the protein level. This is in sharp contrast to a recent demonstration that two mouse PG3 genes are functionally interchangeable (Greer *et al.*, 2000).

How can we explain the observation that two zebrafish duplicates of *hoxb1* play the same developmental roles as *Hoxa1* and *Hoxb1* in mouse? This does not seem to be a simple case of sub-functionalization, as we have no reason to expect that the ancestral *Hoxb1* gene had a *Hoxa1*-like function before its duplication. Rather, it seems a 'function shuffling' has occurred, such that a duplicate from the B cluster has taken on the role originally played by a gene from the A cluster. While zebrafish *hoxa1a* appears to have retained an ancestral function in midbrain patterning (as discussed earlier), it has lost any role in rhombomere patterning. Indeed, while one of the *Hoxa1* sub-functions was parceled out to *hoxa1a*, another function jumped to another cluster entirely. How might *hoxb1b* have taken on this *Hoxa1*-like role? Although the answer is not obvious, a consideration of the *cis*-regulation of PG1 genes may provide some clues. The murine *Hoxa1* and *Hoxb1* genes share distinct similarities in the way they are regulated, presumably reflecting their shared ancestry. For example, the early expression of both is under control of equivalent retinoic acid response elements (reviewed by Marshall *et al.*, 1994). Following the duplication event that produced two *hoxb1* duplicates, loss of a hox/pbx autoregulatory element from one of the duplicates would render its regulation very similar to that of *Hoxa1*. As paralagous Hox coding sequences can be functionally interchangeable (Greer *et al.*, 2000), and in fact paralogue group 1 genes have been substituted between organisms with some success (Lutz *et al.*, 1996), this change alone could conceivably have been sufficient to allow *hoxb1b* to take on the *Hoxa1*-like role in proper formation of rhombomeres. Once functional redundancy became established between *hoxb1b* and *hoxa1a*, the *hoxa1a* gene would have been free to eventually lose its hindbrain patterning role. Alternatively, a physical exchange of regulatory sequences between a *hoxa1* orthologue and *hoxb1b*, via a recombination event, could also explain our observations. In the future, analysis of the zebrafish PG1 regulatory elements may allow these ideas to be tested.

The kind of function shuffling we have demonstrated for the Hox PG1 genes of zebrafish may prove to be common amongst zebrafish paralogues. For example, it has recently been shown using morpholino-based knock-down that the zebrafish *eng2* and *eng3* genes have early developmental roles equivalent to that of the non-orthologous mouse *EN1* gene (Scholpp and Brand, 2001). Furthermore, function shuffling may not be limited to transcription factor genes: the secreted signaling molecule bmp2a from zebrafish appears to play an equivalent functional role to the non-orthologous *Xenopus* Bmp4 during dorsoventral patterning of gastrula stage embryos (Nguyen *et al.*, 1998). Thus, in cases where orthology relationships are unclear, it may not help to assume that common function can help with assignments – synteny relationships are more likely to be a reliable tool.

Conclusions

Hox gene functions are intimately associated with patterning of the body plan. Thus, changes in Hox genes are likely to play a key role in the evolution of new body plans. Consistent with this idea, the large-scale gene duplications at the origin of vertebrates provided many additional Hox genes which correlate with the innovations that characterize the vertebrates (reviewed by Holland *et al.*, 1994). It has been suggested that the additional duplication event in the teleost lineage provided yet more raw genetic material for selection to act upon, and that this may have facilitated the broad radiation of teleosts (see Van de Peer *et al.*, this volume).

The zebrafish has retained at least 10 duplicated Hox genes, opening up the possibility that in some cases duplicates were fixed because one of them attained a novel function. In the two cases that have been investigated so far this appears not to be the case. The *hoxb5* duplicates have sub-functionalized, in accordance with the DDC model (Force *et al.*, 1998), whereas the PG1 genes have gone through an interesting function shuffling, while still not undergoing any obvious neo-functionalization. However, it should be noted that neo-functionalization may prove difficult to recognize, especially in the absence of a representative of the primitive condition. Important changes could be subtle – for example minor but critical changes in timing of gene expression, the dose of the expressed gene or amount of gene product, or origin of a new late expression pattern that would not be detected within the usual time frame of developmental expression analysis. Alternatively, the 10 retained duplicate Hox genes may all prove to have undergone some variation on the sub-functionalization theme. This would not undermine the hypothesis that gene duplication was important in the teleost radiation, but would rather suggest that other genes, perhaps the downstream mediators of Hox function, were the ones to gain novel functions.

Acknowledgements

We thank Ed Stellwag and Ashley Bruce for helpful comments on the manuscript.

References

Amores, A., Force, A., Yan, Y-L., Amemiya, C., Fritz, A., Ho, R.K., Joly, L., Langeland, J., Prince, V., Wang, Y-L., Westerfield, M., Ekker, M. and Postlethwait, J.H. (1998) Genome duplications in vertebrate evolution: evidence from zebrafish *Hox* clusters. *Science*, **282,** 1711–1714.

Aparicio, S. (2000) Vertebrate evolution: recent perspectives from fish. *Trends Genet.*, **16,** 54–56.

Bruce, A., Oates, A., Prince, V.E. and Ho, R.K. (2001) Additional *hox* clusters in the zebrafish: Divergent expression belies conserved activities of duplicate *hoxB5* genes. *Evol Dev.*, **3,** 127–144.

Carpenter, E.M., Goddard, J.M., Chisaka, O., Manley, N.R. and Capecchi, M.R. (1993) Loss of *Hox-A1* (*Hox–1.6*) function results in the reorganization of the murine hindbrain. *Development*, **118,** 1063–1075.

Carroll, R.L. (1988) *Vertebrate Paleontology and Evolution*, W.H. Freeman and Co., New York, NY.

Chisaka, O., Musci, T.E. and Capecchi, M.R. (1992) Developmental defects of the ear, cranial nerves and hindbrain resulting from targeted disruption of the mouse homeobox gene Hox 1.6. *Nature*, **355,** 516–520.

de Rosa, R., Grenier, J.K., Andreeva, T., Cook, C.E., Adoutte, A., Akam, M., Carroll, S.B. and Balavoine, G. (1999) Hox genes in brachiopods and priapulids and protostome evolution. *Nature*, **399,** 772–776.

Di Rocco, G., Mavilio, F. and Zappavigna, V. (1997) Functional dissection of a transcriptionally active, target-specific Hox-Pbx complex. *EMBO J.*, **16,** 3644–3654.

Ferrier, D.E., Minguillon, C., Holland, P.W. and Garcia-Fernandez J. (2000) The amphioxus Hox cluster: deuterostome posterior flexibility and Hox 14. *Evol Dev.*, **2,** 284–293.

Force, A., Lynch, M., Pickett, F.B., Amores, A., Yan, Y.L. and Postlethwait, J. (1999) Preservation of duplicate genes by complementary, degenerative mutations. *Genetics*, **151,** 1531–1545.

Force, A., Amores, A. and Postlethwait, J. (2002) Hox cluster organization in the jawless vertebrate *Petromyzon marinus*. *J. Exp. Zool. (Mol. Dev. Evol.)* **294**, 30–46.

Frohman, M.A. and Martin, G.R. (1992) Isolation and analysis of embryonic expression of Hox–4.9, a member of the murine labial-like gene family. *Mech. Dev.*, **38,** 55–67.

Gaufo, G.O., Flodby, P. and Capecchi, M.R. (2000) Hoxb1 controls effectors of sonic hedgehog and Mash1 signaling pathways. *Development*, **127,** 5343–5354.

Gavalas, S., Studer, M., Lumsden, A., Rijli, F. M., Krumlauf, R. and Chambon, P. (1998) *Hoxa1* and *Hoxb1* synergize in patterning the hindbrain, cranial nerves and second pharyngeal arch. *Development*, **125,** 1123–1136.

Goddard, J.M., Rossel, M., Manley, N.R. and Capecchi, M.R. (1996) Mice with targeted disruption of *Hoxb1*fail to form the motor nucleus of the V11th nerve. *Development*, **122,** 3217–3216.

Godsave, S., Dekker, E.J., Holling, T., Pannese, M., Boncinelli, E. and Durston, A(1994) Expression patterns of Hoxb genes in the Xenopus embryo suggest roles in anteroposterior specification of the hindbrain and in dorsoventral patterning of the mesoderm. *Dev. Biol.*, **166,** 465–76.

Greer, J.M., Puetz, J., Thomas, K.R. and Capecchi, M.R. (2000) Maintenance of functional equivalence during paralogous Hox gene evolution. *Nature*, **403,** 661–665.

Holland, P.W., Garcia-Fernandez, J., Williams, N.A. and Sidow, A. (1994) Gene duplications and the origins of vertebrate development. *Development* .Suppl., 125–133.

Irvine, S.Q., Carr, J.L., Bailey, W.J., Kawasaki, K., Shimizu, N., Amemiya, C.T. and Ruddle , F.H. (2002) Genomic analysis of Hox clusters in the sea lamprey *Petromyzon marinus*. *J. Exp. Zool. (Mol. Dev. Evol.)* **294**, 47–62.

Kappen, C., Schughart, K. and Ruddle, F.H. (1989) Two steps in the evolution of Antennapedia-class vertebrate homeobox genes. *Proc. Natl. Acad. Sci. USA*, **86,** 5459–5463.

Kim, C.B., Amemiya, C., Bailey, W., Kawasaki, K., Mezey, J., Miller, W., Minoshima, S., Shimizu, N., Wagner, G., and Ruddle F. (2000) Hox cluster genomics in the horn shark, Heterodontus francisci. *Proc. Natl. Acad. Sci. USA*, **97,** 1655–1660.

Kmita-Cunisse, M., Loosli, F., Bierne, J. and Gehring, W.J. (1998) Homeobox genes in the ribbonworm *Lineus sanguineus*: evolutionary implications. *Proc. Natl. Acad. Sci. USA*, **95,** 3030–3035.

Kolm, P.J. and Sive, H.L. (1995) Regulation of the Xenopus labial homeodomain genes, HoxA1 and HoxD1: activation by retinoids and peptide growth factors. *Dev. Biol.*, **167,** 34–49.

Long, M., de Souza, S.J., Rosenberg, C., and Gilbert, W. (1996) Exon shuffling and the origin of the mitochondrial targeting function in plant cytochrome c1 precursor. *Proc. Natl. Acad. Sci. USA*, **93,** 7727–7731.

Lufkin, T., Dierich, A., LeMeur, M., Mark, M. and Chambon, P. (1991) Disruption of the Hox–1.6 homeobox gene results in defects in a region corresponding to its rostral domain of expression. *Cell*, **66,** 1105–1119.

Lutz, B., Lu, H.C., Eichele, G., Miller, D., Kaufman, T.C. (1996) Rescue of Drosophila *labial* null mutant by the chicken ortholog Hoxb–1 demonstrates that the function of Hox genes is phylogenetically conserved. *Genes Dev.*, **10,** 176–184.

Lynch, M. and Conery, J.S. (2000) The evolutionary fate and consequences of duplicate genes. *Science*, **290,** 1151–1155.

Lynch, M. and Force, A. (2000) The probability of duplicate gene preservation by subfunctionalization. *Genetics*, **154,** 459–473.

Mark, M., Lufkin, T., Vonesch, J.L., Ruberte, E., Olivo, J.C., Dolle, P., Gorry, P., Lumsden, A. and Chambon, P. (1993) Two rhombomeres are altered in *Hoxa1* mutant mice. *Development*, **119,** 319–338.

Marshall, H., Morrison, A., Studer, M., Popperl, H., Krumlauf, R. (1996) Retinoids and Hox genes. *FASEB J.*, **10,** 969–78.

McGinnis, W. and Krumlauf, R. (1992) Homeobox genes and axial patterning. *Cell*, **68,** 283–302.

McClintock, J.M., Carlson, R., Mann, D.M. and Prince, V.E. (2001) Consequences of Hox gene duplication in the vertebrates: an investigation of the zebrafish Hox paralogue group 1 genes. *Development*, **128,** 2471–2484.

McClintock, J., Kheirbek, M. and Prince, V.E. (2002). Knock-down of duplicated zebrafish *hoxb1* genes reveals distinct roles in hindbrain patterning and a novel mechanism of duplicate gene retention. *Development*, **129,** 2339–2354. .Misof, B.Y. and Wagner, G.P. (1996) Evidence for four Hox clusters in the killifish Fundulus heteroclitus (teleostei) *Mol. Phylogenet. Evol.*, **5,** 309–322.

Misof, B.Y., Blanco, M.J. and Wagner, G.P. (1996) PCR-survey of Hox-genes of the zebrafish: new sequence information and evolutionary implications. *J. Exp Zool.*, **274,** 193–206.

Moens, C. and Prince, V. (2002) Constructing the hindbrain: insights from the zebrafish. *Dev. Dyn.*, **224**, 1–17.

Murphy, P. and Hill, R.E. (1991) Expression of the mouse labial-like homeobox-containing genes, Hox 2.9 and Hox 6.1, during segmentation of the hindbrain. *Development*, **111,** 61–74.

Nadeau, J.H. and Sankoff, D. (1997) Comparable rates of gene loss and functional divergence after genome duplications early in vertebrate evolution. *Genetics*, **147,** 1259–1266.

Naruse, K., Fukamachi, S., Mitani, H., Kondo, M., Matsuoka, T., Kondo, S., Hanamura, N., Morita, Y., Hasegawa, K., Nishigaki, R., Shimada, A., Wada, H., Kusakabe, T., Suzuki, N., Kinoshita, M., Kanamori, A., Terado, T., Kimura, H., Nonaka, M. and Shima, A. (2000) A detailed linkage map of medaka, *Oryzias latipes*. Comparative genomics and genome evolution. *Genetics*, **154,** 1773–1784.

Nasevicius, A. and Ekker, S.C. (2000) Effective targeted gene 'knockdown' in zebrafish. *Nat Genet.*, **26,** 216–220.

Nguyen, V.H., Schmid, B., Trout, J., Connors, S.A., Ekker, M. and Mullins, M.C. (1998) Ventral and lateral regions of the zebrafish gastrula, including the neural crest progenitors, are established by a bmp2b/swirl pathway of genes. *Dev. Biol.*, **199,** 93–110.

Ohno, S. (1970) *Evolution by Gene Duplication*, Springer-Verlag, Heidelberg, Germany.

Pöpperl, H., Bienz, M., Studer, M., Chan, S.K., Aparicio, S., Brenner, S., Mann, R.S. and Krumlauf, R. (1995) Segmental expression of Hoxb1 is controlled by a highly conserved autoregulatory loop dependent uponexd/pbx. *Cell*, **81,** 1031–1042.

Postlethwait, J.H., Yan, Y-L., Gates, M.A., Horne, S., Amores, A., Brownlie, A., Donovan, A., Egan, E.S., Force, A., Gong, Z., Goutel, C., Fritz, A., Kelsh, R., Knapik, E., Liao, E., Paw, B., Ransom, D., Singer, A., Thomson, M., Abduljabbar, T., Yalick, P., Beier, D., Joly, J-S., Larhammar, D., Rosa, F., Westerfield, M., Zon, L.I., Johnson, S.L. and Talbot, W,S. (1998) Vertebrate genome evolution and the zebrafish gene map. *Nat. Genet.*, **18,** 345–349.

Scholpp, S. and Brand, M. (2001) Morpholino-induced knockdown of zebrafish engrailed genes eng2 and eng3 reveals redundant and unique functions in midbrain-hindbrain boundary development. *Genesis*, **30,** 129–133.

Shih, L., Tsay, H., Lin, S. and Hwang, S.L. (2001) Expression of zebrafish Hoxa1a in neuronal cells of the midbrain and anterior hindbrain. *Mech Dev.*, **101,** 279–281

Sidow, A. (1996) Gen(om)e duplications in the evolution of early vertebrates. *Curr. Opin. Genet. Dev.*, **6,** 715–722.

Slack, J.M., Holland, P.W. and Graham, C.F. (1993) The zootype and the phylotypic stage. *Nature*, **361,** 490–492.

Studer, M., Lumsden, A., Ariza-McNaughton, L., Bradley, A. and Krumlauf, R. (1996) Altered segmental identity and abnormal migration of motor neurons in mice lacking *Hoxb1*. *Nature*, **384,** 630–634.

Studer, M., Gavalas, A., Marshall, H., Ariza-McNaughton, L., Rijli, F., Chambon, P. and Krumlauf, R. (1998) Genetic interactions between *Hoxa1* and *Hoxb1* reveal new roles in regulation of early hindbrain patterning. *Development*, **125,** 1025–1036.

A. Meyer, Y. Van de Peer (eds.), Genome Evolution, 195-199.

Phylogenetic analysis of the mammalian *Hoxc8* non-coding region

Chang-Bae Kim[1†], Cooduvalli S. Shashikant[2], Kenta Sumiyama[1], Wayne C.H. Wang[2], Chris T. Amemiya[3] & Frank H. Ruddle[1*]

[1]*Department of Molecular, Cellular, and Developmental Biology, Yale University, PO Box 208103, New Haven, CT 06520, USA*; [2]*Department of Dairy and Animal Science, College of Agricultural Sciences, The Pennsylvania State University, 324 Henning Building, University Park, PA 16802–3503, USA*; [3]*Department of Molecular Genetics, Virginia Mason Research Center, 1201 9th Avenue, Seattle, WA 98101, USA*; [†]*present address: Genetic Resources Center, Korea Research Institute of Bioscience and Biotechnology, Taejon 305–333, Korea.*

[*]*Author for correspondence: E-mail: frank.ruddle@yale.edu*

Received 15.07.2002; Accepted in final form 30.09.2002

Key words: comparative genomics, *Hox* genes, mammalian phylogeny

Abstract

The non-coding intergenic regions of *Hox* genes are remarkably conserved among mammals. To determine the usefulness of this sequence for phylogenetic comparisons, we sequenced an 800-bp fragment of the *Hoxc9–Hoxc8* intergenic region from several species belonging to different mammalian clades. Results obtained from the phylogenetic analysis are congruent with currently accepted mammalian phylogeny. Additionally, we found a TC mini satellite repeat polymorphism unique to felines. This polymorphism may serve as a useful marker to differentiate between mammalian species or as a genetic marker in feline matings. This study demonstrates usefulness of a comparative approach employing non-coding regions of *Hox* gene complexes.

Introduction

Hox gene complexes show remarkable conservation in their sequence, structure and function among diverse species. The degree of sequence similarity in the non-coding regions of *Hox* genes between human and mouse is remarkably higher compared to the similarity of the non-coding regions of other gene clusters, e.g., globin gene clusters. This level of conservation may reflect structural and regulatory interactions critical for preserving the integrity of the *Hox* clusters. Embedded in the non-coding regions are the *cis*-acting elements that may regulate the expression of multiple *Hox* genes within a cluster. The position of these enhancers may be critical in maintaining spatial and temporal collinearity of *Hox* gene expression. Variations in the sequence of the *cis*-acting elements provide useful information on how gene expression can be subtly altered to create divergent morphologies (Shashikant and Ruddle, 1996; Belting *et al.*, 1998; Shashikant *et al.*, 1995, 1998). Increasingly, sequence data representative of regulatory regions of genes such as promoter and enhancer elements are being reported in order to determine their adaptive role in evolution (Yamamoto *et al.*, 1992; Margarit *et al.*, 1998; Spek *et al.*, 1998). The question may be asked: Can non-coding regulatory sequence data provide phyletic information as well as provide insight into adaptive evolution? To test this possibility, we have sequenced an 800-bp fragment of the non-coding region of *Hoxc8* from 26 species representing major mammalian clades. This non-coding region contains a previously characterized *Hoxc8* early enhancer region, variations in which have been associated with differential regulatory strategies in diverse species (Belting *et al.*, 1998; Shashikant *et al.*, 1995, 1998; C.S. Shashikant, unpublished observations). The distribution of nucleotide variation within this region using a small dataset of eight mammalian species was previously reported (Sumiyama *et al.*, 2001). The results described in this study demonstrate the usefulness of non-coding regions of developmentally important genes for constructing phylogenetic trees.

Material and methods

PCR amplification and DNA sequencing

Genomic DNA samples for PCR amplification were the same as those used in our previous paper (Shashikant *et al.*, 1998). Feline DNA which was provided by Dr. Steven O'Brien. In most cases, the primer set, Jstabv (5'-CCCACCTCTCCTCTGCTCCTTTGCTG GAATCACAAAACCCTAAAG-3') and Lcns2 (5'-GCCTCTAACATTGAGCAACAGCGCCACCTCGC GT-3') were used for PCR amplification. In certain cases, enh5P (5'-TCCCACCTCTCCTCTGCTCCTT TGTCGGAATCACAAAACC-3') and enh3Q'A (5'-CTGCCTCTAACATTGAGCAACAGCGCCACCTC GCGT-3') were used. PCR amplification was done for 30–40 cycles using a Mastercycler Gradient cycler (Eppendorf) with different combinations of annealing temperatures (55–70 °C) and $MgCl_2$ concentrations (1–3 mM) to optimize conditions. PCR fragments were cloned in pGEM-T Easy Vector (Promega) and sequenced manually by the dideoxynucleotide method or using automated sequencer (Applied Biosystems 373A LICOR 4200L) following the manufacturer's protocols. To check cross-contamination, we sequenced multiple clones from each species and analyzed diagnostic sequences of each species. To assure accuracy, DNA strands were sequenced in both directions for each clone using forward and reverse primers.

Sequence analyses

Sequences were aligned by the CLUSTAL X program (Thompson *et al.*, 1997). The *Hoxc8* enhancer area is highly conserved while the more downstream area is more variable and alignment required additional manual adjustment. Aligned sequences were analyzed by maximum parsimony, neighbor-joining, and maximum-likelihood methods. Maximum parsimony trees were reconstructed by the branch and bound search in PAUP* (version 4.0; Swofford, 1998) and bootstrap estimates were obtained with 1000 replications. Distance analysis was performed by neighbor-joining (Saitou and Nei, 1987) with distance estimates according to the method of Jukes and Cantor (1969) employing DNADIST and NEIGHBOR programs of the PHYLIP 3.5c package (Felsenstein, 1995). Confidence of internal nodes was tested by bootstrap with 1000 replications. Maximum-likelihood analysis was performed with FASTDNAML (Olsen *et al.*, 1994).

Results and discussion

DNA sequences were obtained for an 800-bp fragment of DNA including the *Hoxc8* early enhancer region from species representing different orders of mammals. The phyletic signal of these sequences was tested by reconstructions of evolutionary relationships among mammalian orders. The neighbor-joining tree is shown in Fig. 1. The trees suggested by three different analytical methods (Maximum parsimony, maximum likelihood, and neighbor-joining) were largely identical. The reconstructed trees suggest that great sperm whale is more closely related to the toothed whale than to the baleen whales. This result is consistent with studies which place the sperm whale with the toothed whales (Arnason and Gullberg, 1994; Ohland *et al.*, 1995; Messenger and McGuire, 1998). However, it should be noted that the neighbor-joining bootstrap value supporting the toothed/sperm whale relationship was not very high. The branching arrangement inferred from our data is consistent with those inferred from mitochondrial genome sequences (Janke and Arnason, 1997; Janke *et al.*, 1997) with respect to interordinal relationships (Fig. 1). In the case of carnivores, the tree strongly supports subordinal relationships. On the other hand, there is insufficient phylogenetic signal to define relationships between feline species. The present data support strongly the association of whales and artiodactyls (Shimamura *et al.*, 1997) as well as the monophyly of whales (Arnason and Gullberg, 1996). To further test hypotheses on the ancestry of whales, more artiodactyls, such as the hippopotamus, should be included in the dataset. Within the cetaceans, the position of great sperm whale is not well resolved.

One potentially significant region unique to the feline lineage was identified after exclusion of ambiguous regions in the sequence alignment (Table 1). All feline species examined showed TC mini-satellite repeats. The length of the repeats ranged between 29–68bp and the number of TC repeats was variable between species. Short tandem repeats (STRs), such as TC repeats, are highly polymorphic, abundant, and near randomly distributed in eukaryotic genomes. They can provide a valuable mapping resource for the

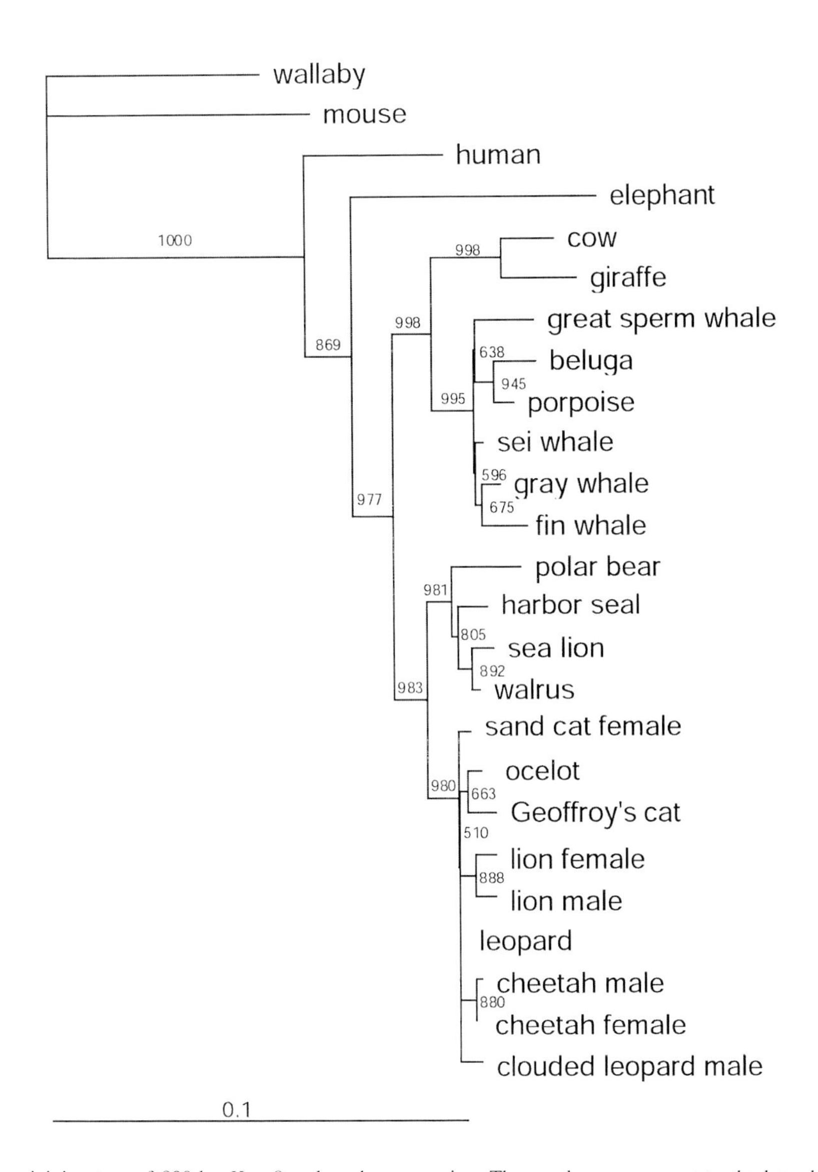

Figure 1. The neighbor-joining tree of 800-bp *Hoxc8* early enhancer region. The numbers on or next to the branch are bootstrap values calculated from 1000 replications. Bootstrap values of more than 500 are shown.

identification of genes associated with hereditary disease. The TC mini-satellite repeats may be useful for population genetic studies within and between feline species. Systematically, these TC insertions are potentially unique sequence markers that characterize the feline clade and differentiate it from other mammalian clades. Moreover, insertions that are unique to the felines may be significant variants in terms of regulation of *Hoxc8* gene expression in different clades.

We show that the *Hoxc8* non-coding region yields phylogenetic branching arrangements for the mammalian orders and suborders that are consistent with those inferred using mitochondrial DNA coding regions. Strikingly, sequence differences of the *Hoxc8* enhancer between some mammalian species were observed. It is possible that these sequences contribute to variation in body plans and could serve as criteria for ordinal assignment. Mitochondrial and nuclear gene sequences have traditionally been used in studies that address phyogenetic relationships (Springer *et al.*, 1997; De Jong, 1998). In this report, we demonstrate the utility of Hox regulatory regions for inferring phylogenetic relationships. We suggest that this approach can serve to elucidate evolutionary mechanisms and provide a new means to bridge investiga-

Table 1. TC di-nucleotide repeats variations in flanking region of Hoxc8 early enhancer

Species	Sequence
Wallaby	$AAC_2(TC)_3CT_3G$
Mouse	$ATC_3(TC)_2CT_3G$
Human	$A(TC)_4TTGAG$
Giraffe	$AC_4(TC)_2GTGCG$
Cow	$AC_4(TC)_2GTGCG$
Elephant	$AATC_3(TC)_3CG$
Sei whale	$AC_4(TC)_2ACGCG$
Gray whale	$AC_4(TC)_6ACGCG$
Fin whale	$AC_4(TC)_2ACGCG$
Great sperm whale	$ATC_3(TC)_2ACGCG$
Beluga	$AC_4(TC)_2ACGCG$
Porpoise	$AC_4(TC)_2ACGCG$
Polar bear	$AC_4(TC)_6G$
Harbor seal	$AC_4(TC)_5G$
Walrus	$AC_4(TC)_4G$
Sea lion	$AC_4(TC)_4G$
Ocelot	$AAC_{10}ATA(TC)_5CTCTG(TC)_{15}TG(TC)_4TT(TC)_2G$
Geoffroy's cat	$AAC_{10}ATA(TC)_5C(TC)_9TG(TC)_2YCTCTT(TC)_2G$
Cheetah (male)	$AC_9ATA(TC)_5C(TC)_5TG(TC)_5TTTCTT(TC)_2G$
Cheetah (female)	$AC_{10}ATA(TC)_5C(TC)_5TG(TC)_5TTTCTT(TC)_2G$
Lion (male)	$AAC_{16}ATA(TC)_5CC(TC)_3TG(TC)_7TT(TC)_2G$
Lion (female)	$AAC_{11}ATA(TC)_5CC(TC)_3TG(TC)_7TT(TC)_2G$
Sand cat (female)	$AAC_{10}ATA(TC)_4T(TC)_{26}TT(TC)_2G$
Leopard	$AAC_9ACA(TC)_5CTCCC(TC)_5(TG)_2(TC)_7TT(TC)_2G$
Clouded leopard (male)	$AAC_9ATA(TC)_5CC(TC)_4CCTT(TC)_2G$

tions of body plan evolution within a phylogenetic framework.

Acknowledgments

We thank Drs. S. O'Brien and U. Arnason for sending genomic DNAs, and Christina Ledje and Chi-Hua Chiu for reading the manuscript. The study was supported by grants from the National Science Foundation (IBN–9630567 and IBN–9905408) and the National Institutes of Health (GM09966 and R24-RR14085) (to FHR and CTA, respectively) and faculty start-up funds for CSS.

References

Arnason, U. and Gullberg, A. (1994) Relationship of baleen whales established by cytochrome b gene sequence comparison. *Nature*, **367**, 726–728.

Arnason, U. and Gullberg, A. (1996) Cytochrome b nucleotide sequences and the identification of five primary lineages of extant cetaceans. *Mol. Biol. Evol.*, **13**, 407–417.

Belting, H.G., Shashikant, C.S. and Ruddle, F.H. (1998) Multiple phases of expression and regulaiton of mouse *Hoxc8* during early embryogenesis. *J. Exp. Zool.*, **282**, 196–222.

De Jong, W.W. (1998) Molecules remodel the mammalian tree. *Trends Ecol. Evol.*, **13**, 270–275.

Felsenstein, J. (1995) PHYLIP (phylogeny inference package). Version 3.5c. Department of Genetics, University of Washington, Seattle.

Janke, A. and Arnason, U. (1997) The complete mitochondrial genome of *Alligator mississippiensis* and separation between recent archosauria (birds and crocodiles). *Mol. Biol. Evol.* **14**, 1266–1272.

Janke, A., Xu, X. and Arnason, U. (1997) The complete mitochondrial genome of the wallaroo (*Macropus robustus*) and the phylogenetic relationship among Monotremata, Marsupialia, and Eutheria. *Proc. Natl. Acad. Sci. USA*, **94**, 1276–1281.

Jukes, T.H. and Cantor, C.R. (1969) Evolution of protein molecules. In *Mammalian Protein Metabolism* (Ed. H.N. Munro), Academic Press, New York, pp. 21–132.

Margarit, E., Gullen, A., Rebordosa, C., Vidal-Taboada, J., Sanchez, M., Ballesta, F. and Oliva, R. (1998) Identification of conserved potentially regulatory sequences of the SRY gene from 10 different species of mammals. *Biochem. Biophys. Res. Commun.*, **245,** 370–377.

Messenger, S.L. and McGuire, J.A. (1998) Morphological molecules, and phylogenetics of cetaceans. *Syst. Biol.*, **47,** 90–124.

Ohland, D.P., Harley, E.H. and Best, P.B. (1995) Systematics of cetaceans using restriction site mapping of mitochondrial DNA. *Mol. Phylogenet. Evol.*, **4,** 10–19.

Olsen, G.J., Matsuda, H., Hagstrom, R. and Overbeek, R. (1994) FastDNAml: a tool for construction og phylogenetic trees of DNA sequences using maximum-likelihood. *Comput. Appl. Biosci.*, **10,** 41–48.

Saitou, N. and Nei, M. (1987) The neighbor-joining method: a new method for reconstructing phylogenetic trees. *Mol. Biol. Evol.*, **4,** 406–425

Shashikant, C.S. and Ruddle, F.H. (1996) Combinations of closely situated *cis*-acting elements determine tissue-specific patterns and anterior extent of early *Hoxc8* expression. *Proc. Natl. Acad. Sci. USA*, **93,** 12364–12369.

Shashikant, C.S., Bieberich, C.J., Belting, H.G., Wang, J.C., Borbely, M.A. and Ruddle, F.H. (1995) Regulation of *Hoxc-8* during mouse embryonic development: identification and characterization of critical eleements involved in early neural tube expression. *Development*, **121,** 4339–4347.

Shashikant, C.S., Kim, C.B., Borbely, M.A., Wang, W.C.H. and Ruddle, F.H. (1998) Comparative studies on mammalian *Hoxc8* early enhancer sequence reveal a baleen whale-specific deletion of a *cis*-acting element. *Proc. Natl. Acad. Sci. USA*, **95,** 15446–15451.

Shimamura, M., Yasue, H., Ohshima, K., Abe, H., Kato, H., Kishiro, T., Goto, M., Munechika, I. and Okada, N. (1997) Molecular evidence from retroposons that whales form a clade within even-toed ungulates. *Nature*, **388,** 666–670.

Spek, C.A., Bertina, R.M. and Reitsma, P.H. (1998) Identification of evolutionarily invariant sequences in the protein C gene promoter. *J. Mol. Evol.*, **47,** 663–669.

Springer, M.S., Cleven, G.C., Madsen, O., De Jong, W.W., Waddell, V.G., Amrine, H.M. and Stanhope, M.J. (1997). Endemic African mammals shake the phylogenetic tree. *Nature*, **388,** 61–64.

Sumiyama, K., Kim, C.B. and Ruddle, F.H. (2001) An efficient *cis*-element discovery method using multiple sequence comparisons based on evolutionary relationships. *Genomics*, **71,** 260–262.

Swofford, D.L. (1998) PAUP*. Phylogenetic Analysis Using Parsimony (* and other methods). Version 4. Sinauer Associates, Sunderland, MA.

Thompson, J.D., Gibson, T.J., Plewniak, F., Jeanmougin, F. and Higgins, D.G. (1997) The CLUSTAL_X windows interface: flexible strategies for multiple sequence alignment aided by quality analysis tools. *Nucleic Acids Res.*, **25,** 4876–4882.

Yamamoto, H., Kudo, T., Masuko, N., Miura, H., Sato, S., Tanaka, M., Tanaka, S., Takeuchi, S., Shibahara, S. and Takeuchi, T. (1992) Phylogeny of regulatory regions of vertebrate tyrosinase genes. *Pigment Cell Res.*, **5,** 284–294.

A. Meyer, Y. Van de Peer (eds.), Genome Evolution, 201-212.

Maximum likelihood methods for detecting adaptive evolution after gene duplication

Joseph P. Bielawski* & Ziheng Yang
Department of Biology, University College London, Darwin Building, Gower Street, London WCIE 6BT, United Kingdom
**Author for correspondence (e-mail: j.bielawski@ucl.ac.uk)*

Received 24.01.2002; Accepted in final form 29.08.2002

Key words: codon model, ECP, EDN, gene family, maximum likelihood, positive selection, Troponin C

Abstract

The rapid accumulation of genomic sequences in public databases will finally allow large scale studies of gene family evolution, including evaluation of the role of positive Darwinian selection following a duplication event. This will be possible because recent statistical methods of comparing synonymous and nonsynonymous substitution rates permit reliable detection of positive selection at individual amino acid sites and along evolutionary lineages. Here, we summarize maximum-likelihood based methods, and present a framework for their application to analysis of gene families. Using these methods, we investigated the role of positive Darwinian selection in the ECP-EDN gene family of primates and the Troponin C gene family of vertebrates. We also comment on the limitations of these methods and discuss directions for further improvements.

Introduction

Duplication of genetic material is generally accepted as an important precursor of functional divergence (Ohno, 1970; Ohta, 1988a, 1988b; Hughes, 1999). Indeed, the majority of genes in higher organisms are members of multigene families or superfamilies (Hughes, 1999). There is often an acceleration of the non-synonymous rate following gene duplication (Li, 1985; Lynch and Conery, 2000), but the mechanism for this acceleration is not clear. Studies of several gene families indicated that natural selection accelerated the fixation rate of nonsynonymous substitutions shortly after a duplication event, presumably to adapt those proteins to a new or modified function (Zhang *et al*., 1998; Schmidt *et al*., 1999; Duda and Palumbi, 1999; Rooney and Zhang, 1999; Bielawski and Yang, 2000). However, an accelerated nonsynonymous rate also could be driven by a relaxation, but not complete loss, of selective constraints. Here, duplicated proteins evolve under relaxed functional constraints for some period of time, after which functional divergence occurs when formerly neutral substitutions convey a selective advantage in a novel environment or genetic background; this model was named the "Dykhuizen-Hartl effect" by Zhang *et al.* (1998). The rapid accumulation of genomic sequences in public databases will allow large scale studies of functional divergence in gene families.

For protein coding genes, the most compelling evidence for positive Darwinain selection is derived from comparison of nonsynonymous (amino acid replacement) and synoymous (silent) substitution rates, d_N and d_S, respectively. The difference between these two rates, measured as the ratio $\omega = d_N/d_S$, reflects the effect of selection on the protein product of the gene (Kimura, 1983). For example, if nonsynonymous mutations are deleterious, purifying selection will reduce or prevent their fixation rate and d_N/d_S will be less than 1, whereas if nonsynonymous mutations are neutral then they will be fixed at the same rate as synonymous mutations and $d_N/d_S = 1$. Only under positive Darwinian selection can nonsynonymous mutations be fixed at a rate higher than that of synonymous substitutions, with $d_N/d_S > 1$. Traditionally, to demonstrate positive Darwinian selection models of neutral evolution and purifying selection must be rejected, *i.e.*, the d_N/d_S ratio must be shown to be significantly greater than 1 (Hughes and Nei, 1988; Yang, 1998).

Models of adaptive evolution by gene duplication (Ohta, 1988a, 1988b; Hughes, 1999) make clear predictions about patterns of genetic changes. After duplication, natural selection favours the fixation of mutations in one or both copies that adapt them to divergent functions. Once new or enhanced functions become established, positive selection ceases and purifying selection acts to maintain the new functions. For protein coding genes, this means nonsynonymous substitutions will be accelerated following the duplication, and then slow down due to increased effects of purifying selection. Furthermore, many amino acids in a protein are under strong structural constraints, and adaptive evolution following a duplication event will most likely operate at only a subset of sites. An ω ratio measured as an average over all sites and evolutionary time will rarely be greater than one, and is too stringent a criterion for detecting positive selection (*e.g.*, Ward *et al.*, 1997; Crandall *et al.*, 1999; Bielawski and Yang, 2001). Hence, the apparent period of relaxed selection following gene duplication (*e.g.*, Lynch and Conery, 2000) might also reflect adaptive divergence of duplicate genes at just a subset of amino acid sites.

Recent ML models of codon substitution relax the assumption of a single ω ratio for all branches of a phylogeny (Yang, 1998). Theses models can provide a framework for constructing likelihood ratio tests of changes in selective pressure following gene duplication (Bielawski and Yang, 2001). Other codon models allow the ω ratio to vary among amino acid sites (Nielsen and Yang, 1998; Yang *et al.*, 2000). Very recently, Yang and Nielsen (2002) developed a third type of model that can simultaneously account for variation in selective constraints among sites and lineages. Here, we summarize all three types of models, and present a framework for their application to gene families. We also comment on limitations of current methods and discuss directions for further improvements.

ML estimation of the d_N/d_S ratio

Markov model of codon evolution

A markov process is used to describe substitutions between 61 of the 64 sense codons. The three stop codons are excluded because they aren't allowed within a protein. Independence among the codon sites of a gene is assumed, and hence the substitution process can be considered one codon site at a time. For any single codon site, the model describes the instantaneous substitution rate from codon i to codon j, q_{ij}.

In the following, we describe the basic model of Goldman and Yang (1994). A similar, but simpler model was described by Muse and Gaut (1994). Because transitional substitutions are known to occur more often than transversions, the rate is multiplied by the κ parameter when the change involves a transition; this is the transition/transversion rate ratio. Usage of codons within genes also can be highly biased, and consequently, the rate of change from i to j is multiplied by the equilibrium frequency of codon j (π_j). Finally, selective constraints acting on substitutions at the amino acid level affect the rate of change when that change represents a nonsynonymous substitution. The rate is multiplied by the ω parameter if the change is nonsynonymous; the ω parameter is the nonsynonymous/synonymous rate ratio (d_N/d_S).

The substitution model is specified by the instantaneous rate matrix, $Q = \{q_{ij}\}$, where

$$q_{ij} = \begin{cases} 0, & \text{if } i \text{ and } j \text{ differ at two or three codon positions} \\ \mu\pi_j & \text{if } i \text{ and } j \text{ differ by a synonymous transversion} \\ \mu\kappa\pi_j, & \text{if } i \text{ and } j \text{ differ by a synonymous transition} \\ \mu\omega\pi_j, & \text{if } i \text{ and } j \text{ differ by a nonsynonymous transversion} \\ \mu\omega\kappa\pi_j, & \text{if } i \text{ and } j \text{ differ by a nonsynonymous transition.} \end{cases}$$

The diagonal elements of the matrix Q are defined by the mathematical requirement that the row sums are equal to zero. Because separate estimation of the rate (μ) and time (t) is not possible, the rate (μ) is fixed so that the expected number of nucleotide substitutions per codon is equal to one. This scaling allows us to measure time (t) by the expected number of substitutions per codon, *i.e.*, genetic distance. The probability that codon i is substituted by codon j after time t is $p_{ij}(t)$, and $\mathrm{P}(t) = \{p_{ij}(t)\} = e^{Qt}$.

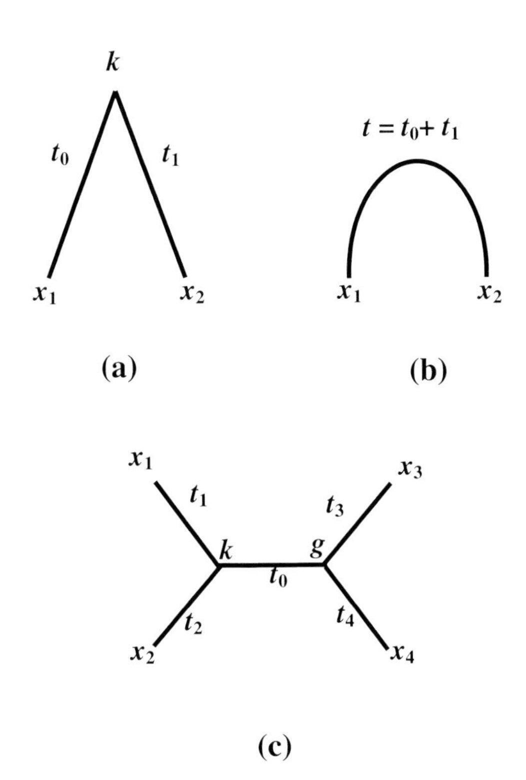

Figure 1. Trees (a) and (b) are for two sequences; (a) is rooted and (b) is unrooted. Codon models are time-reversible; hence, the root cannot be identified and only the sum of branch lengths, *t*, is estimable. Tree (c) is an unrooted tree for four sequences.

ML estimation on a phylogeny

Let us consider a simple case, the likelihood of observing the data of a pair of species. Suppose there are *n* codon sites in a gene, and a certain site (*h*) has codons CCC and CTC. The data at site *h*, denoted $\mathbf{x}_h = \{CCC, CTC\}$ are related to an ancestor with codon *k* by branch lengths t_0 and t_1 (Fig. 1a). The probability of site *h* is

$$L(\mathbf{x}_h) = \sum_k \pi_k p_{k,CCC}(t_0) p_{k,\mathrm{CTC}}(t_1).$$

Since the ancestral codon is unknown the summation is over all 61 possible codons for *k*.

The log likelihood is a sum over all codon sites (*n*) in the sequence

$$l(t,\kappa,\omega) = \sum_{h=1}^{n} \log\{\mathrm{L}(\mathbf{x}_h)\}.$$

Codon frequencies (π_i's) are usually estimated using observed base or codon frequencies. Parameters ω, κ, and *t* are estimated by maximizing the likelihood function. Because the root cannot be identified, t_0 and t_1 cannot be estimated individually, and only $t_1 + t_2 = t$ is estimated (Fig. 1b). Since an analytical solution is not possible, numerical optimisation algorithms are used.

Likelihood calculation for multiple lineages on a phylogeny (*e.g.*, Fig. 1c) is an extension of the calculation for two lineages. As in the case of two sequences, the root cannot be identified and is fixed at one of the ancestral nodes arbitrarily. For example, given an unrooted tree with four species and two ancestral codons, *k* and *g*, the probability of observing the data at codon site *h*, $\mathbf{x}_h = \{x_1, x_2, x_3, x_4\}$ (Fig. 1c), is

$$\mathrm{L}(\mathbf{x}_h) = \sum_k \sum_g [\pi_k p_{kx_1}(t_1) p_{kx_2}(t_2) p_{kg}(t_0) p_{gx_3}(t_3) p_{gx_4}(t_4)]$$

The quantity in the brackets is the contribution to the probability of observing the data by ancestral codons *k* and *g* at the two ancestral nodes. For an unrooted tree of *N* species, with *N*–2 ancestral nodes, the data at each site will be a sum over $61^{(N-2)}$ possible combinations of ancestral codons. The log likelihood is a sum over all codon sites in the alignment

$$l = \sum_{h=1}^{n} \log\{L(\mathbf{x}_h)\}.$$

As in the two-species case, numerical optimisation is used to estimate ω, κ, and the (2*N*–3) branch length parameters (*t*'s).

Detecting lineage-specific changes in selective pressure

Models of functional divergence in gene families emphasize the episodic nature of substitution rates (Ohta, 1988a, 1988b; Hughes, 1999). Under these models, most evolution in gene families will be by purifying selection, with episodes of either reduced selective constraints or positive Darwinian selection following duplication events. If divergence is driven by positive Darwinian selection, nonsynonymous mutations might be fixed at a much higher rate than synonymous mutations immediately following the duplication event. However, if the gene family evolves under purifying selection at other times, comparisons among distantly related pairs of sequences are unlikely to yield d_N/d_S ratios greater than 1. For this reason, adaptive evolution will be very difficult to detect in gene families by using a pairwise approach to estimating d_N/d_S ratios.

We need tools to (1) estimate selective pressures at different time points in the phylogenetic history of a gene, and (2) test the hypothesis that those estimates differ significantly. Codon models that allow independent ω ratios in different parts of a phylogeny provide a framework for analysing changes in selective pressure over time (Yang, 1998; Bielawski and Yang, 2001). The advantage of this likelihood-based approach is that it does not depend on the accuracy of hypothetical ancestral sequences, as in other approaches, although it does incur greater computational costs (Yang, 2001).

Models of variable selective pressures among branches

The null model (the one-ratio model) assumes the same ω ratio for all branches (Fig. 2; R1). Nested models are constructed based on the assumption that selective constraints change following a gene duplication (Fig. 2). Here, the likelihood calculation is modified so that independent ω's are used to calculate rate matrices (Q) and transitions probabilities for different branches. In a simple case where there is only a single duplication event in a phylogeny, model R2 (two-ratio model) assumes two independent ω ratios: one ratio for all branches predating a duplication event and a second for all branches postdating a duplication event (Fig. 2; R2). A likelihood ratio test (LRT) of the one-ratio model with model R2 examines the difference between average selective constrains before and after a duplication event. A more complex model assumes three independent ω ratios: one for all branches predating the duplication event, a second for the branches immediately following the duplication event, and a third for all subsequent branches (Fig. 2; R3). An LRT comparing model R2 and R3 examines the difference between selective constraints at branches immediately after the duplication and those in subsequent branches. Finally, R4 extends R3 to allow selective pressure to differ between paralogous genes (Fig. 2; R4). Different hypotheses, perhaps involving multiple duplication events, also can be constructed; we used this example (Fig. 2) to illustrate the general framework of the ML approach.

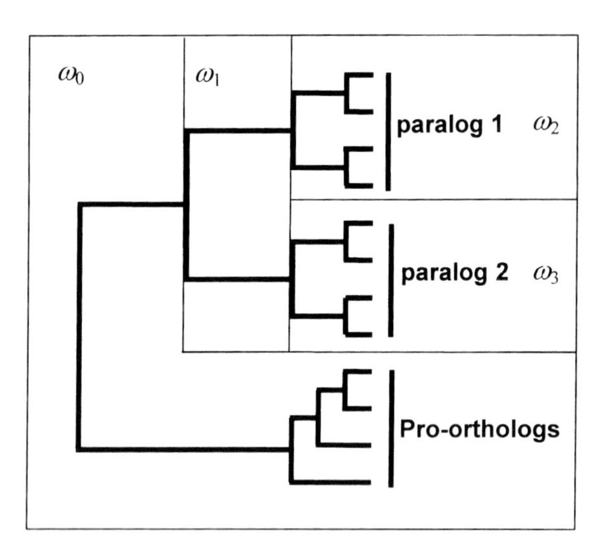

One ratio (R1): $\omega_0 = \omega_1 = \omega_2 = \omega_3$
Two ratios (R2): $\omega_0 \neq \omega_1 = \omega_2 = \omega_3$
Three ratios (R3): $\omega_0 \neq \omega_1 \neq \omega_2 = \omega_3$
Four ratios (R4): $\omega_0 \neq \omega_1 \neq \omega_2 \neq \omega_3$

Figure 2. Phylogeny for a hypothetical gene family. Pro-orthologs are the single-copy genes that predate the duplication event. Paralogs 1 and 2 refer to the two sets of genes that derived from the duplication event. The one-ratio model (R1) assumes all branches have the same ω parameter. The R2 model assumes one ω is for all branches that predate the duplication event, and a second ω for all branches that follow the duplication event. Model R3 assumes one ω for all branches that predate the duplication event, a second ω for the branches that immediately follow the duplication event, and a third ω is for subsequent branches. Model R4 is an extension of R3, and allows selective constraints to differ in the two paralogs.

Changes in nonsysnonymous substitution rates following gene duplication in the ECP-EDN gene family

The ECP-EDN gene family of primates is comprised of the eosiophil-derived neurotoxin (EDN) and the eosinophil cationic (ECP) genes. ECP and EDN are ribonucleases present in the large specific granules of eosinophilic leukocytes (Rosenberg and Domachowske, 1999). While both ECP and EDN have host-defence roles, their specific functions differ. ECP is a cationic toxin, apparently functioning as a non-specific toxin to parasites and bacteria (Rosenberg and Domachowske, 1999). EDN is considerably less cationic, but has strong ribonucleic activity making it a potent antiviral agent via ribonucleic degradation of viral RNA (Rosenberg and Domachowske, 1999). Zhang *et al.* (1998) studied the evolution of this gene

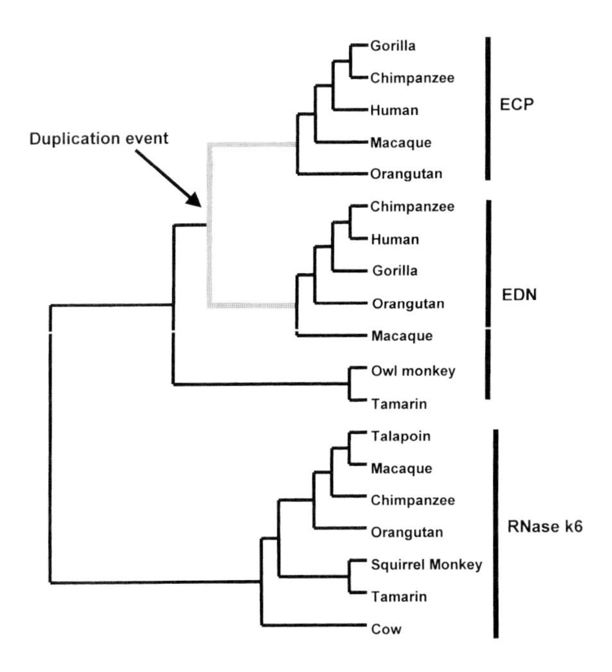

Figure 3. A phylogenetic tree for the ECP-EDN gene family. The tree is rooted with RNase k6 genes. All analyses were conducted using unrooted topologies; this topology is rooted for convenience. Note that branch lengths are not to scale.

family and found an excess of nonsynonymous substitutions over synonymous substitutions in the branch leading to the ECP gene. They suggested that the anti-parasitic function of ECP evolved shortly after the duplication that gave rise to ECP and EDN. We use the ECP-EDN family to demonstrate the application of codon models to studies of evolution by gene duplication. A phylogenetic hypothesis for the ECP-EDN gene family is presented in Figure 3.

We estimated ω as an average over all sites and branches (Fig. 3) and the ratio was substantially less than 1 (one ratio model; $\omega = 0.51$). The one-ratio model was compared with model R2, and the LRT indicated that model R2 provided a significantly better fit to these data ($2\delta = 14.92$, d.f. = 1, $P = 0.0001$). Estimate of ω under model R2 indicated a significant increase in the average rate of nonsynonymous substitution following the duplication event (Table 1; $\omega_0 = 0.38$ vs. $\omega_1 = 0.87$). To test if more recent selective pressure in ECP and EDN differed from that immediately following the duplication event, we compared model R2 with R3. The likelihood of model R3 was significantly better than R2 ($2\delta = 4.2$, d.f. = 1, $P = 0.04$), with estimates of ω's indicating that adaptive evolution had occurred subsequent to the duplication event (Table 1; $\omega_1 = 2.00$). Lastly, we

Table 1. Parameter estimates and likelihood scores for the ECP-EDN gene family under different branch-specific models

Model	NP	Parameters for branches [a]	*l*
One-ratio (R1)	1	$\omega_0 = 0.51$ $\omega_1 = \omega_0$ $\omega_2 = \omega_0$ $\omega_3 = \omega_0$	−3109.63
Two-ratios (R2)	2	$\omega_0 = 0.38$ $\omega_1 = 0.87$ $\omega_2 = \omega_1$ $\omega_3 = \omega_1$	−3102.17
Three-ratios (R3)	3	$\omega_0 = 0.38$ $\boldsymbol{\omega_1 = 1.99}$ $\omega_2 = 0.72$ $\omega_3 = \omega_2$	−3100.06
Four-ratios (R4)	4	$\omega_0 = 0.38$ $\boldsymbol{\omega_1 = 2.00}$ $\omega_2 = 0.45$ $\boldsymbol{\omega_3 = 1.56}$	−3095.50

[a] Models R1 to R4 are presented in Figure 2. The topology for the ECP-EDN gene family is presented in figure 3: ω_0 is for branches that predate the duplication event, ω_1 for the branches that immediately follow the duplication event (grey branch in Fig. 3), ω_2 is for branches in the EDN clade excluding Owl monkey and Tamarin, and ω_3 is for all branches in the ECP clade. NP is the number of freely estimated ω ratios.

investigated the assumption of the same selective pressures for both EDN and ECP by comparing model R3 with R4 . Likelihood of model R4 was significantly better than R3 ($2\delta = 9.64$, d.f. = 1, $P = 0.002$), indicating different selective pressures in ECP and EDN. Interestingly, parameter estimates indicated positive Darwinian selection for the ECP clade ($\omega_3 = 1.56$) and purifying selection in the EDN clade ($\omega_2 = 0.45$).

Our findings indicate there was a significant increase in the rate of fixation of nonsynonymous substitutions following the duplication that gave rise to the ECP-EDN gene family. This rate increase was partially due to adaptive evolution immediately following the duplication. Presumably, natural selection increased the fixation rate for nonsynonymous mutations that adapted ECP for greater anti-parasitic activity, and nonsynonymous mutations that enhanced anti-viral ribonuclease activity in EDN. Our findings also suggest that ECP has continued to evolve under positive Darwinian selection long after the initial period of function divergence. Rosenberg and Domachowske (1999) have speculated that both EDN and ECP might have acquired specialized anti-viral activity, perhaps against respiratory viral pathogens. Perhaps subsequent adaptive evolution in the ECP clade reflects long term selective pressure for effective anti-viral activity against respiratory viral pathogens.

Identification of amino acid sites under adaptive evolution

In general, most amino acid sites are subject to strong functional constraints, with d_N, and consequently ω, close to zero (Sharp, 1997). Most studies of molecular evolution indicate adaptive changes occur at only a subset of sites (*e.g.*, Golding and Dean, 1998); hence, use of ω averaged over all sites has little power to detect positive selection (*e.g.*, Endo *et al.*, 1996). In the previous section we assumed that all amino acid sites were subject to identical selective pressure, with a single ω ratio applied to all sites. By allowing an independent ω for specific intervals of time, such as for those branches that immediately postdate a duplication event, we greatly increased the power to detect positive selection. However, finding adaptive evolution during a certain time interval provided no information about variation in selective constraints among sites. Hence, we also need tools (1) to test for evolution by positive Darwinian selection at a subset of codons, and (2) to identify such sites when they exist.

Two strategies can be taken. In the first, amino acid sites are classified into several independent ω ratio classes based on prior knowledge of structural and functional domains. Likelihood calculation in this case is similar to that under the model of one ω ratio for all sites, except that different ω parameters are used to calculate the transition probabilities at different sites (Yang, 2001; Yang and Swanson 2002). Under the second strategy, information of structural and functional domains of the protein is unknown, or not used, and a statistical distribution is used to account for variation of the ω ratio among sites (Nielsen and Yang, 1998; Yang *et al.*, 2000). After ML estimates of parameters in the ω distribution are obtained, an empirical Bayes approach is used to predict the most likely ω class for each site, with the posterior probability providing a measure of reliability (Nielsen and Yang, 1998; Yang *et al.*, 2000). It is important to point out that these codon models assume a constant selection pressure along the tree and are thus conservative in detecting positive selection. Despite this assumption, they proved useful for identifying sites under positive selection and studying the process of adaptive molecular evolution (*e.g.*, Zanotto *et al.*, 1999; Bishop *et. al.*, 2000; Haydon *et al.*, 2001; Swanson *et al.*, 2001).

Models of variable selective pressure among sites

Usually, the structural and functional domains of a protein are not well understood or are completely unknown. For this reason we focus on models that use a statistical distribution to account for variation of the ω ratio among sites. Collectively, Nielsen and Yang (1998) and Yang *et al.* (2000) implemented 12 such models; here we discuss five of them (M1, M2, M3, M7, and M8) because they comprise the set generally recommended for data analysis (Yang *et al.*, 2000; Anisimova *et al.*, 2001).

The first two models, introduced by Nielsen and Yang (1998), specify just a few discrete ω classes. The 'neutral' model (M1) assumes two classes of sites with fixed ω values; in one class of sites nonsynonymous mutations are completely selected against, with $\omega_0 = 0$, and the other class is comprised of neutral sites, with $\omega_1 = 1$. The proportion of sites is estimated via ML for only one ω class (p_0), as $p_1 = 1 - p_0$. The 'selection' model (M2) adds a third class of sites with the underlying ω ratio freely estimated from the data. M2 has parameters p_0, p_1, and ω_2. These models appear too simple to capture the complexity of the substitution process of various proteins (Yang, 2001). They are conservative for the purpose of testing and identifying sites under positive selection (Anisimova *et al.*, 2001).

Yang *et al.* (2000) introduced more general models M3, M7, and M8. M3 (discrete) assumes k site classes, with corresponding ω_i and p_i for each class ($i = 1, \ldots k$) estimated as parameters. M7 (beta) assumes that ω ratios are distributed among sites according to a beta distribution. Note that the beta distribution is very flexible, taking a variety of shapes within the interval (0,1) depending on the parameters p and q. M8 (beta&ω) is an extension of M7, having an extra class of sites with an independent ω ratio freely estimated from the data.

An LRT of the one-ratio model with M3 is a test of variable selective pressures among sites. To spe-

Table 2. Parameter estimates and likelihood scores for the ECP-EDN gene family under different models of variable ω ratios among sites.

Model	Parameter estimates	Positively selected sites	l
One-ratio	$\omega = 0.79$	None	−1912.05
Neutral (M1)	$(\omega_0 = 0), f_0 = 0.36$ $(\omega_1 = 1), (f_1 = 0.64)$	Not allowed	−1890.38
Selection (M2)	$(\omega_0 = 0), f_0 = 0.36$ $(\omega_1 = 1), f_1 = 0.38$ $\boldsymbol{\omega_2 = 2.22}, (\boldsymbol{f_2 = 0.26})$	37 sites[a]	−1886.59
Discrete (M3)	$\omega_0 = 0.17, f_0 = 0.56$ $\boldsymbol{\omega_1 = 1.9}$, $(\boldsymbol{f_1 = 0.44})$	70 sites[b]	−1885.49
Beta (M7)	$p = 0.011, q = 0.05$	Not allowed	−1891.10
Beta&ω (M8)	$p = 20, q = 99$ $f_0 = 0.56$ $\boldsymbol{\omega_1 = 1.9}, (\boldsymbol{f_1 = 0.44})$	70 sites[b]	−1885.50

[a] None of the positive selection sites identified under model M2 had posterior probabilities >0.95.

[b] Positive selection sites identified under M3 and M8 were identical; sites with posterior probabilities ≥0.95 under M8 were as follows (sites with posterior probabilities ≥ 0.99 are in bold): **19M**, 27A, **28R**, **30P**, **34R**, **39A**, **44S**, **45L** , **48P**, **49R**, **52I**, 55R, **62W**, 72R, **77N**, **85Q**, 86S, **88R**, **93R**, 94T, 96N, **100R**, 102R, **103F**, 117A, 118Q, **124T**, 126A, **127D**, **129P**, 130G, 143P, 144?, 159T. Letters refer to the amino acid residue found in the human ECP gene.

cifically test for a portion of sites evolving by positive Darwinian selection, LRTs are conducted to compare M1 with M2 and M7 with M8. Positive selection is indicated when a freely estimated ω parameter is >1 and the LRT is significant.

Detection of positive selection sites in the EDN-ECP gene family

We now re-examine the EDN-ECP dataset using models of variable ω ratios among sites. Because previous analysis indicated positive selection followed the duplication event, and because the RNase-k6 sequences were highly divergent, we excluded the outgroups from these analyses, leaving twelve sequences in the dataset. Note that a very high divergence can reduce power to detect positive selection under models of variable ω ratios among sites (Anisimova *et al.*, 2001). Results obtained under several models are presented in Table 2.

Averaging ω over all sites and branches for these sequences gives an ω ratio of 0.818, an average which indicates evolution by weak purifying selection. However, a LRT of the one-ratio model with M3 ($k = 3$) indicates that selective pressure is not uniform among sites ($2\delta = 53.12$, d.f. = 4, $P < 0.0001$). Furthermore, all three models which permit an ω parameter to exceed 1 (M2, M3, and M8) indicated a large fraction of sites evolving under positive Darwinian selection (Table 2). LRTs indicate that models M2 and M8 fit the data significantly better than models M1 and M7, in which positively selected sites are not allowed (M1 v. M2: $2\delta = 7.58$, d.f. = 2, $P = 0.02$; M7 v. M8: $2\delta = 11.2$, d.f. = 2, $P = 0.004$). Clearly, some variation in selective pressure is due to evolution by positive selection. It is worth noting that parameter estimates also suggest a large fraction of sites evolving under strong purifying selection (Table 2). Clearly, estimation of the single ω parameter, as an average over sites, did not provide a sensible measure of selective constraints for the ECP-EDN gene family.

Zhang *et al.* (1998) reported that the number of arginines increased substantially along the branch leading to the ECP clade. We wanted to evaluate the pattern of arginine evolution in the ECP clade at sites subject to positive selection. The Bayes theorem was used to identify candidate positive selection sites. There was no major increase in the number of argenines at those sites within the ECP clade. This suggests that more recent adaptive pressure in ECP might differ from the early selection pressure to enhance cationic toxicity. Interestingly, residues critical to ribonucleic activity in ECP are evolutionary conserved despite their making no contribution to the cationic toxicity of ECP (Rosenberg and Domachowske, 1999).

It is important to note that the current Baysean analysis did not distinguish between sites related to adaptive divergence of ECP and EDN, and those sites that have been the target of continual adaptive pressure in ECP. Additional sequences of ECP will be needed to infer positive selection sites specific to that gene. Bayes identification of positively selected sites in EDN and ECP, combined with reconstruction of ancestral amino acid sequences and site-directed mutagenesis, could help identify the key sites associated with EDN and ECP activities.

Lineage-specific changes in selective pressure at specific amino acid sites

Yang and Nielsen (2002) recently extended previous codon models to permit the ω ratio to vary both among lineages and among sites. These new models are called "branch-site" models. Because they were developed to identify adaptive evolution in pre-specified branches, they are suited to studying the role of positive selection in gene families.

We are interested in knowing if adaptive evolution has occurred in the lineages immediately following a duplication event. The branch, or branches, of interest are called the "foreground" branches, and all others are called "background" branches. In the branch-site models, selective constraints are assumed to vary among sites, but at a subset of sites selective constraints are permitted to change in the foreground lineage. Sites changing in the foreground lineage are permitted to have $\omega > 1$. The basic model has four ω site classes. The first two classes of sites, with ω_0 and ω_1, are uniform over the entire phylogeny. The other two classes allow some sites with ω_0 and ω_1 to change to positive selection (ω_2) in the foreground lineage, *i.e.*, sites where $\omega_0 \rightarrow \omega_2$ and $\omega_1 \rightarrow \omega_2$.

Yang and Nielsen (2002) implemented two versions of branch-site models (A and B). In model A, ω_0 is fixed to 0 and ω_1 is fixed to 1; hence positive selection is permitted only in the foreground branch. This model is compared with model M1 (neutral) with d.f. = 2. In model B, ω_0 and ω_1 are free parameters. Thus some sites could evolve by positive selection across the entire phylogeny, whereas other sites could evolve by positive selection in just the foreground branch. Model B is compared with M3 (discrete) with $k = 2$ site classes and d.f. = 2.

Application to the Troponin C gene family

We use the Troponin C gene family to demonstrate the application of the branch-site models (Yang and Nielsen, 2002) to studies of evolution by gene duplication. Through association with actin and tropomyosin, Troponin C inhibits actomysin interaction at submicromolar Ca^{2+} concentrations, and stimulates the interaction at micromolar concentrations (Farah and Reinach, 1995). In vertebrates there was a single duplication in this family that initiated evolution of the two distinct muscle isoforms for Troponin C: (i) the fast skeletal-muscle isoform, and (ii) the cardiac and slow skeletal-muscle isoform. Ohta (1994) examined estimates of d_N and d_S within and between the different isoforms of Torponin C and found a higher rates of nonsynonymous substitution between isoforms than within an isoform. Ohta (1994) considered this as evidence for an accelerated nonsynonymous rate following the duplication event and suggested divergence by positive selection. However, in no comparisons were d_N/d_S ratios greater than 1, and the results also seemed compatible with reduced purifying selection after gene duplication. Here we investigate the evolutionary forces associated with the functional divergence of these two isoforms using branch models, sites models, and branch-site models. The phylogeny for the Troponin C gene family is presented in Figure 4. We treat the branch labelled A as the foreground branch and all other branches in the phylogeny as background branches.

Averaging ω over all sites and branches yielded the estimate $\omega = 0.017$ (one ratio model; Table 3), indicating the overwhelming role of purifying selection in this gene family. We relax the assumption of a single ω for all branches in the two-ratios model. Here we assign one ω ratio for the branch separating the two isoforms (ω_0) and a second for all other branches (ω_1). The branch separating the two isoforms (Fig. 4, branch A) immediately postdates the Troponon C duplication event. Parameter estimates suggested a higher relative rate of nonsynonymous substitution in branch A (Table 3), but the two-ratio model did not provide a significantly better fit to these data (Table 4).

We also applied sites models (Nielsen and Yang, 1998; Yang *et al.*, 2000) to Troponoin C (Table 3).We employed the following model pairs: (i) one-ratio (M0) and M3; (ii) M1 and M2; and (iii) M7 and M8. M3 fitted these data significantly better than the one-ratio model (Table 4), but no parameter estimate

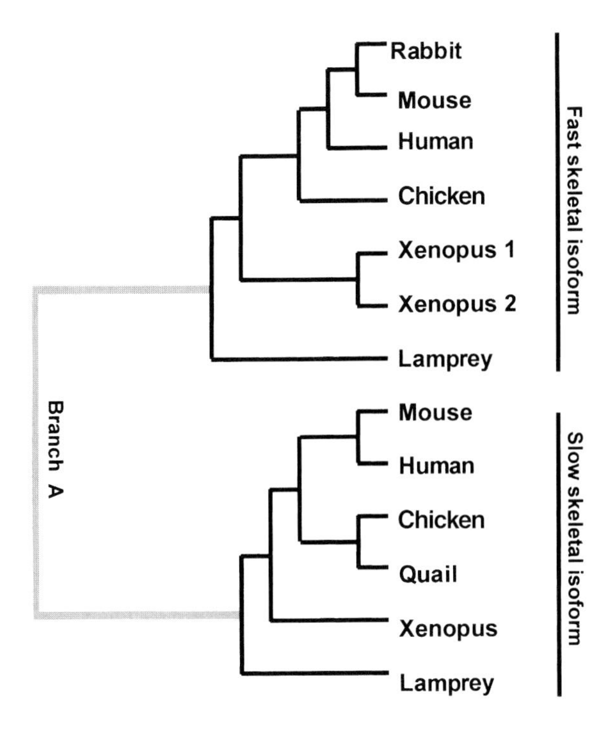

Figure 4. A phylogenetic tree for the Troponin C gene family. Note branch lengths are not to scale. Branch A, indicated in grey, represent the lineage that immediately follows the duplication event in Troponin C. All analyses were conducted using unrooted topologies; this topology is rooted for convenience.

indicated positive selection (Table 3). This finding indicates significant variation in selective constraints among sites, but not by positive Darwinian selection. Comparisons of M1 with M2 and M7 with M8 are consistent with this interpretation, in that both models that permit an ω ratio to exceed 1 (M2 and M8) also failed to indicate sites under positive selection (Table 3).

Branch-site models were applied to Troponin C with branch A (Fig. 4) specified as the foreground branch. In contrast, the new branch-site models indicated evolution by positive Darwinian selection in the Troponin C gene family. Both models A and B indicated a relatively large fraction of sites (model A: 14%; model B: 22%) evolving under positive selection in the foreground branch. Comparison of model A with its null model (M1), and model B with its null model (M3; $k = 2$) indicated that the branch-site models provided a significantly better fit to these data (Table 4). Positively selected sites identified by the Bayes method were listed in Table 3. Model B identified fewer sites than model B (Table 3).

Table 4. Likelihood ratio test statistics (2δ) for the Troponin C dataset.

	2δ	df	*P*-value [a]
LRT of ω at branch A(Fig. 4)			
One ratio vs. two ratios	1.02	1	0.31
LRTs of variable ω's among sites			
One ratio vs. M3	88.62	2	<0.0001
M1 vs. M2	679.14	2	<0.0001
M7 vs. M8	2.70	2	0.26
LRTs of variable ω's along branch A (Fig. 4)			
M1 vs. Model A	152.54	2	<0.0001
M3 vs. Model B	13.86	2	0.001

Initial application of the branch-sites models to two single copy genes (primate lysozme and tumour suppressor BRCA1) yielded mixed results, and Yang and Nielsen (2002) speculated that branch-site models might not provide a significant improvement over previous models in many single copy genes. In contrast, our analysis of the Troponin C gene family illustrates the utility of these new models. Previous codon models did not have the power to detect adaptive evolution in Troponin C because adaptation appears to have occurred at a subset of sites for a limited period of time. This mode of molecular adaptation is likely to be very common in gene families. Indeed, these new models also were successful in detecting adaptive evolution after the gene duplication of the cone opsin that gave rise to rhodopsin genes of vertebrates, a gene family in which the sites-models and branch-models were not effective (B.S.W. Chang, personal communication).

Power and accuracy of ML methods

Initial application of these codon based models led to the detection of positive selection in many genes for which it had not previously been known. In particular, relaxing the assumption of a constant selective pressure over sites appears to increase the power of the detection method considerably. For example, Yang *et al.* (2000) analysed 10 genes from a variety of genomes and identified six of them to be under positive selection. Moreover, previous studies using average d_N and d_S over all sites and branches indicated that evolution in the *nef* gene of HIV1 (Pilkat *et al.*, 1997) and the *DAZ* gene in primates (Agulnick *et al.*, 1998) was dominated by genetic drift, yet sub-

Table 3. Parameter estimates and likelihood scores for the Troponin C gene family under different branch models, sites models, and branch-site models.

Model	p	Estimates of ω parameters	Positive selection	l
One-ratio	1	$\omega = 0.02$	None	−2945.08
Branch specific				
Forground/background	2	$\omega_0 = 0.15$ (branch A; Fig 4) $\omega_1 = 0.02$ (background)	None	−2944.57
Site specific				
Neutral (M1)	1	$(\omega_0 = 0)$, $f_0 = 0.45$ $(\omega_1 = 1)$, $f_1 = 0.55$	Not allowed	−3245.91
Selection (M2)	3	$(\omega_0 = 0)$, $f_0 = 0.38$ $(\omega_1 = 1)$, $f_1 = 0.02$ $\omega_2 = 0.03$, $(f_2 = 0.60)$	None	−2906.34
Discrete (M3), $k = 2$		$\omega_0 = 0.01$, $f_0 = 0.76$ $\omega_1 = 0.08$, $(f_1 = 0.24)$	None	−2900.77
Beta (M7)	2	$p = 0.42$, $q = 17.08$	Not allowed	−2898.82
Beta&ω (M8)	4	$p = 0.25$, $q = 8.22$ $f_0 = 0.69$ $\omega = 0.015$, $(f_1 = 0.31)$	None	−2897.47
Branch-site				
Model A	3	$(\omega_0 = 0)$, $f_0 = 0.43$ $(\omega_1 = 1)$, $f_1 = 0.33$ $\boldsymbol{\omega_2 = 89}$, $\boldsymbol{(f_{2+3} = 0.14)}$	**Foreground:** **22 sites**[a] Background: Not allowed	−3169.63
Model B	5	$\omega_0 = 0.01$, $f_0 = 0.56$ $\omega_1 = 0.07$, $f_1 = 0.22$ $\boldsymbol{\omega_2 = 76}$, $\boldsymbol{(f_{2+3} = 0.22)}$	**Foreground:** **12 sites**[b] Background: None	−2893.84

[a] Positive selection sites identified under Model A with posterior probabilities ≥ 0.95 (sites with posterior probabilities ≥ 0.99 are in bold): **15V**, 16E, 17Q, **23K**, **32I**, **34V**, 35L, 37A, 39D, **41C**, **64Q**, **65E**, **66M**, **68D**, **78V**, **90C**, **95S**, **109M**, **126M**, 157Y, **163F**, **165K**. Letters refer to the amino acid residue found in the slow skeletal-muscle isoform of Human.

[b] Positive selection sites identified under Model B with posterior probabilities ≥ 0.95 (sites with posterior probabilities ≥ 0.99 are in bold): 11Y, **15V**, 17Q, 23K, 24N, 35L, **41C**, **60P**, **64Q**, **90C**, 95S, **163F**. Letters refer to the amino acid residue found in the slow skeletal-muscle isoform of Human.

sequent ML analyses indicated a small fraction of sites evolving under positive selection and the remaining sites under purifying selection (*nef*, Nielsen and Yang, 1998; Zanotto *et al.*, 1999; *DAZ*, Bielawski and Yang, 2000).

Although successful in those cases, the accuracy and power of the LRT and Bayesian site identification were largely unknown. Anisimova *et al.* (2001) used computer simulation to address the performance of the site models (Nielsen and Yang, 1998; Yang *et al.*, 2000), leading to the following observations: (i) the LRT statistic (2δ) does not follow the χ^2 distribution because some model parameters are fixed at the boundary of parameter space, and use of the χ^2 distribution made the test conservative; *i.e.*, false positives always occurred less frequently than expected under the specified significance level of the test. (ii) Despite being conservative, the LRT still provided a powerful means of detecting positive selection given enough variation in the data; power was 100% with datasets of 17 taxa, but quite low for datasets of only 5 or 6 taxa. (iii) Sequence length, sequence divergence, and strength of positive selection all influence power of the LRT, but the number of sampled taxa is the most important. (iv) Power is low for both highly similar and highly divergent sequences, and is highest at intermediate or moderately high divergence. (v) The LRT is robust to the assumed distribution for ω across sites.

Anisimova *et al.* (2002) also used computer simulation to study the power and accuracy of Bayesian identification of sites evolving under positive selection. The findings can be summarized as follows: (i) identification of sites is not feasible from a few highly similar sequences; (ii) accuracy of site identification is most dependent on the numbers of lineages. The number of lineages and the sequence divergence are the most important factors affecting power. (iii) Unlike results for LRT, the assumed distribution for ω across sites can affect both accuracy and power of site identification. Anisimova *et al.* (2002) recommended using multiple models to identify positively selected sites.

Limitations and Future prospects

The site models are computationally complex and intensive. Models M2 and M8 are known to have multiple local optima in some datasets. Because accuracy is sensitive to parameter estimation, suboptimal parameter estimates, based on a local optimum, could increase the probability of a type-I error. To avoid being trapped at a local optimum, it is important to run M8 multiple times; minimally it should be run with one initial $\omega > 1$ and one initial $\omega < 1$. Results corresponding to the highest likelihood value should be used.

Codon models are limited in effectiveness by the sample of sequences. Typically, the power to detect adaptive evolution will be low for sequences less than one hundred codons in length. Furthermore, the analysis is impacted by the requirement that information is drawn from both synonymous and nonsynonymous substitutions. Hence, the power can be quite low for highly similar and highly divergent sequences (Anisimova *et al.*, 2001). Although sampling additional lineages can improve power when sequence divergence is low or high, the window of suitable sequence divergence is expected to be narrower than for phylogeny reconstruction.

Gu (1999, 2001) developed a maximum likelihood approach to testing for functional divergence that is based on amino acid substitution rates. The method can detect specific amino acid residues that contribute to functional divergence following gene duplication (Gu, 1999, 2001). Although amino acid substitution rate might not be as sensitive a measure of selective pressure as the d_N/d_S ratio, amino acid evolution is much slower than nucleotide evolution, and Gu's (1999, 2001) approach should provide an excellent tool for studying evolution by gene duplication in gene families too divergent for analysis with codon models.

An interesting avenue for future development of codon models is the "covarion-like" model (Galtier, 2001). Covarion-like models allow site specific rates to vary among lineages, and can be achieved by adding as few as two additional parameters to non-covarion models (Galtier, 2001). Galtier's (2001) ML implementation for nucleotide models is promising, as it revealed a significant amount of site-specific rate variation in ribosomal RNA sequences. Covarion-like codon models might provide an alternative framework for investigating episodic adaptive evolution acting on a few amino acid sites (Galtier, 2001). This approach would have the advantage of not requiring *a priori* knowledge of the specific branches where episodes of adaptive evolution have occurred.

References

Agulnik, A.I., Zharkikh, A., Boettger-Tong, H., Bourgeron, T., McElreavey, K. and Bishop, C.E. (1998) Evolution of the DAZ gene family suggests that Y-linked *DAZ* plays little, or a limited, role in spermatogenesis but underlines a recent African origin for human populations. *Hum. Mol. Genet.* **7,** 1371–1377.

Anisimova, M., Bielawski, J.P. and Yang, Z. (2001) Accuracy and power of the likelihood ratio test in detecting adaptive molecular evolution. *Mol. Biol. Evol.* **18,** 1585–1592.

Anisimova, M., Bielawski, J.P. and Yang, Z. (2002) Accuracy and power of Bayesian prediction of amino acid sites under positive selection. *Mol. Biol. Evol.* **19**, 950–958.

Bielawski, J.P. and Yang, Z. (2000) Positive and negative selection in the *DAZ* gene family. *Mol. Biol. Evol.* **18,** 523–528.

Bishop,J.G., Dean, A.M. and Mitchell-Olds, T. (2000) Rapid evolution of plant chitinases: molecular targets of selection in plant-pathogen coevolution. *Proc. Natl. Acad. Sci. USA* **97,** 5322–5327.

Crandall, K.A., Kelsey, C.R., Imanichi, H., Lane, H.C. and Salzman, N.P. (1999) Parallel evolution of drug resistance in HIV: failure of nonsynonymous/synonymous substitution rate ratio to detect selection. *Mol. Biol. Evol.* **16,** 372–382.

Domachowske, J.B., Bonville, C.A., Dyer, K.D. and Rosenberg, H.F. (1998) Evolution of antiviral activity in the ribonuclease A gene superfamily: evidence for a specific interaction between eosinophil-derived neurotoxin (EDN/RNase 2) and respiratory syncytial virus. *Nucleic Acids Res.* **26,** 5327–5332.

Duda, T.F. and Palumbi, S.R. (1999) Molecular genetics of ecological diversification: duplication and rapid evolution of toxin genes of the venomous gastropod *Conus*. *Proc. Natl. Acad. Sci. USA* **96,** 6820–6823.

Dykhuizen, D. and Hartl, D.L. (1980) Selective neutrality of 6PGD allozymes in *E. coli* and the effects of genetic background. *Genetics* **96,** 801–817.

Endo, T., Ikeo, K. and Gojobori, T. (1996) Large-scale search for genes on which positive selection may operate. *Mol. Biol. Evol.* **13,** 685–690.

Farah, C.S. and Reinach, F.C. (1995) The troponin complex and regulation of muscle contraction. *FASEB J.* **9,** 755–767.

Galtier, N. (2001) Maximum-likelihood phylogenetic analysis under a covarion-like model. *Mol. Biol. Evol.* **18,** 866–873.

Golding, G.B. and Dean, A.M. (1998) The structural basis of molecular adaptation. *Mol. Biol. Evol.* **15,** 355–369.

Goldman, N. and Yang, Z. (1994) A codon based model of nucleotide substitution for protein-coding DNA sequences. *Mol. Biol. Evol.* **11,** 725–736.

Gu, X. (1999) Statistical methods for testing functional divergence after gene duplication. *Mol. Biol. Evol.* **16,** 1664–1674.

Gu, X. (2001) Maximum-likelihood approach for gene family evolution under functional divergence. *Mol. Biol. Evol.* **18,** 453–464.

Haydon, D.T., Bastos, A.D., Knowles, N.J. and Samuel, A.R. (2001) Evidence for positive selection in foot-and-mouth-disease virus genes from field isolates. *Genetics* **157,** 151–154.

Hughes, A.L. (1999) *Adaptive Evolution of Genes and Genomes*, Oxford University Press, Oxford, UK.

Hughes, A.L. and Nei, M. (1988) Pattern of nucleotide substitution at major histocompatibility complex class I loci reveals overdominant selection. *Nature* **335,** 167–170.

Kimura, M. (1983) *The Neutral Theory of Molecular Evolution*, Cambridge University Press, Cambridge, UK.

Leigh Brown, A.J. (1997) Analysis of HIV–1 *env* gene reveals evidence for a low effective number in the viral population. *Proc. Natl. Acad. Sci. USA* **94,** 1862–1865.

Li, W.-H. (1985) Accelerated evolution following gene duplication and its implications for the neutralist-selectionist controversy. In *Population Genetics and Molecular Evolution* (Eds., Otha, T. and Aoki, K.), Japan Scientific Press, Tokyo, pp. 333–352.

Lynch, M. and Conery, J.S. (2000) The evolutionary fate and consequences of duplicate genes. *Science* **290,** 1151–1155.

Muse, S.V. and Gaut, B.S. (1994) A likelihood approach for comparing synonymous and nonsynonymous nucleotide substitution rates, with applications to the chloroplast genome. *Mol. Biol. Evol.* **11,** 715–725.

Nielsen, R. and Yang, Z. (1998) Likelihood models for detecting positively selected amino acid sites and applications to the HIV–1 envelope gene. *Genetics* **148,** 929–936.

Ohno, S. (1970) *Evolution by Gene Duplication*, Springer-Verlag, Berlin.

Ohta, T. (1988a) Further simulation studies on evolution by gene duplication. *Evolution* **42,** 375–386.

Ohta, T. (1988b) Multigene and supergene families. *Oxf. Surv. Evol. Biol.* **5,** 41–65.

Ohta, T. (1993) Pattern of nucleotide substitution in growth hormone-prolactin gene family: a paradigm for evolution by gene duplication. *Genetics* **134:** 1271–1276.

Ohta, T. (1994) Further examples of evolution by gene duplication revealed through DNA sequence comparisons. *Genetics* **138,** 1331–1337.

Plikat, U., Nieselt-Struwe, K. and Meyerhans, A. (1997) Genetic drift can dominate short-term human immunodeficiency virus type 1 *nef* quasispecies evolution in vivo. *J. Virol.* **71,** 4233–4240.

Rooney, A.P. and Zhang, J. (1999) Rapid evolution of primate sperm protein: relaxation of functional constraint or positive Darwinian selection? *Mol. Biol. Evol.* **16,** 706–710.

Rosenberg, H.F. and Domachowske, J.B. (1999) Eosinophils, riobnucleases and host defence: solving the puzzle. *Immunol. Res.* **20,** 261–274.

Rosenberg, H.F., Dyer, K.D., Tiffany, H.L. and Gonzalez, M. (1995) Rapid evolution of a unique family of primate ribonuclease genes. *Nature Genet.* **10,** 219–223.

Schmidt, T.R., Goodman, M. and Grossman, L.I. (1999) Molecular evolution of the *COX7A* gene family in primates. *Mol. Biol. Evol.* **16,** 619–626.

Sharp, P.M. (1997) In search of molecular Darwinism. *Nature* **385,** 401–404.

Swanson, W.J., Yang, Z., Wolfner, M.F. and Aquadro, C.F. (2001) Positive Darwinian selection in the evolution of mammalian female reproductive proteins. *Proc. Natl. Acad. Sci. USA* **98,** 2509–2514.

Ward, T.J., Honeycutt, R.L. and Derr, J.N. (1997) Nucleotide sequence evolution at the kappa-casein locus: evidence for positive selection within the family Bovidae. *Genetics* **147,** 1863–1872.

Yang, Z. (1997) PAML: a program package for phylogenetic analyses by maximum likelihood. *Cabios* **13,** 555–556.

Yang, Z. (1998) Likelihood ratio tests for detecting positive selection and application to primate lysozyme evolution. *Mol. Biol. Evol.* **15,** 568–573.

Yang, Z. (2001) Adaptive molecular evolution. In *Handbook of Statistical Genetics* (Eds., Balding, D.J., Bishop, M. and Cannings, C.), Wiley & Sons, New York, NY, pp. 327–350.

Yang, Z. and Nielsen, R. (2002) Codon-substitution models for detecting molecular adaptation at individual sites along specific lineages. *Mol. Biol. Evol.* **19,** 908–917.

Yang, Z. and Swanson, W.J. (2002) Codon-substitution models to detect adaptive evolution that account for heterogeneous selective pressures among site classes. *Mol. Biol. Evol.*, **19,** 49–57.

Yang, Z., Nielsen, R., Goldman, N. and A.-M.K. Pederson, W.J. (2000) Codon-substitution models for heterogeneous selection pressure at amino acid sites. *Genetics* **155,** 431–449.

Zanotto, P.M. de A., Kallas, E.G., de Souza, R.F. and Holmes, E.C. (1999) Genealogical evidence for positive selection in the *nef* gene of HIV–1. *Genetics* **153,** 1077–1089.

Zhang, J., Rosenburg, H.F. and Nei, M. (1998) Positive Darwinian selection after gene duplication in primate ribonuclease genes. *Proc. Natl. Acad. Sci. USA* **95,** 3708–3713.

A. Meyer, Y. Van de Peer (eds.), Genome Evolution, 213-224.

Approach of the functional evolution of duplicated genes in *Saccharomyces cerevisiae* using a new classification method based on protein-protein interaction data

Christine Brun[1], Alain Guénoche[2] & Bernard Jacq[1,*]
[1]*Laboratoire de Génétique et Physiologie du Développement, IBDM and* [2]*Institut de Mathématiques de Luminy, Parc Scientifique de Luminy, CNRS, Case 907, F–13288 Marseille Cedex 9, France*
[*] *Author for correspondence: E-mail: jacq@lgpd.univ-mrs.fr*

Received 21.01.2002; Accepted in final form 29.08.2002

Key words: duplicated genes, functional evolution, protein classification, protein-protein interactions, regulation network

Abstract

The concept of protein function is widely used and manipulated by biologists. However, the means of the concept and its understanding may vary depending on the level of functionality one considers (molecular, cellular, physiological, etc.). Genomic studies and new high-throughput methods of the post-genomic era provide the opportunity to shed a new light on the concept of protein function: protein-protein interactions can now be considered as pieces of incomplete but still gigantic networks and the analysis of these networks will permit the emergence of a more integrated view of protein function. In this context, we propose a new functional classification method, which, unlike usual methods based on sequence homology, allows the definition of functional classes of protein based on the identity of their interacting partners. An example of such classification will be shown and discussed for a subset of *Saccharomyces cerevisiae* proteins, accounting for 7% of the yeast proteome. The genome of the budding yeast contains 50% of protein-coding genes that are paralogs, including 457 pairs of duplicated genes coming probably from an ancient whole genome duplication. We will comment on the functional classification of the duplicated genes when using our method and discuss the contribution of these results to the understanding of function evolution for the duplicated genes.

Duplications in the yeast genome

Gene duplication is a general phenomenon, which concerns from 7 to more than 30% of the ORFs, depending on the genome (Coissac *et al.*, 1997). This conclusion was drawn from sequence analysis performed on genomes from procaryotes, archea and eucaryotes. In the yeast *Saccharomyces cerevisiae*, early studies concluded that as much as 40% of the genome was duplicated (Dujon *et al.*, 1994; Feldmann *et al.*, 1994). More recent studies demonstrated that 55 to 57 duplicated blocks each spanning 55 kb and containing 6.9 genes on an average, can be found (Wolfe and Shields, 1997; Seoighe and Wolfe, 1999). All together, they account for 50% of the genome and they are supposedly the remnants of an ancient whole genome duplication likely to have occurred 100 million years ago.

The abundant structural and functional genomic data now available for yeast, in combination with the 'classical' molecular, genetic, biochemical and cellular knowledge which has accumulated on this organism since several decades, makes it a good choice for studying the gene duplication phenomenon in eucaryotes. An important and as yet unresolved question in this field is to understand how the function of duplicated genes has evolved since the duplication event. Clues are at the moment very partial due to a general lack of information on the actual function of duplicated genes.

In this article, we will first discuss the notion of gene (or protein) function and illustrate the need for generic and powerful methods for comparing functions. Based on recent work in our laboratory, we will propose a simple method that uses protein-protein interaction data to compare gene/protein functions in the absence of any other information. This method has been applied to 429 proteins of yeast and leads to several functional clusters, some of them corresponding to already known complexes, while some others are at least partially new. Then, considering the proteins, which correspond to duplicated genes in our test sample, we have been able to compare them from a functional point of view and to correlate functional evolution to sequence evolution.

How can we measure and compare gene/protein functions?

Determining the function of the 'new' genes revealed by the sequencing of many procaryotic and euc aryotic complete genomes is certainly one of the most important tasks of the biology of this beginning 21st century. At least three types of problems are associated with function discovery and description. The first one is a semantic problem: what do we exactly mean by 'function' of a gene or a protein and how do we describe it? A second point is: how can we compare functions and how different is this from comparing sequences or structures? The third and most important one is technical: do we have powerful, high throughput, methods allowing determination of the function of thousands of genes in one or a few standardized ways? We are addressing these three points below.

Function definition

The term 'gene function' (or 'protein function') is certainly one of the most widely used in biology, but is probably also (and unfortunately so) the one for which there is the most severe lack of common accepted definition. In fact, the notion of function is a complicated one with regards to the facts that (i) determining a function implies defining the structural level for which this function is valid and (ii) a gene/protein has often more than one function.

(i) When trying to describe the function of a biological macromolecule, it is probably important to realize that several different structural levels can be considered and that specific functions are associated to each level. One can find at least six structural levels of increasing biological complexity in multicellular organisms: molecules, sub-cellular structures, cells, tissues and organs, organisms and populations. Each of these six structural levels can be distinguish by specific components, structures, processes and concepts, not valid when describing other levels. Importantly, each structural level can be associated to a specific functional level: respectively molecular complexes and interaction networks, cell trafficking, cell migrations and inter-cellular communications, physiological regulations, behaviour, and finally inter-species relationships and ecological equilibriums. Therefore, the function of a protein can be described at different levels, for instance a transcription factor (molecular) is involved in the establishment of cell polarity (cellular) and in peripheral nervous system morphogenesis (organismal).

(ii) We have already discussed how a structural biologist and a geneticist could describe the function of the same protein in completely different terms, although both correctly capture a part of the 'complete' function of the protein (Jacq, 2001). It is essential to note that the function of a protein is generally pleiotropic (for a recent review, see Jeffery, 1999). This pleiotropy can be observed within and between organisational levels. For instance, a protein has often more than one function at the molecular level: *Xenopus* TFIIIA (Romaniuk, 1985), *Drosophila* bicoid (Rivera-Pomar *et al.*, 1996) and modulo (Perrin *et al.*, 1999) proteins are all DNA-binding proteins which are also RNA-binding proteins. But the function of a protein can rarely be described at only one structural level: the *Drosophila* bicoid protein is a DNA- RNA-binding protein and a transcription factor at the molecular level, and it is an essential determinant of the formation of anterior structures (head, thorax) at the level of the organism.

At this stage, two important conclusions are drawn. First, it is always necessary to specify what structural level is being examined when defining the function of a given gene/protein. Second, it is probably more appropriate and accurate to speak about 'the functions' rather than 'the function' of a gene or a protein. All these issues are discussed in more details in (Jacq, 2001). Within the context of the present paper, we will further discuss function at the two first levels only, i.e. molecular and cellular.

How can we measure and compare functions?

Since at least three decades, biologists have developed several powerful methods to compare structures (i.e. representations of biological objects with a physical existence). This is particularly true for nucleic acid and protein sequences, which are represented by successions of letters, and for which many different powerful alignment methods and associated tree representations for sequence proximities have been developed. The situation is completely different when function is concerned. Indeed, a function is not a structure but rather corresponds to a description of properties displayed by this biological structure. Function descriptions are usually found as sentences in plain text and/or as series of keywords in databases. Since comparison of sentences is difficult, no serious attempt to compare gene/protein functions was made until recently. Despite these difficulties, people have nevertheless tried for a long time to predict functions using sequence or structure comparisons: a popular way to predict the function of an unknown protein is to infer it from sequence alignments with proteins of known functions. The central assumption of function inference is that the more two sequences are similar, the more their functions are likely to be related. Although there is a positive correlation between sequence and function similarities, inferring the function of an unknown protein from sequence remains a risky operation, which could lead to wrong conclusions (Devos and Valencia, 2000). There are at least three main limitations to functional inferences from sequences. First, subtle sequence differences in key residues of an enzyme active site for instance, can change or abolish a function. Second, due to the mosaic nature of proteins, sequence similarities limited to some domains point towards similar molecular functions but give little information on the precise cellular functions of the unknown protein. Finally, we have many examples of proteins, which do not exhibit any detectable sequence similarity, but either have similar three-dimensional structures or are nevertheless engaged in the same cellular function, being components of the same protein complex for instance. It is only very recently that people became interested in developing new tools for function comparison (Devos and Valencia, 2000; Todd *et al.*, 2001).

A general conclusion is that protein function prediction, using the methods presently available (including function inference from sequence alignments) is very limited at the moment (Devos and Valencia, 2000). We will describe in the two next sections how the results brought by the new high-throughput proteomics methods can be used to set up a new generic and powerful method to predict function on a large-scale basis.

Protein function from the perspective of molecular interactions and genetic networks

At all the previously discussed structural levels, functions are achieved by interactions between biological structures. At the lower structural level of informational macromolecules, interactions are either molecular or genetic. Only molecular interactions will be considered here. They involve DNA, RNA and proteins and play an essential role in all known biological processes. Indeed, protein-DNA, protein-RNA and protein-protein interactions account for the great majority of biological macromolecular interactions. Consequently, many different types of experimental approaches have been developed in the last years to identify interactions and to test their biological relevance. Understanding molecular and genetic interactions and describing them in terms of complex networks of interacting molecules is a current concern. Such interaction networks are capable to respond to both external stimuli and stresses, as well as to internal changes occurring within components of the network. Being able to formally describe interactions and networks, to query and manipulate them is now largely recognized as essential to the study of gene regulation and function. Importantly, the simulation of network behaviour from mathematical models (Thomas, 1973; Thomas *et al.*, 1995; Hlavacek and Savageau, 1996; Sharp and Reinitz, 1998; for recent reviews, see Smolen *et al.*, 2000; von Dassow *et al.*, 2000) started to receive a practical confirmation when small networks have been engineered and their predictable behaviour experimentally tested in procaryotes (Becskei and Serrano, 2000; Elowitz and Leibler, 2000; Gardner *et al.*, 2000).

In the context of our study, we will focus only on protein-protein interactions and on network structure rather than dynamics. The structure of any protein network can be represented as a graph, in which nodes are proteins and edges the relationships linking them (their physical interaction in this case). More precisely, for the purpose of our study, we will systematically examine the vicinity relationships of components of the protein network, and infer functional comparisons from these relationships.

A new method to compare and predict the functions of proteins

Principle

Defining new ways of comparing proteins from a functional point of view would be, as already discussed, a very desirable goal. Considering proteins as individual members of an gigantic network, in which each protein has a finite number of interactions with several other specific molecular partners, could represent a new and (as far as we know) as yet unexplored means by which to compare proteins at a functional level. Basically, the idea is not to compare proteins themselves but instead to compare the list of their partners: the more they share partners, the more they should be functionally related.

The feasibility of this idea was already tested on a small set of protein-protein interactions from *Saccharomyces cerevisiae* (Jacq, 2001). Although preliminary results were obtained on a very small sample only (14 different proteins and their 47 interactors), they indeed show that proteins with common interaction partners tend to cluster together and that these clusters correspond to proteins with related functions. This prompted us to develop a general method, which allows the calculation of a functional distance between proteins from lists of pairs of interacting proteins. Such a distance should provide a direct measurement of the functional relationships between proteins, independently of their primary sequence homology. A functional relationship could mean either that proteins are involved in the same biological process or that they share common functional features (molecular or cellular function, belonging to the same functional family, etc.).

Program

We first calculate a distance on a set of proteins that share common interactors. Then, using only the distance values, we build classes of proteins. 'Relevant proteins' to be clustered are defined as having at least p interactions ($p > 1$), the other proteins being there only to measure distance values. From all possible pairs of distances, a neighbour joining algorithm is then applied which finds related proteins and finally the results of the clustering method are displayed as a tree.

The program first builds a rectangular table of interactions T in which rows corresponds to relevant proteins and columns to proteins, relevant or not. This table permits to define a relation (or a graph) between proteins: pair (i, j) is 'in relation' (or is an edge) if and only if they interact together ($T(i, j) = 1$ or $T(j, i) = 1$), or if they both act on the same protein, that is there exists k such that $T(i, k) = T(j, k) = 1$. Resulting from the transitive closure of this relation, a large class appears, denoted X, containing almost all the relevant proteins, on which we compute a symmetrical difference distance D. Let Int(i) be the list of proteins interacting with i, extended with i itself. The distance between i and j is equal to the number of interactors of i that do not act on j, plus the number of interactors of j that do not act on i, this sum being divided by the number of proteins belonging to one or the other list:

$$D(i, j) = |\mathrm{Int}(i) \Delta \mathrm{Int}(j)| \;/\; |\mathrm{Int}(i) \cup \mathrm{Int}(j)| \;.$$

In doing so, only the related proteins have a distance smaller than 1. This distance is classical on graph. Overall, the obtained distances were very closed to ultrametric ones and consequently, protein clusters can appear in a tree representation.

Constitution of the dataset and computational treatment

Interactions were extracted from the Physical Interaction Table of the MIPS site (http://mips.gsf.de/proj/yeast/CYGD/db/index.html; Mewes *et al.*, 2000) and from the YPD database (http://www.proteome.com/data bases/index.html; Costanzo *et al.*, 2000, 2001). Only direct interactions were selected based on the method used to identify them: when the method could lead to a bridging effect (co-immunoprecipitation, affinity chromatography, copurification, co-sedimentation, crosslink, mass spectrometry, etc.), the interaction was not taken into account. Consequently, the data collection was then essentially composed of results coming from two-hybrid experiments, *in vitro* binding assays, far western experiments, gel retardation assays and biochemical experiments.

We have chosen 3 as p value: only proteins involved in at least 3 different interactions have been processed. This parameter selected 429 proteins participating in 3692 interactions. The program was then applied to this sample, a tree representation was computed by BIONJ (Gascuel, 1997) and the resulting class was drawn using NJPlot (Perriere and Gouy, 1996).

A functional tree for 429 yeast proteins (7% of the yeast proteome)

After computation of the dataset, the obtained tree, represented in Figure 1, is mainly composed of very long branches and a large number of nodes are very close to the root. At a first glance, one then would

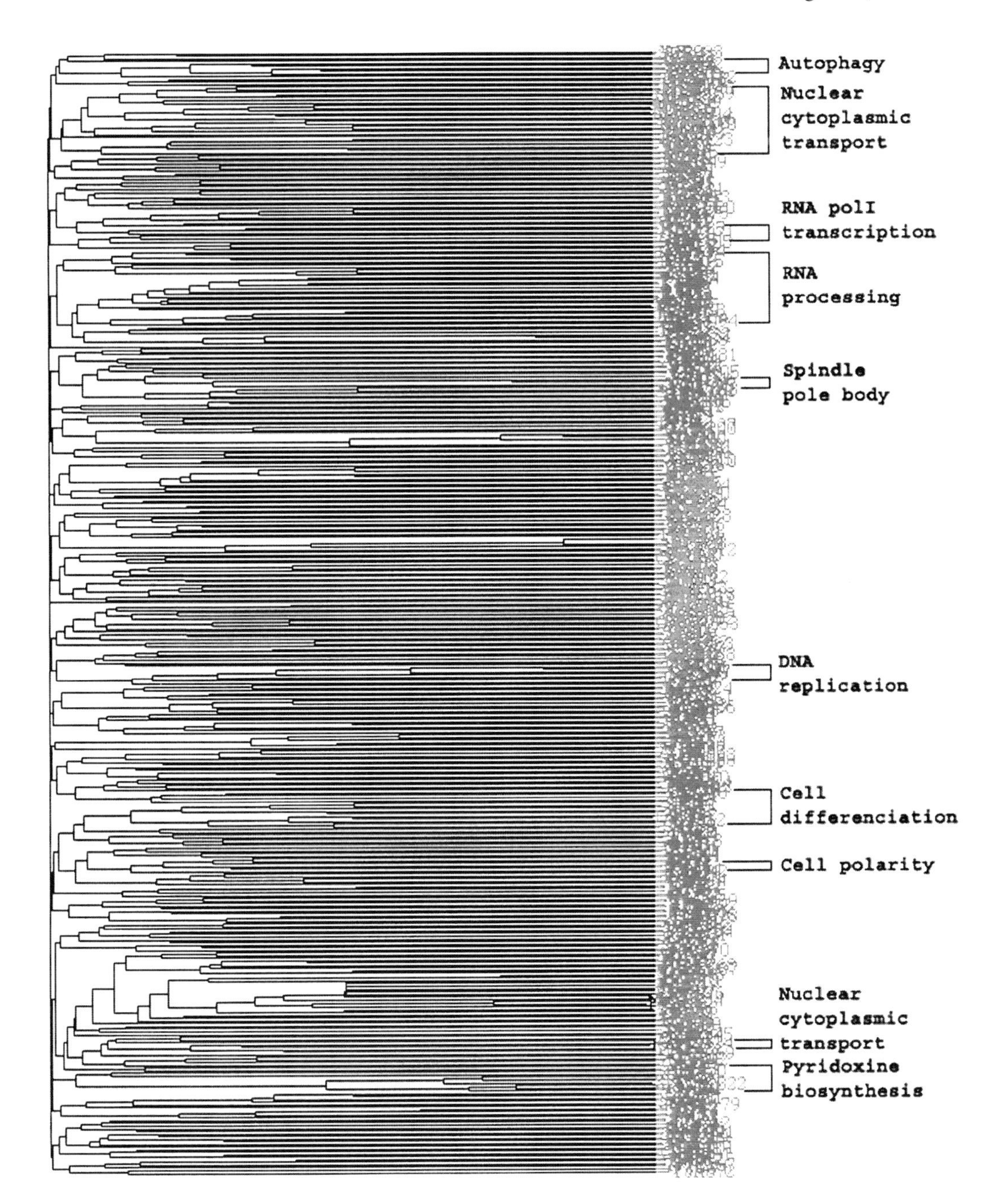

Figure 1. A functional tree for 429 yeast proteins. NJPlot was used to represent the computed classification. Names of the proteins were coloured in light grey for sake of clarity. Several relevant clusters and subtrees are annotated in the margin.

consider clusters appearing in this tree with caution. However, it should be emphasized that this is not a phylogenetic tree and that deep branching may simply reflect the pleiotropy of proteins. Therefore, in the present work, more attention was given to the analysis of the elements found within the different clusters than to the relative arrangement of the clusters. For future analysis, using other distance measurements may improve the tree general aspect by displacing a large number of nodes away from the root (work in progress).

An analysis of the tree depicted in Figure 1 showed at least 10 different clusters composed of proteins interacting either in a same physical complex (LSM proteins which participate in mRNA decay and pre-mRNA splicing (reviewed in He and Parker, 2000), MCM proteins which are involved in the initiation of DNA replication (reviewed in Lei and Tye, 2001), in a same pathway (SNZ and SNO proteins are involved in a pyridoxine biosynthesis pathway (Padilla *et al.*, 1998; Osmani *et al.*, 1999), or in a same biological process (nuclear-cytoplasmic transport, cell differentiation, etc.). A more detailed analysis (work in progress) is revealing additional biologically meaningful clusters.

Thus our classification method, based solely on interaction data, allowed the clustering of proteins already known to be functionally related and this ability participates in the validation of the method. However, we are aware that the robustness of the method still needs to be tested by statistical means (bootstrap and jacknife tests, work in progress).

Furthermore, the detailed analysis of several clusters described below will illustrate three other important contributions of our method: the grouping of some clusters into a larger protein ensemble biologically meaningful at the cellular level, the clustering of some known proteins into new functional groups leading to unforeseen functional relationships between proteins and finally, the proposal of a functional assignment for some unknown proteins in search of a function.

Detailed subtree analysis

The MCM subtree (Figure 2)

The hexameric complex MCM2–7 is an essential component of the prereplication chromatin assembled at replication origins during G1 phase. It is also the

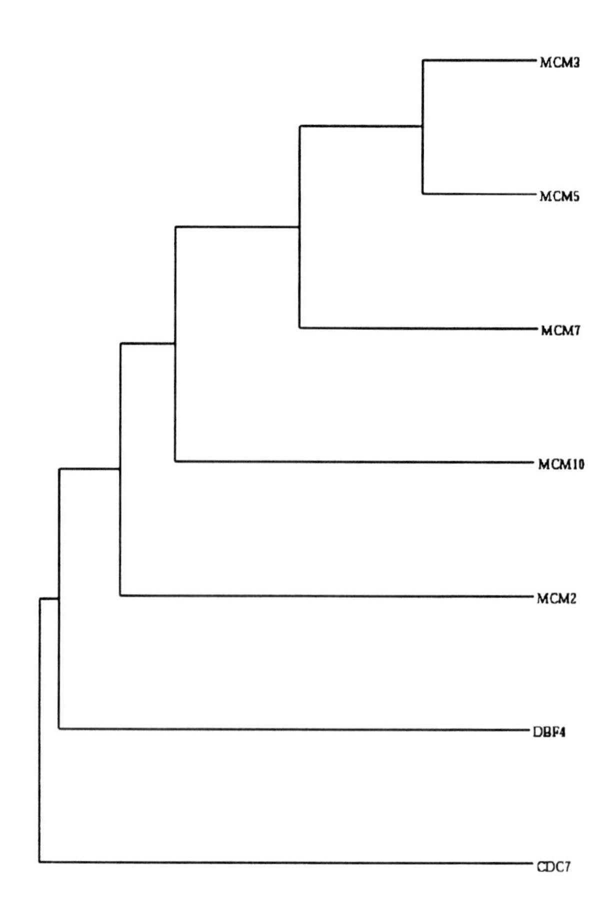

Figure 2. The MCM subtree. Detail of the DNA replication cluster from Figure 1.

presumed helicase of the growing replication forks in S phase. Four out of six MCMs are clustered into the subtree: the two missing ones are either not included into the dataset (MCM4) or clustered in another subtree (MCM6). Other proteins in the tree are directly related to the MCM functions: MCM10 is involved in the recruitment of the MCM complex onto the replication origin and the CDC7-DBF4 protein kinase, a protein complex on its own, triggers the switch from an inactive origin-bound MCM complex into an enzymatically active helicase, by phosphorylating the complex (reviewed in Lei and Tye, 2001).

Subtrees in our present analysis often show incomplete physical complexes. This is due first, to the fact that only proteins displaying at least 3 interactors (p value = 3) are taken into account in the analysis and second, to the limited extent of our present knowledge on yeast interactions.

Finally, as exemplified above, the method is able to group an enzymatic complex (CDC7-DBF4) with

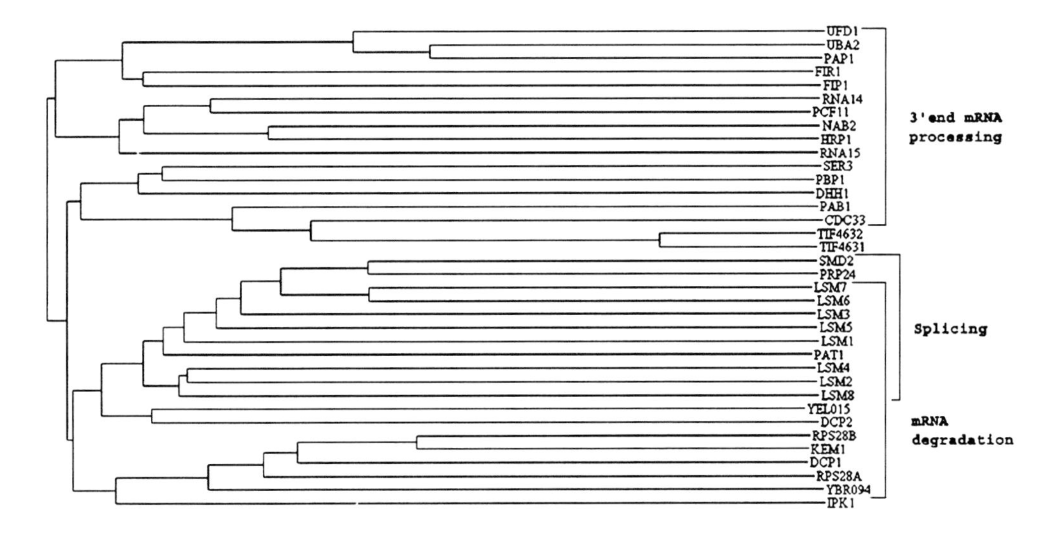

Figure 3. The RNA processing subtree. Annotated detail of the RNA processing cluster from Figure 1.

its substrate (MCM2–7). Such significant clustering then tends to validate the method we use.

The RNA processing subtree (Figure 3)

Strikingly, certain specialised molecular mechanisms are clustered within the tree to form a broader cellular process as illustrated for RNA processing. This deeply rooted subtree is subdivided into several subtrees: two of them cluster essentially proteins involved in the processing of the mRNA 3′ end and the two others group proteins having a role in RNA splicing and degradation.

Detailing these two last subtrees, at least two heptameric LSM complexes exist in yeast cells: LSM1–7 affects mRNA decapping and LSM2–8 functions in pre-mRNA splicing through interactions with U6 snRNA (reviewed in He and Parker, 2000). All these proteins are clustered in the subtree. As expected, other known splicing factors interacting with the LSMs are also clustered (SMD2, PRP24), as well as the proteins with which LSMs act in decapping: PAT1, DCP1, DCP2, and KEM1/XRN1.

Large-scale two-hybrid screens brought novel interactions. Among them, interactions of all the LSM proteins with two proteins belonging to the small ribosomal subunit, RPS28A and B, were demonstrated (Ito *et al.*, 2001; Uetz *et al.*, 2000). It was then proposed that these interactions suggest either an involvement of the LSMs in translation or ribosomal biogenesis or an unforeseen role of the ribosomal proteins in splicing (30). The clustering of both ribosomal proteins with the partners of the LSMs in mRNA degradation demonstrated herein, may suggest an unexpected role for RPS28A and B in mRNA degradation. Finally, the clustering of an additional protein of still unknown function, YBR094W, interacting with RPS28A and B and with the decapping and degradation enzymes DCP1 and KEM1, respectively, further supports this hypothesis.

The predictive chromatin tree (Figure 4)

Interestingly, our classification method points out clusters of proteins appearing intuitively related, even though no biochemical or genetical results corroborate such a clustering. For instance, the cluster shown in Figure 4 contains five proteins that can be grouped into several partially overlapping categories: MCM6 and EST1 are involved in DNA replication, NOG1 and EST1 are found in the nucleolus, MCM6, EST1 and SPT2 are associated to chromatin and finally YAL028 is a protein of unknown function. We expect that this clustering has a potential predictive value and consequently points towards new functional relationships between these yet unrelated proteins.

Finally, this detailed analysis shows that proteins of unknown function are found clustered with well known proteins, therefore leading to the proposal of a putative function for them. Among others, an example of functional assignment is found in the RNA processing tree: the YBR094W protein (noted as protein

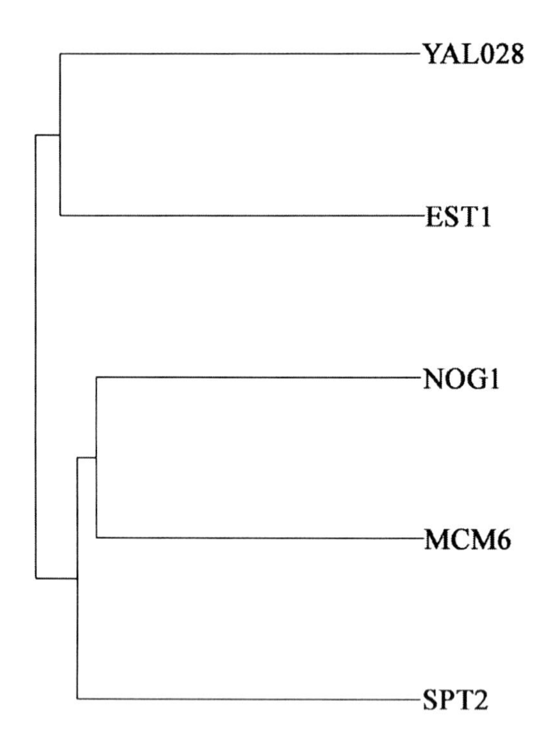

Figure 4. The predictive chromatin subtree.

of unknown function in the YPD database) can now unambiguously be assigned to the mRNA decay group.

Can the interaction network study bring new insights for the understanding of the function of duplicated genes?

From a theoretical point of view, duplicated genes could have evolved towards three distinct fates: neofunctionalization, subfunctionalization or nonfunctionalization (Lynch and Force, 2000). Unfortunately enough, only a few examples of duplicated genes have been studied from a functional point of view, therefore precluding any general evaluation of functional evolutionary fates in the case of duplicated genes. We describe below how the study of interaction networks may represent a new and useful tool for studying the phenomenon leading to functional divergence among duplicated genes.

In order to approach the functional evolution of yeast duplicated genes, we considered 3 different classes of duplicated genes (Table 1) representing a total of 64 proteins: ten pairs of paralogs supposedly resulting from a whole yeast genome duplication (Wolffe and Shields, 1997; Seoighe and Wolfe, 1999), ten pairs of genes considered to be duplicated because displaying 21 to 100% of sequence identity on more than 75% of their total length (but for which duplication dates are unknown), and finally twelve pairs of genes belonging to two distinct families of duplicated genes (eight pairs of LSM genes displaying a low sequence identity, four pairs of SNO-SNZ genes with a high sequence identity). These gene pairs, as well as their position in the tree, are described in Table 1. The number of nodes separating them gives the distance in the tree, closest relatives being separated by one node only. In addition, a set of information helping to understand the positioning of gene pairs within the tree is reported in the same table: the percentage of shared interactors between genes of a same pair and their sequence identity determined on at least 75% of their total length. Finally, a functional status aimed to point out the putative functional divergence of duplicated genes, was determined for each pair, on the basis of textual information extracted from the YPD database (http://www.proteome.com/databases/index.html; Costanzo *et al.*, 2000, 2001). A change in functional status is reported mainly when proteins from a pair are found with different cellular roles but still identical biochemical/molecular function. It is to note that other subtle changes in function have also been taken into account: for example, when only one gene out of the two from the same pair is able to complement a deficient allele of another gene. All the structural and functional information is reported in Table 1.

The results of the functional analysis on the small subset of yeast duplicated proteins are depicted in Table 1 and Figure 5 (where the sequence identity between genes of a same pair is plotted against the percentage of shared interactors). From table 1, it can be seen that eleven out of the thirty-three pairs of studied proteins (33%) are found tightly clustered in the tree (noted as 1 node away), indicating a strong or total functional conservation (independently from the 'Functional status' reported in column 6). Ten pairs of proteins (31%) separated by two to five nodes away only are also considered to be clustered, indicating a significative functional conservation. Finally, twelve pairs of proteins (36%) are found more than 9 nodes away that we interpret as a notable change in function. On the basis of the tree analysis, we therefore consider that 21 pairs of proteins (11 + 10, i.e. 64%) are conserved in function whereas 36% exhibit a detectable change in function. It is satisfactory to

Table 1. Duplicated gene pairs and their structural and functional associated variables

Gene 1	Gene 2	Position in the tree	% shared interactors	% sequence identity	Functional status[a]
Ancient Duplication					
CNA1	CMP2	**1 node away**	50	67	C
YCK1	YCK2	**1 node away**	20	66	U
MKK1	MKK2	**1 node away**	50	59	U
YNL047C	YIL105C	**1 node away**	21.40	53	ND
TIF4631	TIF4632	**1 node away**	80	50	U
NUP100	NUP116	**1 node away**	13.80	36	U
ASM4	SWI5	3 nodes away	13.63	29	C
EBS1	EST1	10 nodes away	0	27	C
YJL058C	YBR270C	Not clustered	4	44	ND
NUP170	NUP157	Not clustered	14	42	U
TUB4	TUB1	Not clustered	0	30	C
Other Duplications					
SOR1	YDL246	**1 node away**	80	100	U
FUS3	SS1	**1 node away**	46.15	54	C
YPT31	YPT1	**1 node away**	16.66	46	C
PHO85	CDC28	Not clustered	0	53	C
SLT2	KSS1	Not clustered	5.30	50	C
RHO1	CDC42	Not clustered	12	49	C
RHO4	RHO1	Not clustered	14.28	47	C
RAD51	DMC1	Not clustered	12.50	46	C
FUS3	HOG1	Not clustered	0	45	C
STE5	FAR1	Not clustered	13.33	21	C
Gene Families					
LSM2	LSM4	**1 node away**	20	31	U
LSM1	LSM5	2 nodes away	18.64	29	C
LSM3	LSM5	2 nodes away	18.60	27	U
LSM1	LSM3	3 nodes away	15.70	32	C
LSM3	LSM7	3 nodes away	27	31	U
LSM7	LSM5	4 nodes away	25.70	29	U
LSM1	LSM7	5 nodes away	15.20	32	C
LSM4	LSM7	9 nodes away	16.66	35	U
SNZ1	SNZ3	**1 node away**	50	81	U
SNZ2	SNZ3	2 nodes away	45.45	100	U
SNZ1	SNZ2	2 nodes away	44.44	80	U
SNO1	SNO2	2 nodes away	75	67	U

[a] U, unchanged function; C, changed function; ND, not determined because the function of the proteins are unknown.

Table 1. Duplicated gene pairs and their structural and functional associated variables

[a] U, unchanged function; C, changed function; ND, not determined because the function of the proteins are unknown.

note that these conclusions established only on an interaction criteria (column 3 in table 1) correlate well with what is known from other sources (column 6). Indeed, there is an agreement of at least 74% between columns 3 and 6 (no functional information was available in two instances for one protein of a pair).

How is change in function correlated with change in sequence and with the percentage of shared interactors? Surprisingly, the eleven pairs of strongly related proteins display a broad range of percentage of shared interactors (from 13.80 to 80%) as well as of sequence identity (from 31 to 100%). As expected however, the majority of the 11 pairs of tightly clus-

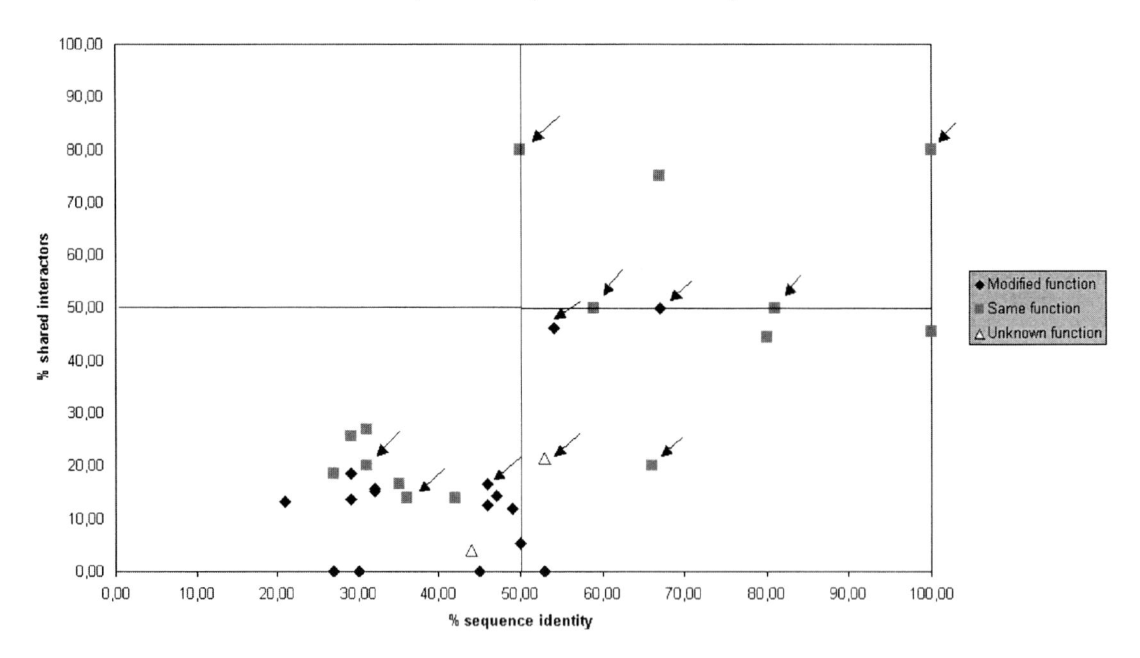

Figure 5. Sequence identity vs. Interactors identity. Values reported in Table 1 were plotted and a shape code was used to pinpoint the functional status of the each gene pairs. In addition, black arrows show the 11 gene pairs tightly clustered in the functional tree.

tered proteins (73%) are more than 50% identical in sequence, but a few pairs of proteins highly identical in sequence are not completely clustered. They however all belong to the same gene family (SNO-SNZ family) and are found within the same subtree. Conversely, as expected, proteins that are not clustered at all in the tree display less than 50% sequence identity.

A change in functional status within a pair of duplicated genes appears to happen preferentially among pairs with low sequence identity (81% of pairs in which a functional change occurred share less than 50% of sequence identity). As expected, the majority of these pairs share only a few interactors (less than 20%) and are not clustered in the tree.

When the percentage of shared interactors vs. the sequence identity is studied (Figure 1), it is striking to note that as soon as sequences diverge between the genes of a given pair, the percentage of shared interactors drops very rapidly. This illustrates the fact that interactions are very sensitive to sequence changes.

Interestingly enough, a low percentage of shared interactors is not always associated with a change in functional status among genes of a same pair (see gene pairs with less than 50% of shared interactors). It needs to be associated with a low sequence identity in order to lead to functional divergence. Similarly, a low percentage of sequence identity is not the only parameter to change function: it should be coupled with a loss of shared interactors. Consequently, it is tempting to speculate that two critical thresholds exist to modify the function of duplicated genes, one for the percentage of shared interactors (~20%), the other for the percentage in sequence identity (~45%). For both variables, values below these thresholds are necessary to observe a functional divergence. Strikingly, a few genes pairs have reached both thresholds without yet changing their functional status and therefore constitute an exception to the above conclusion. Interestingly, the majority of these gene pairs belong to the LSM family and act together in the same protein complexes. It could then be argued that each protein of the complex, obviously participating to the same cellular process, assumes a distinct set of interactions necessary for the complex function.

How can this study contribute to the understanding of the functional fate of duplicated genes?

If one considers the theoretical evolution of duplicated genes towards nonfunctionalization, subfunctionalization, neofunctionalization (Lynch and Force, 2000) and functional redundancy, variables we measured in the study should allow to distinguish between these different cases. Independently of the percentage of sequence identity, the percentage of shared interactors should permit to discriminate between functional redundancy on one hand and sub- and neofunctionalization on the other hand. Indeed, in the case of functional redundancy, the percentage of shared interactors is expected to be high whereas sub- and neofunctionalization would imply a low percentage of shared interactors.

Among gene pairs which are candidates to have evolved towards subfunctionalization and neofunctionalization, considering the percentage of sequence identity should allow to discriminate between sub- and neofunctionalization. Indeed, a low percentage of sequence identity is required together with a low percentage in shared interactors in order to induce a functional change. Interestingly enough, this last parameter does not appear to influence subfunctionalization. Indeed, the YCK1/YCK2 pair, which displays 66% of sequence identity as well as the LSMs pairs which are only ~30% identical are all good candidates to have evolved through subfunctionalization.

Conclusions

We have presented in this paper and in another study (Brun, Guénoche and Jacq, submitted) a new generic method allowing a functional classification of proteins, based on the identity of their interactors. With this information only, the method is able to point out already known protein complexes, pathways and broad cellular process. It also has a predictive value when it permits functional assignment of protein of unknown function through its clustering with well known proteins and when it proposes the existence of new functional groups of proteins. Hopefully, this method will become a valuable tool, helping to analyse results broughtby the new high throughput techniques such as large-scale yeast two-hybrid experiments.

We asked whether this method may help the understanding of functional divergence among duplicated genes through the study of a small subset of duplicated yeast genes. For a pair of duplicated genes, considering 3 different variables, (the clustering in the tree resulting from the application of the functional classification method, the percentage of shared interactors between duplicated genes and their sequence identity), we were able to pinpoint the existence of at least two thresholds which need to be both reached in order to obtain a functional divergence. In addition, it appears that a careful analysis of the 3 variables may give a hint for the assignment of an evolutionary fate to duplicated gene pairs.

Although rigorous statistical tests still need to be applied, other types of distance tested and a larger set of proteins analysed, the present results underline the fact that systematic studies of protein interactors can already be successfully used to bring new insights into biological processes. Overall, this work supports the idea that the study of interactors represents a new and powerful tool to answer biological questions related to function.

Acknowledgements

We thank Laurence Röder and Denis Thieffry for critical reading of the manuscript. Christine Brun thanks Valigen S.A. for financial support. The research of the group of Bernard Jacq is supported by the CNRS genome program and by a grant from the French inter-organisms bioinformatics program.

References

Becskei, A. and Serrano, L. (2000) Engineering stability in gene networks by autoregulation. *Nature*, **405**, 590–593.

Coissac, E., Maillier, E. and Netter, P. (1997) A comparative study of duplications in bacteria and eucaryotes: the importance of telomeres. *Mol. Biol. Evol.*, **14**, 1062–1074.

Costanzo, M.C., Crawford, M.E., Hirschman, J.E., Kranz, J.E., Olsen, P., Robertson, L.S., Skrzypek, M.S., Braun, B.R., Hopkins, K.L., Kondu, P., Lengieza, C., Lew-Smith, J.E., Tillberg, M. and Garrels, J.I. (2001) YPD, PombePD and WormPD: model organism volumes of the BioKnowledge library, an integrated resource for protein information. *Nucleic Acids Res.*, **29**, 75–79.

Costanzo, M.C., Hogan, J.D., Cusick, M.E., Davis, B.P., Fancher, A.M., Hodges, P.E., Kondu, P., Lengieza, C., Lew-Smith, J.E., Lingner, C., Roberg-Perez, K.J., Tillberg, M., Brooks, J.E. and Garrels, J.I. (2000) The yeast proteome database (YPD) and

Caenorhabditis elegans proteome database (WormPD): comprehensive resources for the organization and comparison of model organism protein information. *Nucleic Acids Res.*, **28,** 73–76.

Devos, D. and Valencia, A. (2000) Practical limits of function prediction. *Proteins*, **41,** 98–107.

Dujon, B., Alexandraki, D., Andre, B., Ansorge, W., Baladron, V., Ballesta, J.P., Banrevi, A., Bolle, P.A., Bolotin-Fukuhara, M., Bossier, P. *et al.* (1994) Complete DNA sequence of yeast chromosome XI. *Nature*, **369,** 371–378.

Elowitz, M.B. and Leibler, S. (2000) A synthetic oscillatory network of transcriptional regulators. *Nature*, **403,** 335–338.

Feldmann, H., Aigle, M., Aljinovic, G., Andre, B., Baclet, M.C., Barthe, C., Baur, A., Becam, A.M., Biteau, N., Boles, E. *et al.* (1994) Complete DNA sequence of yeast chromosome II. *EMBO J.*, **13,** 5795–5809.

Gardner, T.S., Cantor, C.R. and Collins, J.J. (2000) Construction of a genetic toggle switch in Escherichia coli. *Nature*, **403,** 339–342.

Gascuel, O. (1997) BIONJ: an improved version of the NJ algorithm based on a simple model of sequence data. *Mol. Biol. Evol.*, **14,** 685–695.

He, W. and Parker, R. (2000) Functions of Lsm proteins in mRNA degradation and splicing. *Curr. Opin. Cell Biol.*, **12,** 346–350.

Hlavacek, W.S. and Savageau, M.A. (1996) Rules for coupled expression of regulator and effector genes in inducible circuits. *J. Mol. Biol.*, **255,** 121–139.

Ito, T., Chiba, T., Ozawa, R., Yoshida, M., Hattori, M. and Sakaki, Y. (2001) A comprehensive two-hybrid analysis to explore the yeast protein interactome. *Proc. Natl. Acad. Sci. USA*, **98,** 4569–4574.

Jacq, B. (2001) Protein function from the perspective of molecular interactions and genetic networks. *Brief Bioinform.*, **2,** 38–50.

Jeffery, C.J. (1999) Moonlighting proteins. *Trends Biochem. Sci.*, **24,** 8–11.

Lei, M. and Tye, B.K. (2001) Initiating DNA synthesis: from recruiting to activating the MCM complex. *J. Cell Sci.*, **114,** 1447–1454.

Lynch, M. and Force, A. (2000) The probability of duplicate gene preservation by subfunctionalization. *Genetics*, **154,** 459–473.

Mewes, H.W., Frishman, D., Gruber, C., Geier, B., Haase, D., Kaps, A., Lemcke, K., Mannhaupt, G., Pfeiffer, F., Schuller, C., Stocker, S. and Weil, B. (2000) MIPS: a database for genomes and protein sequences. *Nucleic Acids Res.*, **28,** 37–40.

Osmani, A.H., May, G.S. and Osmani, S.A. (1999) The extremely conserved pyroA gene of Aspergillus nidulans is required for pyridoxine synthesis and is required indirectly for resistance to photosensitizers. *J. Biol. Chem.*, **274,** 23565–23569.

Padilla, P.A., Fuge, E.K., Crawford, M.E., Errett, A. and Werner-Washburne, M. (1998) The highly conserved, coregulated SNO and SNZ gene families in Saccharomyces cerevisiae respond to nutrient limitation. *J. Bacteriol.*, **180,** 5718–5726.

Perriere, G. and Gouy, M. (1996) WWW-query: an on-line retrieval system for biological sequence banks. *Biochimie*, **78,** 364–369.

Perrin, L., Romby, P., Laurenti, P., Berenger, H., Kallenbach, S., Bourbon, H.M. and Pradel, J. (1999) The Drosophila modifier of variegation modulo gene product binds specific RNA sequences at the nucleolus and interacts with DNA and chromatin in a phosphorylation-dependent manner. *J. Biol. Chem.*, **274,** 6315–6323.

Rivera-Pomar, R., Niessing, D., Schmidt-Ott, U., Gehring, W.J. and Jackle, H. (1996) RNA binding and translational suppression by bicoid. *Nature*, **379,** 746–749.

Romaniuk, P.J. (1985) Characterization of the RNA binding properties of transcription factor IIIA of Xenopus laevis oocytes. *Nucleic Acids Res.*, **13,** 5369–5387.

Seoighe, C. and Wolfe, K.H. (1999) Updated map of duplicated regions in the yeast genome. *Gene*, **238,** 253–561.

Sharp, D.H. and Reinitz, J. (1998) Prediction of mutant expression patterns using gene circuits. *Biosystems*, **47,** 79–90.

Smolen, P., Baxter, D.A. and Byrne, J.H. (2000) Mathematical modeling of gene networks. *Neuron*, **26,** 567–580.

Thomas, R. (1973) Boolean formalization of genetic control circuits. *J. Theor. Biol.*, **42,** 563–585.

Thomas, R., Thieffry, D. and Kaufman, M. (1995) Dynamical behaviour of biological regulatory networks--I. Biological role of feedback loops and practical use of the concept of the loop-characteristic state. *Bull. Math. Biol.*, **57,** 247–276.

Todd, A.E., Orengo, C.A. and Thornton, J.M. (2001) Evolution of function in protein superfamilies, from a structural perspective. *J. Mol. Biol.*, **307,** 1113–1143.

Uetz, P., Giot, L., Cagney, G., Mansfield, T.A., Judson, R.S., Knight, J.R., Lockshon, D., Narayan, V., Srinivasan, M., Pochart, P., Qureshi-Emili, A., Li, Y., Godwin, B., Conover, D., Kalbfleisch, T., Vijayadamodar, G., Yang, M., Johnston, M., Fields, S. and Rothberg, J.M. (2000) A comprehensive analysis of protein-protein interactions in *Saccharomyces cerevisiae*. *Nature*, **403,** 623–627.

von Dassow, G., Meir, E., Munro, E.M. and Odell, G.M. (2000) The segment polarity network is a robust developmental module. *Nature*, **406,** 188–192.

Wolfe, K.H. and Shields, D.C. (1997) Molecular evidence for an ancient duplication of the entire yeast genome. *Nature*, **387,** 708–713.

A. Meyer, Y. Van de Peer (eds.), Genome Evolution, 225-234.

Development gene networks and evolution

Jonathan P. Rast
Division of Biology 156–29, California Institute of Technology, Pasadena, CA 91125, USA
Correspondence: E-mail: jprast@its.caltech.edu

Received 01.07.2002; Accepted in final form 29.08.2002

Key words: *brachyury*, gene network, sea urchin, transcription

Abstract

Animal development relies on complex programs of gene regulation that are likely to account for a significant fraction of the information carried in genomes. The evolution of these regulatory programs is a major contributor to the diversity of animal forms, yet the architecture of the transcriptional networks that comprise developmental programs is only beginning to come into focus. The sea urchin offers an uncomplicated system in which to study transcriptional regulation and the networks that direct embryogenesis. This review describes the approaches that we are taking towards this problem using the sea urchin embryo and some comparative methods that will eventually lead to a deeper understanding of the evolution of developmental networks.

Introduction

Animal genomes encode the sequences of proteins and noncoding RNAs, and also the genetic programs needed to regulate gene expression. In the course of development the spatial and temporal dimensions of expression and the rates of transcription for each gene are controlled by these regulatory programs. The *cis*-regulatory sequences that encode these programmed interconnections account for much of the genetic information that is passed between generations. In complex metazoans the raw quantity of sequence dedicated to these control functions probably approaches or exceeds that used in gene coding regions. This regulatory information is arguably the prime target of evolutionary change that accounts for much of the diversity of animal life. Nonetheless the structure of these programs and the mechanisms by which they evolve are not well understood. A grasp of the nature of regulatory networks and how they are encoded in DNA is essential to a full understanding of genome function and evolution.

Information on the surroundings and the internal state of a cell is brought to particular genes in the nucleus by DNA binding proteins. The specificity of transcriptional control is mediated by sites that bind these proteins to the genes under regulation. While this genomic "sensory" process may entail a variety of signaling molecules and membrane receptors, and cascades of cytoplasmic mediators, the final effectors of regulation are the activities of those transcription factors which transduce each signal in the nucleus. The control region of each gene integrates the information carried to it by these DNA binding proteins and determines an appropriate output. The output of a control region will always be different than any particular one of its inputs, and two differently organized *cis*-regulatory DNA sequences can interpret the same input information in different ways. An important point here is that the DNA sequence of control regions actively contributes to the control process, beyond specifying the transcription factors to be brought to it, though this is indeed the fundamental function of *cis*-regulatory DNA sequence. The DNA itself scaffolds and structures the DNA-binding proteins and determines their effect on the rate of gene transcription. In this way *cis*-regulatory DNA serves as an information-processing unit that interprets the nuclear (and cellular) environment and effects an appropriate response, by modulating the transcriptional rate of the gene that it regulates. In the course of evolution, the DNA sequence of these control regions can bring more or less information into the decision process and the ways in which this information is used to produce a transcriptional outcome can be modified.

Some of the genes being regulated encode transcription factors, so the wider regulatory scheme comprises a web of interactions linking each gene into a complex network. Signaling systems allow cells to affect the activity of transcription factors in neighboring cells, thereby linking the network among the nuclei of a multicellular organism. Within the structure of such networks lies the information needed to establish and differentiate the regulatory environment of each cell in a developing embryo. Ultimately it is within the context of gene networks that DNA sequence changes are coupled to the gross morphological and physiological changes that are the direct targets of evolutionary selection.

Evolutionary analysis can focus on different levels of biological organization. The field of molecular evolution has been largely concerned with alterations of gene coding sequence. The evolution of gene regulatory mechanism has received much less attention, though it has long been recognized that evolution in this domain is likely to be of great importance (Britten and Davidson, 1971). Recently, however, more effort is being devoted to the evolutionary analysis of the regulatory regions of individual genes (e.g., Ludwig *et al.*, 2000; Sumiyama *et al.*, 2002). This becomes increasingly feasible as the regulatory mechanisms of an increasing number of genes are deciphered, and as comparative methods for efficient regulatory sequence analysis mature. It is an interesting twist of priorities that evolutionary analysis of regulatory regions is quickly being established as a tool to decipher control region function.

The biological phenomena that are the ultimate targets of evolutionary explanation (such as beak morphology in a finch or the appearance of the adaptive immune system of vertebrates) lie at a higher level of organization. They emerge from the coordinated activities of many genes. Prediction of how these systems evolve requires knowledge of the regulatory systems of their individual component genes and an understanding of how those genes function in the context of entire regulatory networks. The process of animal development is enormously complex and is likely to make use of a large majority of the regulatory information that is contained in the genomes. Nonetheless because development is accomplished in a highly regulated and stereotypic manner, untangling the genetic programs that underlie it is a tractable problem with modern molecular techniques. Additionally development is generally unidirectional and the mechanisms that confer directionality ensure that gene expression proceeds through a series of stable states that are amenable to this type of investigation.

This review presents our approach toward a network level explanation for development in the sea urchin embryo. An explanation of why this system is suitable for this type of research is briefly presented, along with a description of the tools that we use to analyze network architecture. An analysis of transcript factor function in gastrulation is employed to illustrate this approach. Finally the use of comparative methods to analyze these networks from an evolutionary viewpoint is discussed.

Tools for understanding gene regulation on a network scale

Although it has been possible to establish regulatory connections for some time, technological innovations have increased the efficiency of this work to the point where systems-scale network analysis is now becoming feasible. These technologies include global gene expression analysis based on cDNA arrays, large insert bacterial artificial chromosome (BAC) cloning, efficient large-scale sequencing and an expanding arsenal of tools with which to specifically disrupt gene function. This review focuses on the methodology and findings in an investigation of genetic programs that regulate development. The work described is taken from an ongoing project to discover the architecture of the transcriptional network that regulates endomesodermal specification and differentiation in the embryo of the purple sea urchin, *Strongylocentrotus purpuratus*. This work and its supporting methodology are described in a series of recent publications (Bolouri and Davidson, 2002; Brown *et al.*, 2002; Davidson *et al.*, 2002a; Oliveri *et al.*, 2002; Ransick *et al.*, 2002; Rast *et al.*, 2002; Yuh *et al.*, 2002) and in a recent review (Davidson *et al.*, 2002b). The work is in progress and the most recent version of the network is posted on the web site: http://www.its.caltech.edu/~mirsky/endomes.htm.

Sea urchin embryogenesis as a model for transcriptional analysis

The ability to understand *cis*-regulatory function for individual genes must underlie any deep knowledge of transcriptional networks. The sea urchin embryo offers an exceptional experimental system for studying developmental *cis*-regulation. Its advantages stem from the ease with which transgenic reporter constructs can be tested, the relative simplicity of the embryo and the larval form into which it develops, and the well characterized tools that are available for disrupting gene function. Staged sea urchin embryos are also available in huge quantities enabling investigations of rare embryonic proteins and mRNAs. Additionally, the phylogenetic position of sea urchins among the deuterostomes strengthens its relevance as a simple model for comparison to vertebrate development.

Detailed work on the sea urchin embryo has shown that, in the process of developmental specification, even genes with seemingly simple expression patterns can be controlled by cis-regulatory systems that are complex and counterintuitive. The regulation of the *endo16* gene during sea urchin embryogenesis illustrates this point. This gene encodes a secreted protein that is ultimately expressed in the larval midgut. The expression of *endo16* is initiated early in the process of endomesodermal specification before there is any morphological hint of the structure in which it will ultimately function. The control region of this gene has been modeled in great detail, and many insights into how developmental control systems operate emerge from its analysis (Yuh *et al.*, 1998, 2001). While the expression pattern of *endo16* appears on first glance to be unremarkable, the regulatory machinery by which this takes place is surprisingly elaborate. This is unexpected until one considers the complexity of the decision process that must be encoded in the control region of this gene.

A 2.3 kb region of DNA taken from upstream of *endo16* is sufficient to drive accurate expression of a reporter gene in transgenic embryos. This region can be divided into five modules that regulate subfunctions of the overall transcription pattern. Thirteen different proteins bind within the control region at more than 50 separate binding sites and each of these thirteen species of interactions has a demonstrable effect on qualitative or quantitative aspects of transcription. Much of the regulatory machinery is negative in function, devoted to maintaining the gene in the off state in embryonic cells where it is never expressed or turning it off in mesenchymal cells that migrate away from the expression domain. Notably, there is a transition late in embryogenesis that passes the positive function of the regulatory region from the module that activates it early in embryogenesis to a module that is dedicated to its later activation, when it runs in response to a positive regulator probably localized in the gut.

The *endo16* study illuminates two characteristics of systems that regulate the transcription of genes expressed during embryogenesis: (1) These *cis*-regulatory mechanisms can be complex. They operate in ways that are not easily predicted simply by observing expression patterns, or even by scanning the region surrounding a gene with reporter constructs. A deep understanding requires iterative testing of reporter constructs, modeling, and experimental confirmation. (2) Gene regulatory systems that operate in the course of development are complex because they are required to function in a regulatory landscape that is itself evolving. They accomplish this by dividing their regulatory machinery among specialized modules and sometimes use molecular switches to change between modes of regulation as specification proceeds. These regulatory systems also tend to rely heavily on negative regulation for spatial restriction during specification, then to switch to spatially restricted positive regulation in their terminal expression phases. All of this complexity is encoded directly in the DNA surrounding the gene, and this is where its function is experimentally most accessible.

To move beyond individual genes and to begin to explain the programs that underlie complex developmental phenomena requires an understanding of the functions of many genes. These phenomena emerge from gene interactions and their causal explanations can only be found in the structure of gene networks. Models of developmental *cis*-regulatory systems, as exemplified by the *endo16* gene, can be extended to include entire sets of genes that are used in the process of cell specification and differentiation. The complexity of these broader systems dictates that the details for particular genes must be reduced from the level of explanation that is achieved in the *endo16* model. *Cis*-regulatory interactions involving ubiquitous factors are ignored and only those interactions that directly and significantly affect the spatial distribution and amplitude of developmental genes are considered. A genetic regulatory model of this type is described below.

Endomesodermal development in the sea urchin embryo

The endoderm and mesoderm of the sea urchin larva are specified in a group of cells at the vegetal pole of the early embryo. The endoderm eventually gives rise to a three-part gut, including an esophagus, stomach and hindgut, and to some endoderm derived contractile cells that form the gut sphincters. The mesoderm differentiates into at least five cell types, including the primary mesenchyme cells (PMCs) which lay down the larval skeleton, the small micromeres which are incorporated into coelomic pouches that will later contribute to the adult rudiment, the mesodermal pharyngeal muscle cells, and two independent migratory cell types, the pigment and blastocoelar cells (Ruffins and Ettensohn, 1996). The latter four mesodermal cell-types are designated secondary mesenchyme cells (SMCs). The transcriptional processes that lead to the differentiation of these cells begin early in cleavage long before any hint of the structures that they will form emerges. Early molecular events depend on an asymmetry along the animal-vegetal 'axis', that is laid down maternally in the egg. This initial cytoarchitectural polarity is transduced into a signaling cascade, which originates from the vegetal-most blastomeres, the micromeres. These central cells organize progressive transcriptional domains in the cells that surround them. A second tier of transcriptional autoregulation and cross-regulation lock these signaling patterns into place such that they persist after the original signal has disappeared. The cells that participate in the process of endomesodermal specification lie in a thickened region at the bottom of the bastula called the vegetal plate. A fate map of the vegetal plate is shown in Figure 1. This map is based on the lineage tracings (Ruffins and Ettensohn 1996), and is supported by the expression patterns of a number of marker genes that we and others observe in these cells.

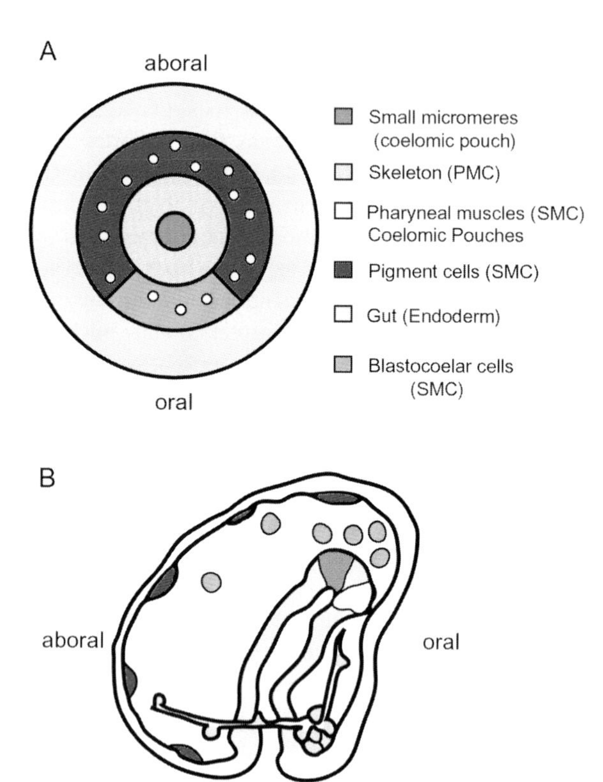

Figure 1. The distribution and development of endoderm and mesoderm in the sea urchin embryo. **A**. A hypothetical view from the vegetal pole of the embryo showing the location of precursors for endoderm and mesodermal cell types. The view is hypothetical because not all of the cell types are yet specified at a stage when they are in this location. For example specification of the blastocoelar cells probably occurs after the primary mesenchyme cells have ingressed. The ectodermal cells are excluded from the diagram. The information used in this diagram is taken from lineage tracing data (Ruffins and Ettensohn, 1996) and is supported by the expression patterns of genes discovered in the network analysis described in the text. **B**. The distribution of endoderm and mesodermal cell types in the late gastrula shown in cross-section. The pharyngeal muscles and coelomic pouches are not yet elaborated at this stage. Cells are color coded as in A.

The transcriptional network that regulates endomesoderm

The purpose of the approach that we take is to describe the transcriptional network that underlies endomesodermal specification. That is, we wish to define the connections among *cis*-regulatory elements and transcription factors that participate in the division of the vegetal plate into transcriptional territories and finally into differentiated endodermal and mesodermal cell types. Of course, in one way or another, thousands of genes are probably involved in this process. There is surely a wealth of interconnectivity that provides the feedback control and partial redundancy to ensure that even in the face of perturbation, development will proceed to the same end. The complexity of this interconnectivity is of great interest, but initially we are concerned with a subset of genes that can be identified as necessary components of the decision making process. Thus the number of genes in the network can be limited to a manageable set.

We must deal also with another dimension of complexity, that of the number of *cis*-regulatory connections into each gene. *Endo16* is probably not unusual in terms of regulatory complexity. Many developmental genes that are studied in detail integrate information from large numbers of transcription factors using multimodule *cis*-regulatory control regions (for reviews, see Arnone and Davidson, 1997; Davidson, 2001). Many of these interactions, however, involve ubiquitous factors that, though necessary for function, do not play pivotal roles in the developmental decision process. Therefore the approach taken here is to first limit the analysis to connections that if perturbed critically alter the course of gene expression. These consist mainly of inputs from spatially and temporally restricted transcription factors and signaling system termini.

Important genes were identified from the considerable resource of past studies, and from a series of differential screens in which the development of endomesoderm was strategically interrupted and analyzed for genes with altered transcription. Additional genes were selected because their orthologs are involved in endomesodermal specification in other organisms. Information of the expression of these genes and the phenomena with which they are associated was then integrated into a transcriptional interaction model. The process of building this first stage network model is outlined below:

1. A minimal logical network was assembled from available information.
2. Genes from the differential screens and genes whose orthologs play roles in endomesoderm in other species were incorporated into the model.
3. The function of these genes was perturbed by several methods including: gene mis-expression in the case of repressors; injection of engrailed-fusion dominant-negative constructs; interference with relevant signaling systems; and mainly, specific translational interference by morpholino oligonucleotide injection.
4. The effects of specific perturbations on the transcription of other network genes were quantified by realtime quantitative PCR (QPCR).
5. The model was revised according to these observations and reanalyzed begining at Step 3.

The result of this analysis is a hypothetical map of the transcriptional interactions that operate in the process of specification. Ultimately these must be confirmed or falsified by reporter gene experiments that manipulate the *cis*-regulatory regions of each gene. To accomplish this, the sequence surrounding each gene in the network is obtained and the region is analyzed for *cis*-regulatory function. Random sub-cloning and comparative methods are used to expedite the identification of *cis*-regulatory modules. Eventually these are tested in reporter constructs and are manipulated to identify specific *cis*-regulatory connections. The perturbation and analysis methods outlined in Steps 3 and 4 above can be enlisted to alter specifically the function of transcription factors and signaling systems, and then to determine their effects on reporter construct function, thus limiting the scope of the analysis to short stretches of DNA.

The purpose of the model, parts of which will be described here, is to explicitly lay out the *cis*-regulatory connections among genes that control the division of the vegetal plate into transcriptional domains. Initially the model includes only the architecture of these connections, but eventually it will describe the flow of regulatory information through these connections. For the latter purpose detailed quantitative information will be required.

A specific case: the function of *brachyury* in the early embryo

We do not wish to review the Endomesodermal Specification Model in detail here but rather to use part of the model to exemplify how this approach can be used to focus investigation on the level of gene interaction (specifically on the level of transcription control networks). We take as an example an analysis of the regulatory connections to and from the transcription factor *brachyury* in terms of the role that these interactions play during endodermal morphogenesis at gastrulation. Here the analysis is centered on the function of a single gene, but the approach is also being applied to complex developmental phenomena, such as cell-type specification and skeletal patterning.

Transcription factors are often associated with particular developmental functions, cell types or structures. However, transcription factors function directly only to control rates of gene expression. While they clearly play specific developmental roles, the level at which they operate is removed from gross morphological phenomena by several layers of causal connections. Among these layers are the transcriptional networks that are the targets of the present studies, and it is at this level that individual molecular functions are integrated into developmental plans. Thus, to understand how transcription factors operate in a concrete sense, it is first necessary to know the inputs that control their own expression and to identify genes that they in turn regulate.

Brachyury is a member of the T-box family of transcription factors that is well known for its role in mesoderm and notochord development in vertebrates (Kispert *et al.*, 1995). However, its primitive function is in gastrulation and hindgut development (Peterson *et al.*, 1999). This is evidenced in the early expression pattern of its orthologs in a variety of protostome and deuterostomes representatives. For example, *brachyury* homologs exhibit a strikingly similar pattern of expression during gastrulation both in indirectly developing deuterostome echinoderms, and in a protostome, a polychaete annelid (Arendt *et al.*, 2001). We have set out to identify the *brachyury* target genes that function during gastrulation. In doing so we hope both to explain *brachyury* function in sea urchin gastrulation and to define a common set of network interactions that may be evolutionarily conserved in *brachyury* function throughout the Bilateria.

The pattern of *Brachyury* expression in sea urchin embryos is based on investigations in three species (Croce *et al.*, 2001; Gross and McClay, 2001; Rast *et al.*, 2002) and is illustrated in Figure 2A. *Brachyury* expression begins at the mid-blastula stage in a subset of the presumptive endoderm, forming a ring around the future mesodermal cells (dark gray in cross section in Figure 2, A1-A3). At gastrulation *brachyury* expression is extinguished just before the cells that will form the archenteron enter the blastopore. Throughout gastrulation it is expressed in this dynamic fashion in all of the endodermal cells that will contribute to the archenteron. A separate phase of *brachyury* expression is initiated in the oral ectoderm (shown in light gray in Figure 2, A2-A3). This zone of *brachyury* expression eventually localizes to a region where the mouth will form, the stomodeum. We are primarily concerned with the blastoporal phase of *brachyury* expression and its endodermal target genes, though some of these may overlap with the genes that Brachury regulates in the stomodeum.

Genes that require *brachyury* for transcription were identified by a subtractive differential screening procedure on arrayed embryo cDNA libraries (Rast *et al.*, 2000; Rast *et al.*, 2002). Some of these genes are expressed coincidently with *brachyury*, and are candidates as direct target genes. Some of the coincident genes are expressed around the blastopore, but are absent from the gut, just as is *brachyury*. Others remain in the gut in cells that have extinguished *brachyury* expression. Transcription of this latter class of genes may be initiated by Brachyury then regulated by other factors that run in the archenteron.

Notably, some of the genes that are expressed coincidently with *brachyury* have homologs in other species that are known to function as cytoskeletal modifiers, suggesting that these genes may be part of the causal linkage between *brachyury* expression and morphogenesis. The next step in this analysis is to determine whether or not the connection between Brachyury and these genes is direct. Of course, ultimately the process of gastrulation is far more complex than can be explained by merely turning on and off a few genes. Other factors should eventually be indentified from the differential screen and eventually the network linkages among these genes will explain *brachyury's* role in gastrulation.

To look upstream of *brachyury*, expression levels were measured in embryos after processes that are known to function in endomesodermal specification were interrupted (Davidson *et al.*, 2002a; R.A. Cameron, unpublished). From these analyses *brachyury* transcription appears to be affected by genes that act in endodermal specification. For example brachyury responds to the nuclearization of ß-catenin and is downstream of *gataE* function. Interestingly brachyury is downregulated by the function of the *foxA* forkhead transcription factor which runs in the gut. This offers a mechanism by which *brachyury* expression is extinguished in cells entering the archenteron. Clearly there must be other negative regulators that maintain *brachyury* in only a subset of the cells that are affected by these upstream factors.

The upstream and downstream connections are illustrated in Figure 2B. Although they are far from complete they point towards an explanation of how *brachyury* functions and how this function is tied to the phenomenon of gastrulation. The emerging view is that *brachyury* responds to the mediators of endodermal specification and in turn activates genes that encode effectors of morphogenesis. It is certain that the complex morphogenetic movements associated with *brachyury* function will require the participation of other transcription factors as well as other downstream effector proteins. Nevertheless, this system is shallow and there is little cell division between the times that Brachyury protein is transiently expressed and when the cells undergo morphogenesis.

In addition there is a non-cell autonomous component to Brachyury function that implies the activity

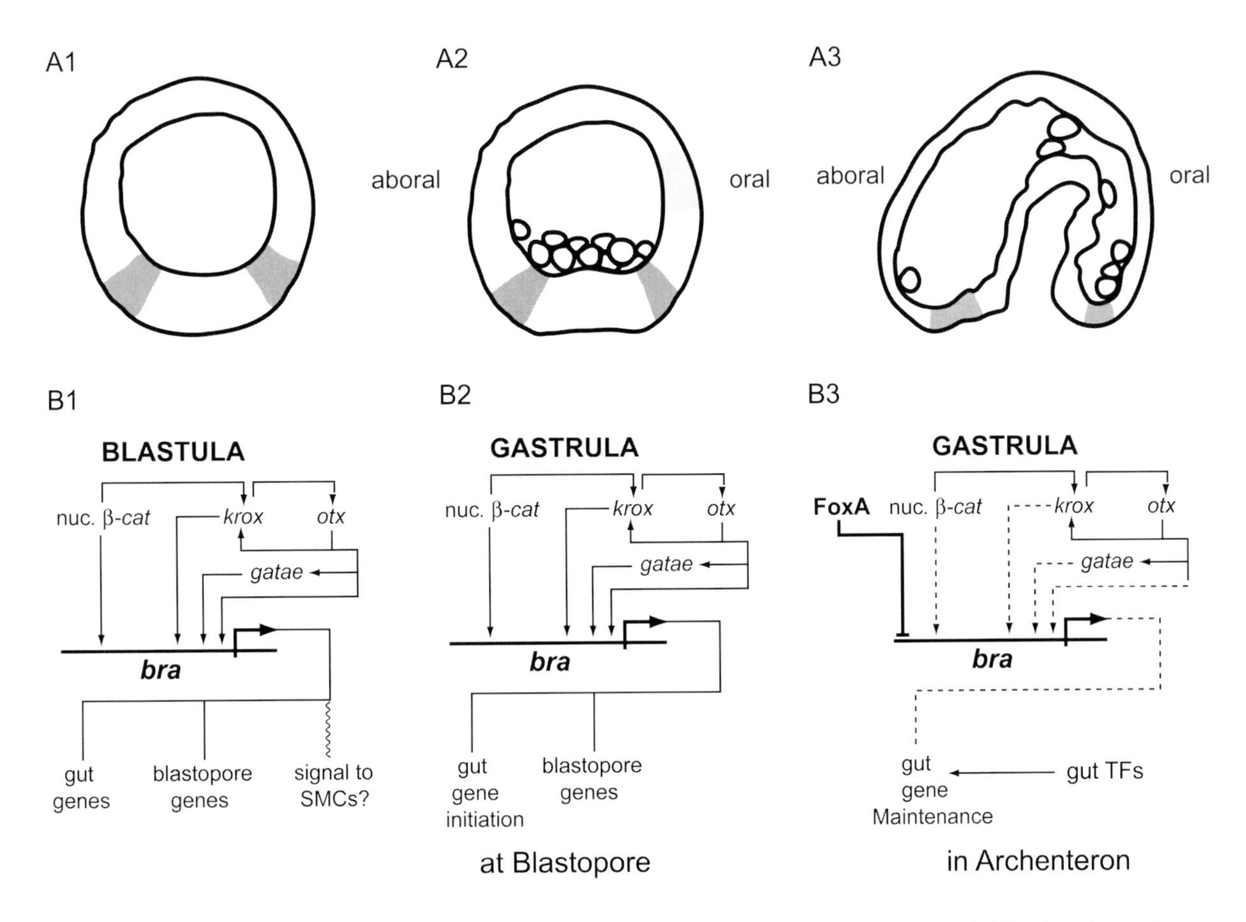

Figure 2. The function of brachyury in the gastrulating sea urchin embryo. **A**. *brachyury* expression pattern in blastula and gastrula stage embryos. A1. Mid-blastula. A2. Late (mesenchyme) blastula. A3. Gastrula. Expression is initiated early in the vegetal plate in the endodermal precursor cells (dark gray shading). In the late blastula a second zone of expression is established in the oral ectoderm (light gray shading). During gastrulation vegetal expression is maintained at the blastopore and turned off in the archenteron. Oral ectoderm expression is progressively restricted to the region where the larval mouth will form. **B**. Upstream and downstream transcriptional connections of *brachyury*. B1. In the blastula *brachyury* is under the control of factors that direct endodermal specification. *krox* and *otx* mutually interact to lock on major positively acting gut factors like *gataE*. *Brachyury* in turn activates genes that are later expressed in the gut and at the blastopore. At this early point *brachyury* affects genes that are transcribed in the mesoderm where *brachyury* itself is not expressed. This may indicate an early *brachyury* dependent signal from endoderm to mesoderm. B2. During gastrulation endoderm specification factors transiently maintain *brachyury* expression in cells just before they enter the blastopore. B3. As cells move into the archenteron, *brachyury* expression is extinguished as is expression of the downstream blastoporal genes. This negative function is downstream of *foxA* expression in the archenteron. Connections that are inactive at this stage are shown as dotted lines. In the absence of *brachyury*, the maintenance of gut gene transcription must be under the control of gut transcription factors. The interactions described are hypothesized from message prevalence measurements after embryo perturbation and other experiments. Whether the connections are direct or indirect can be distinguished by *cis*-regulatory analysis.

of signaling systems. Elements of all of these systems should fall out of the downstream differential screen and the upstream network and *cis*-regulatory work that is ongoing. Eventually when we say that *brachyury* functions in gastrulation we will be able to provide a precise description of the *cis*-regulatory linkages that are the cause of this function.

Comparative approaches toward gene regulation and network architecture

The properties of regulatory sequence evolution can themselves be used as a tool to expedite the analysis of genetic control systems. Genomic sequences containing regulatory regions from appropriately distanced species can be compared to locate evolutionarily conserved regions. Sequence conservation can then reveal *cis*-regulatory elements and their binding sites for transcriptional factors. These binding sites usually occur in clusters and, for reasons that are presently unclear, the sequence surrounding them is often also conserved. Patches of conservation that fit the description of regulatory regions can be tested in reporter constructs. An analysis of this type was performed in sea urchins on the particularly complex regulatory region of the transcription factor *otx*, which is expressed from three different transcriptional start sites in different territories of the sea urchin embryo (Yuh *et al.*, 2002). The sequence surrounding orthologs of this gene were compared from BAC clones isolated from two sea urchin species (*Strongylocentrotus purpuratus* and *Lytechinus variegatus*; divergence time ~50 mya). Conserved noncoding regions were identified using the Family Relations program (Brown *et al.*, 2002). This program identifies short stretches of sequence that are conserved above a threshold level (e.g., 18 of 20 nucleotides or 40 of 50 nucleotides). The comparison is independent of a global alignment between the two sequences and so is immune to problems of small-scale rearrangements. Seventeen patches of conserved sequence were identified in 60 kb of BAC sequence surrounding the *otx* gene. The average size of these patches was about 600 bp. Eleven of these fragments (65%) were able to confer transcriptional activity when linked to a reporter under the control of an unrestricted basal promoter. When assayed by *in situ* hybridization, expression from most of these was spatially restricted. Only about 3% of similar sized DNA fragments taken randomly from the genome and placed into the reporter construct are able to activate transcription (P. Oliveri and R.A. Cameron, unpublished). Some of the 6 conserved regions that did not confer activity may contain repressor elements that can be tested with different reporter configurations.

An interesting observation emerges from this study. This strategy differs from the traditional approaches that generally strive to define the minimal elements that are able to confer an expected expression pattern. Many of the 11 active elements drive expression patterns that are identically restricted and correspond to one or another of the *otx* transcripts. Whether these elements are functionally redundant or subtly different is as yet unclear, but the multiplicity of such regions may tend to be overlooked in searches for minimal sufficient elements. The sequence conservation approach uses a biological criterion in so far as the detection algorithm based on conservation is accurate and able to exclude noise. The sequence that becomes the focus of experimentation is that which nature has seen fit to conserve. In combination with the increased efficiency afforded by this approach, control sequences that may otherwise have gone undetected are likely to be found.

Recent comparison among animal genomes supports a view that has been emerging for some time: on first approximation, the remarkable diversity of animal forms is not so much rooted in differences in their complement of genes as in the way that these genes are used. The primary difference among animals then is in the realm of gene network structure and it is there that evolution plays its most significant role in modifying development. As the architecture of these networks emerges from investigations like the one described here, they will be open to evolutionary analysis. The development of echinoderm larvae offers an excellent comparative system in which to study network evolution. Echinoderms have a solid fossil record and, relative to the purple sea urchin, species with a range of divergence times are available that extend in well-documented increments throughout the Phanerozoic. The larvae of different echinoderms with similar but modified morphology can be compared in terms of developmental network structure. Importantly, the same tools and qualities that make the sea urchin amenable to network analysis (e.g., simplicity of form, ease of transgenesis and efficient morpholino antisense interference) can be utilized in other echinoderm embryos. The network architecture for one species can immediately be used to generate testable evolutionary hypotheses that can be applied to other species.

Conclusions

Much of the work described here initially differs little from some of the more traditional molecular genetic approaches toward development except that the focus, rather than on the effect of gene function on phenotype, is on the effect of gene function on the transcription of other genes. After an initial phase of collecting relevant genes and determining the phenomenological transcriptional connections among genes, the effort turns toward establishing concrete *cis*-regulatory connections in terms of transcription factor binding sites. The effectiveness of this work will be enhanced by the comparative methods described in the last section, computational methods that search the sequence surrounding interesting sets of genes for binding site-like motifs and improved molecular technology that allows the efficient generation of informative reporter constructs. Eventually this work will produce a collection of transcriptional circuitry that will lead to generalizations that will assist in determining further network architectures, both in the sea urchin and in other more complex developmental systems.

The guiding principle of this work is that a causal explanation of developmental phenotype can only be found in the interplay among genes and this is most experimentally accessible at the level of gene transcription. If the evolution of novel structures is primarily accomplished by creating new gene regulatory programs, then the evolutionary changes that account for these programs will be found in the form of sequence changes in gene regulatory regions. The meaning of these changes can only be linked to the higher-order developmental processes in which they work in the context of transcriptional networks.

Acknowledgements

I would like to thank Prof. Eric H. Davidson who is the driving force behind the Endomesodermal Network Project and whose suggestions significantly enhanced the manuscript, and Dr. Michele Anderson for insightful comments that greatly improved the text. Network research is by nature a collaborative and inclusive endeavor, and the success of this project is heavily dependent on the elegant work done throughout the community of sea urchin researchers. This work was supported by grants from the NIH (HD–37105, GM–61005, RR06591, and RR15044) and from the NASA/Ames Fundamental Space Biology Program (NAG2–1368). BAC sequences for comparative analyses were provided by the Joint Genome Institute of the U.S. department of energy and by Lee Rowen and Leroy Hood of the Institute for Systems Biology, Seattle.

References

Arendt, D., Technau, U. and Wittbrodt, J. (2001) Evolution of the bilaterian larval foregut. *Nature* **409,** 81–85.

Arnone, M.I. and Davidson, E.H. (1997) The hardwiring of development: organization and function of genomic regulatory systems. *Development* **124,** 1851–1864.

Bolouri, H. and Davidson E.H. (2002) Modeling DNA sequence-based *cis*-regulatory gene networks. *Dev. Biol.* **246,** 2–13.

Britten, R.J. and Davidson, E.H. (1971) Repetitive and non-repetitive DNA sequences and a speculation on the origins of evolutionary novelty. *Q. Rev. Biol.* **46,** 111–138.

Brown, C.T., Rust, A.G., Clarke, P.J.C., Pan, Z., Schilstra, M.J., De Buysscher, T., Griffin, G., Wold, B.J., Cameron, R.A., Davidson, E.H. and Bolouri, H. (2002). New computational approaches for analysis of cis-regulatory networks. *Dev. Biol.* **246,** 86–102.

Croce, J., Lhomond, G. and Gache, C. (2001) Expression pattern of Brachyury in the embryo of the sea urchin *Paracentrotus lividus*. *Dev. Genes Evol.* **211,** 617–619.

Davidson, E.H. (2001) *Genomic Regulatory Systems: Development and Evolution*, Academic Press, San Diego, CA.

Davidson, E.H., Rast, J.P., Oliveri, P., Ransick, A., Calestani, C., Yuh, C-.H., Minokawa, T. Amore, G., Hinman, V., Arenas-Mena, C., Otim, O., Brown, C.T., Livi, C.B., Lee, P.Y., Revilla, R., Schilstra, M.J., Clarke, P.J.C., Rust, A.G., Pan, Z., Arnone, M.I., Rowen, L., R., Cameron, R.A., McClay, D.R., Hood, L. and Bolouri, H. (2002a) A provisional regulatory gene network for specification of endomesoderm in the sea urchin embryo. *Dev. Biol.* **246,** 162–190.

Davidson, E.H., Rast, J.P., Oliveri, P., Ransick, A., Calestani, C., Yuh, C-.H., Minokawa, T., Amore, G., Hinman, V., Arenas-Mena, C., Otim, O., Brown, C.T., Livi, C.B., Lee, P.Y., Revilla, R., Rust, A.G., Pan, Z., Schilstra, M.J., Clarke, P.J., Arnone, M.I., Rowen, L., Cameron, R.A., McClay, D.R., Hood, L. and Bolouri, H. (2002b) A genomic regulatory network for development. *Science* **295,** 1669–1678.

Gross, J.M. and McClay, D.R. (2001) The role of Brachyury (T) during gastrulation movements in the sea urchin, *Lytechinus variegatus*. *Dev. Biol.* **239,** 132–147.

Kispert, A., Koschorz, B. and Herrmann, B.G. (1995) The T protein encoded by Brachyury is a tissue-specific transcription factor. *EMBO J.* **14,** 4763–4772.

Ludwig, M.Z., Bergman, C., Patel, N.H. and Kreitman, M. (2000) Evidence for stabilizing selection in a eukaryotic enhancer element. *Nature* **403,** 564–567.

Oliveri, P., Carrick, D.M., Davidson, E.H. (2002) A Regulatory gene network that directs micromere specification in the sea urchin embryo. *Dev. Biol.* **246,** 209–228.

Peterson, K.J., Cameron, R.A., Tagawa, K., Satoh, N. and Davidson, E.H. (1999) A comparative molecular approach to mesodermal patterning in basal deuterostomes: the expression pattern of Brachyury in the enteropneust hemichordate *Ptychodera flava*. *Development* **126,** 85–95.

Ransick, A., Rast, J.P., Minokawa, T., Calestani, C. and Davidson, E.H. (2002) New early zygotic regulators of endomesoderm specification in sea urchin embryos discovered by differential array hybridization. *Dev. Biol.* **246,** 132–147.

Rast, J.P., Amore, G., Calestani, C., Livi, C. B., Ransick, A. and Davidson, E.H. (2000) Recovery of developmentally defined gene sets from high-density cDNA macroarrays. *Dev. Biol.* **228,** 270–286.

Rast, J.P., Cameron, R.A., Poustka, A.J. and Davidson, E.H. (2002) *Brachyury* target genes in the early sea urchin embryo isolated by differential macroarray screening. *Dev. Biol.* **246,** 191–208.

Ruffins, S.W. and Ettensohn, C.A. (1996) A fate map of the vegetal plate of the sea urchin (*Lytechinus variegatus*) mesenchyme blastula. *Development* **122,** 253–263.

Sumiyama, K., Irvine, S.Q., Stock, D.W., Weiss, K.M., Kawasaki, K., Shimizu, N., Shashikant, C.S., Miller, W. and Ruddle, F.H. (2002) Genomic structure and functional control of the Dlx3- 7 bigene cluster. *Proc. Natl. Acad. Sci. USA* **99,** 780–785.

Yuh, C-.H., Bolouri, H. and Davidson, E.H. (1998) Genomic cis-regulatory logic: experimental and computational analysis of a sea urchin gene. *Science* **279,** 1896–1902.

Yuh, C-.H., Bolouri, H. and Davidson, E.H. (2001) Cis-regulatory logic in the endo16 gene: switching from a specification to a differentiation mode of control. *Development* **128,** 617–629.

Yuh, C-.H., Brown, C.T., Livi, C., Rowen, L., Clarke, P.J.C. and Davidson, E.H. (2002) Patchy interspecific sequence similarities efficiently identify positive *cis*-regulatory elements in the sea urchin. *Dev. Biol.* **246,** 148–161.

Index of Authors

Index of Keywords